Entomology Monographs

Series Editor

Hideharu Numata, Graduate School of Science, Kyoto University, Kyoto, Japan

Insects are the most diverse group of organisms, and many biological advances have been derived from this group. Although entomology is considered to be a classic science, recent developments in molecular methods, application of modern theoretical concepts and collaboration with related sciences have opened new directions in entomology. Japanese scientists play a significant role in these fields, and this book series will focus on such developments. The book series Entomology Monographs publishes refereed volumes on all aspects of entomology, including ecology, ethology, physiology, taxonomy, systematics, morphology, evolutionary developmental biology, genetics, biochemistry, and molecular biology in insects and related arthropods. Authors are not restricted to Japanese entomologists, and other international experts will also be considered on the basis of their recent contribution to these fields.

More information about this series at http://www.springer.com/series/15687

Junichi Yukawa • Makoto Tokuda
Editors

Biology of Gall Midges

Evolution, Ecology, and Biological Interactions

Springer

Editors
Junichi Yukawa
Entomological Laboratory, Faculty of Agriculture
Kyushu University
Fukuoka, Japan

Makoto Tokuda
Laboratory of Systems Ecology, Faculty of Agriculture
Saga University
Saga, Japan

ISSN 2522-526X ISSN 2522-5278 (electronic)
Entomology Monographs
ISBN 978-981-33-6533-9 ISBN 978-981-33-6534-6 (eBook)
https://doi.org/10.1007/978-981-33-6534-6

This Springer imprint is published by the registered company Springer Nature Singapore Pte Ltd.
The registered company address is: 152 Beach Road, #21-01/04 Gateway East, Singapore 189721, Singapore

Preface

Galls can be unexpectedly found on plants growing in gardens, parks, hills, and fields because of their outstanding feature in shape and color. Some galls exhibit structures like wonderful works of art with curious appearance and vivid color. People who found such swellings for the first time must be naturally surprised and very much wonder what happened to the plants. Once people note that the swellings were caused by tiny organisms such as insects or eriophyoid mites, they may be stupefied by the working of natural creatures. Because galls are so attractive, such new findings and experiences would make people's interests direct toward nature more strongly than before. On successive occasions, people try to find different sorts of gall on different plant species in the field. Thus, huge amounts of information on gall-inducing insects and their host plants have been accumulated until today by many amateur naturalists.

Galls can be easily observed in the field continuously from early to final stage of the development of galls and gall inducers because of their outstanding features and immobility. At the same time, we can obtain important data of host plant such as phenology, abundance as food resources, and the survival of galled organs. By taking these advantages, galls and gall-inducing insects have been used by many biologists as extremely convenient organisms for a wide range of ecological, ethological, evolutionary, and biogeographic studies. A number of our studies would not be intensively and extensively performed without using galls and gall-inducing insects. In other words, the studies include various interesting and important findings or implications that would not be obtained solely from investigations of non-gall inducing insects.

As will be mentioned in Chap. 1, the family Cecidomyiidae including tiny insects called gall midges is the most speciose family in Diptera, and possibly among whole insect orders. As a result, galls induced by cecidomyiids are much more diverse across various plant families and numerous in the number of sorts compared to those caused by other arthropods such as Cynipidae, Tenthredinidae (Hymenoptera), Aphidoidea, Psylloidea (Hemiptera), Sesiidae (Lepidoptera), Curculionidae

(Coleoptera), and Eriophyoidea (Acarina). Therefore, this book focuses particularly on gall-inducing cecidomyiids, their galls, and host plants.

We provide various practical ecological, ethological, evolutionary, and biogeographic data for gall-inducing cecidomyiids based on our field surveys, laboratory experiments, and genetic analysis. We also refer to various papers on gall-inducing insects published by a world of biologists. In some Chapters, practical methods of field surveys and data analysis are presented. We also provide Chapters of parasitoids, invasive pests, and beneficial gall midges for applied entomologists.

This book is primarily intended to demonstrate the appeal of galls and gall-inducing insects as extremely convenient organisms for a wide range of biological studies. In particular, we would like to emphasize that gall-inducing cecidomyiids are one of the most ideal insects to study ecology and evolution. This book will be helpful to open the doors to further cryptic study subjects. In addition, we can expect that an integration of various ecological, ethological, evolutionary, and biogeographic data demonstrated in this book will serve for the further progress of macroevolutionary studies of insects.

Ecological way of thinking is essential under recent conditions of global environment. Readers can learn an ecological way of thinking through diverse interrelations between insects and plants, and the analysis of ecological data from gall-inducing cecidomyiids. Then, they realize importance of actual ecological data to discuss environmental issues such as global warming, a decline of biodiversity, and pest management. We expect applied entomologists to be able to obtain useful ideas from some of the Chapters in establishing an appropriate method for IPM (Integrated Pest Management) and IBM (Integrated Biodiversity Management).

This book would be helpful for young biologists who are seeking an attractive study subject and appropriate target organisms in the field of ecology, ethology, or evolutionary biology. By referring to various Chapters of this book, we hope they can find an interesting study subject with convenient target insects and practice field surveys and data analysis based on the methods demonstrated.

In this book, we provide photographs of cecidomyiid galls as many as possible to introduce diverse structures of gall to readers including amateur naturalists. By showing these photographs, this book possibly contributes to increase people who will pay much more attention to nature and environmental issues and acquire ecological way of thinking.

Fukuoka, Japan
Saga, Japan
January 2021

Junichi Yukawa
Makoto Tokuda

Acknowledgements

First, we gratefully dedicate this book to the late Kiritani K (Emeritus Researcher, NARO Institute for Agro-environmental Sciences, Japan) for his long-term support for our ecological studies of gall midges. We would like to express our sincere gratitude to Gagné RJ (Systematic Entomology Laboratory, USDA) and Harris KM (former Director of the International Institute of Entomology, UK) for their comments and information on the systematics, taxonomy, morphology, and biology of gall midges throughout our research activities. We have been strongly influenced by Price PW (Northern Arizona University, USA) and Craig TP (University of Minnesota Duluth, USA) through their ecological studies of gall-inducing insects, and by Yamamura K through his method of key-factor/key-stage analysis. We appreciate their ecological studies.

We deeply thank many graduate and undergraduate students of Entomological Laboratories of Kagoshima and Kyushu Universities, and Laboratory of Systems Ecology of Saga University for their great help in field surveys of gall midge populations. The filed data are well exploited in this book.

Our thanks are also due to the following persons for giving us information on various gall-inducing arthropods and for providing us with important literatures: Abe Y (Kyushu University, Japan), Ide T (National Museum of Nature and Science, Japan), late Masuda H (Yamanashi Pref.), Wachi N (University of the Ryukyus, Japan), Yang MM (National Chung Hsing University, Taiwan), and Yamazaki K (Osaka City Institute of Public Health and Environmental Science, Japan).

We extend our thanks to the following Japanese colleagues for providing us with photographs of cecidomyiid galls used in this book: Aizawa T (Gunma Pref.), Hashimoto N and Iwasaki A (Central Agricultural Experiment Station, Hokkaido), Matsunaga K (Fukuoka Pref.), Matoba I and Matoba M (Wakayama Pref.), Minami T (Hokkaido), Nagai A (Miyazaki Pref.), Shimizu Y (Okinawa Prefectural Plant Protection Center), and Yafuso M (formerly University of the Ryukyus). In addition to these colleagues, many others have found various sorts of cecidomyiid gall around Japan. Thanks to their contributions, we could obtain important information on the life history traits, host plants, and distribution range of gall-inducing cecidomyiids in

Japan. They are as follows: Aoki M, Gôukon K, Gyotoku N, Hirota S, Ichita T, Ikeda F, Inoue E, Kaburagi K, Kakizaki M, Kanai K, Kidokoro T, Kikuchi H, Kogure I, Kohama T, Koizumi M, Makanai H, Matsumoto S, Mishima M, Mitsui H, Miyakuni S, Miyatake T, Murase M, Nakamura S, Nishi K, Nodu T, Nunoyama Y, Odajima H, Okamoto M, Ogata K, Ogawa H, Oguri Y, Sanui T, Sasaki M, Shiga M, Shirota Y, Sonobe R, Tabuchi K, Tajiri S, Tokioka K, Tominaga A, late Usuba S, Watanabe M, Yamagishi K, Yamauchi S, Yasuda K, and Yoshimura H.

The following publishers or societies allowed us to use photographs that had been used in previous publications: Zenkoku Nôson Kyôiku Kyôkai, Tokyo (Yukawa and Masuda 1996, *Insect and Mite Galls of Japan in Colors*; Kiritani and Yukawa 2010, *Effects of Global Warming on Insects*, through the courtesy of Motomura K), and the Nanki Biological Society (*Nanki Seibutu* 2018, through the courtesy of Doei H). We appreciate their kind courtesies.

Elsayed AK (formerly Saga University, Japan, presently The University of Tokyo) recently described many Japanese gall midge species together with their biological notes, which were quite useful for providing manuscripts, Ohno S (Okinawa Prefectural Agricultural Research Center) newly identified some species of spider mites for Sect. 13.3, Kamitani S (Kyushu University) scanned many old color-slide photographs and sized them up to higher *dpi*, and Yamaguchi D (Kyushu University) downloaded from the Kyushu University Library many important references used in this book. To these persons, we are deeply indebted. The study of fungal symbiont in Sect. 2.7 was partly supported by JSPS KAKENHI Grant No. 18K05682 to Uechi N.

Contents

Part I
Introduction

Chapter 1
Taxonomy, Phylogeny, and Larval Feeding Habits

Makoto Tokuda and Junichi Yukawa

Abstract Cecidomyiidae are currently divided into seven subfamilies. The higher taxonomy and their phylogenetic relationships are presented in this chapter. Some studies estimate that Cecidomyiidae are one of the most speciose families in Diptera or even in whole insects. Recent molecular phylogenetic analyses of Cecidomyiidae were mostly consistent with morphology-based phylogenetic assumptions and suggested to have radiated in the Upper Cretaceous, coincided with the radiation of angiosperms. As a result of radiation, gall midges belonging to the subfamily Cecidomyiinae have become the most dominant gall inducers among various gall-inducing arthropod taxa. Therefore, we take up the gall midges in this book to show their diverse ecological aspects including interrelations with their host plants and associated arthropods. From the contents of this book, we expect readers can learn something about an ecological way of thinking, which is essential under recent conditions of global environment.

Keywords Higher taxonomy · Species number · Phylogenetic relationship · Gall-inducing cecidomyiid

1.1 Introduction

Cecidomyiidae, commonly known as gall midges, are supposed to be the most speciose family in Diptera (Hebert et al. 2016; Borkent et al. 2018; Brown et al. 2018). The family exhibits diverse feeding habits including those saprophagous, mycophagous, phytophagous, and zoophagous (predacious and endoparasitic)

M. Tokuda (✉)
Laboratory of Systems Ecology, Faculty of Agriculture, Saga University, Saga, Japan
e-mail: tokudam@cc.saga-u.ac.jp

J. Yukawa
Entomological Laboratory, Faculty of Agriculture, Kyushu University, Fukuoka, Japan
e-mail: galling-cecidomyiids@kyudai.jp

J. Yukawa, M. Tokuda (eds.), *Biology of Gall Midges*, Entomology Monographs,
https://doi.org/10.1007/978-981-33-6534-6_1

(e.g. Möhn 1955; Harris 1967, 1973; Yukawa 1971; Gagné 1994, 1995; Skuhravá 1997; Roskam 2005; Jaschhof and Jaschhof 2009, 2013; Gagné and Jaschhof 2017). Among them, most phytophagous species have tight associations with their host plants and induce galls with species-specific shape. As shown in the subsequent chapters, gall inducers are suitable materials to take ecological data in the field and interesting study subjects to clarify interaction with the host plants. In this chapter, we introduce higher taxonomy of gall midges, phylogenetic relationship among subfamilies, and feeding habits of larvae including non-gall inducers.

1.2 Taxonomy

Cecidomyiidae are a dipteran family belonging to Sciaroidea, Bibliomorpha in Nematocera (Gagné and Jaschhof 2017). The family was previously divided into three subfamilies, Lestremiinae, Porricondylinae, and Cecidomyiinae (e.g. Skuhravá 1997; Gagné 1989, 1994). However, the family Cecidomyiidae is currently divided into six subfamilies, Catotrichinae, Lestremiinae, Micromyinae, Winnertziinae, Porricondylinae, and Cecidomyiinae (Gagné and Jaschhof 2017) following the taxonomic revisions of genera that were traditionally placed in Lestremiinae (Jaschhof 2000; Jaschhof and Jaschhof 2009) or in Porricondylinae (Jaschhof and Jaschhof 2013). The subfamily Catotrichinae was regarded as the most basal subfamily of Cecidomyiidae by their morphological features (Jaschhof 2000; Roskam 2005). Based on the morphological features, Jaschhof and Jaschhof (2009, 2013) supposed the phylogenetic relationships of Cecidomyiidae as (Catotrichinae + (Lestremiinae + (Micromyinae + (Winnertziinae (Porricondylinae + Cecidomyiinae))))) (Fig. 1.1) (Gagné and Jaschhof 2017). Subfamilies other than

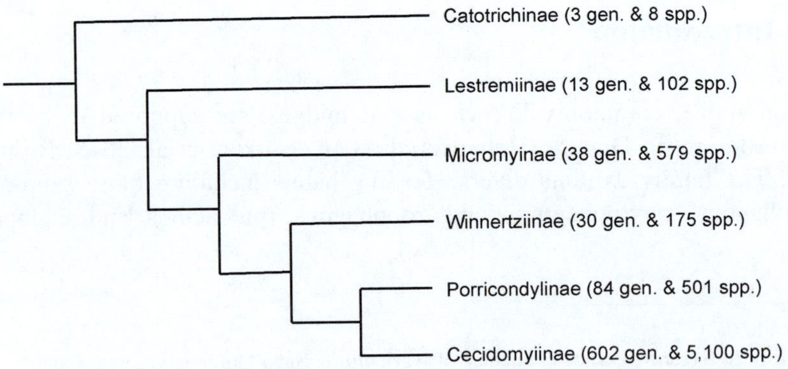

Fig. 1.1 Phylogenetic relationships among different subfamilies of Cecidomyiidae inferred by Jaschhof and Jaschhof (2009, 2013). The number of extant genera and species (Gagné and Jaschhof 2017) were shown in parentheses. Also note that, based on a molecular phylogenetic study by Sikora et al. (2019), Lestremiinae, not Catotrichinae, was situated at the most basal clade of Cecidomyiidae, and Micromyinae, Winnertziinae and Porricondylinae became paraphyletic groups

Cecidomyiinae are fundamentally saprophagous or mycophagous, while Cecidomyiinae, the largest subfamily consisting of approximately 600 genera and 5100 species, exhibits diverse feeding habits including many herbivores (especially gall-inducers) as well as some parasitoids/predators of arthropods and fungus feeders (Gagné and Jaschhof 2017).

Nearly 7000 species and at least 800 genera have been described in Cecidomyiidae (Gagné and Jaschhof 2017). However, the actual number of species seems to be far more abundant. For example, Hebert et al. (2016) investigated Canadian insect species based on DNA barcodes and estimated that Cecidomyiidae comprises 16,000 species among 94,000 Canadian insect species. If Canadian insects account for 1% of the whole insect species in the world, the number of cecidomyiid species can be estimated as 1.8 million worldwide, which exceeds even the number of whole global coleopteran species (estimated as 1.5 million spp. by Stork et al. 2015) (Hebert et al. 2016). The extraordinary diversity of Cecidomyiidae has also been noted in the tropics. Borkent et al. (2018) and Brown et al. (2018) conducted an intensive inventory survey of dipteran species in a tropical cloud forest in Costa Rica, and reported that Cecidomyiidae were the most speciose family in Diptera.

1.3 Phylogeny

In recent years, a couple of comprehensive molecular phylogenetic studies of Cecidomyiidae were published. Sikora et al. (2019) conducted molecular phylogenetic analyses based on nine genetic markers including both nuclear and mitochondrial genes mainly focusing on the relationships among the six subfamilies. Then, they clarified that the present morphology-based concept of higher classification of Cecidomyiidae could be largely supported with some exceptions (Sikora et al. 2019). In the analysis, Lestremiinae, not Catotrichinae, was situated most basally in the family, and Micromyinae, Winnertziinae and Porricondylinae were regarded as paraphyletic groups (Sikora et al. 2019).

Dorchin et al. (2019) focused on the phylogenetic relationships among tribes in Cecidomyiinae and on the evolution of feeding habits in the subfamily. Based on five mitochondrial and nuclear genes, they conducted molecular phylogenetic analyses using 142 species belonging to 88 genera from 13 tribes including species with various feeding habits. As a result, Cecidomyiinae seemed to have radiated in the Upper Cretaceous, coincided with the radiation of angiosperms. Transition from fungus-feeding to plant-feeding habit seemed to have occurred once or twice in the subfamily, and predatory habit seemed to have evolved only once.

1.4 Larval Feeding Habits Other than Gall Induction

Larvae of gall midges that belong to the subfamilies Catotrichinae, Lestremiinae, Micromyinae, Winnertziinae, and Porricondylinae are mycophagous or saprophagous, living in fallen trees or decayed matter. In contrast, larvae of the subfamily Cecidomyiinae exhibit a wide variety of feeding habits as a result of transition from fungus- to plant-feeding habit (Dorchin et al. 2019). For example, some species are mycophagous feeding on rust fungi on living plants (e.g. Yukawa et al. 2018) or on fungal mycelia as a successor in vacated galls of other inducers (e.g. Yukawa and Haitsuka 1994), and some are predators or parasitoids attacking spider mites, aphids, mealybugs, psyllids, and other tiny organisms (e.g. Harris 1967; Abe et al. 2010) (Chap. 13). In addition, larvae of the genus *Cecidomyia* (= *Retinodiplosis*) live in the resin of pines (e.g. Gagné 1978), spruce and fir (Barnes 1951), and those of *Resseliella resinicola* Sanui & Yukawa in the resin of the Japanese cedar (Sanui and Yukawa 1985).

1.5 Gall-Inducing Cecidomyiids

The majority belonging to the subfamily Cecidomyiinae is phytophagous, and many species induce galls on various plant organs and species. Galls are induced by various organisms such as microbes, bacteria, fungi, algae, Nematoda, mites, and insects including Thysanoptera, Hemiptera (Tingidae, Aphidoidea, Psylloidea, Coccoidea), Lepidoptera, Coleoptera, Diptera (Cecidomyiidae, Cyclorrhapha), and Hymenoptera (Tenthredinidae, Chalcidoidea, Cynipidae) (e.g. Mani 1964; Dregger-Jauffret and Shorthouse 1992). Among them, gall midges of the subfamily Cecidomyiinae are the most dominant gall inducers. This book focuses particularly on gall-inducing cecidomyiids, their galls, and host plants.

We provide Chaps. 2–7 with various practical ecological, ethological, evolutionary, and biogeographic data for gall-inducing cecidomyiids based on our field surveys, laboratory experiments, and genetic analyses. Chaps. 9 and 10 refer to the topics of arthropod community and association with host plants, respectively. In Chap. 8, practical methods of field surveys and data analyses are presented. We prepare Chap. 11 for parasitoids, Chap. 12 for invasive pests, and Chap. 13 for beneficial gall midges for applied entomologists. We also refer to various papers on gall-inducing insects published by a world of biologists.

Ecological way of thinking is essential under recent conditions of global environment. We hope readers can learn something about an ecological way of thinking through ecological data obtained from diverse interrelations between gall-inducing cecidomyiids and their host plants. We also expect applied entomologists to be able to obtain useful ideas from some of the Chapters to establish an appropriate method for IPM (Integrated Pest Management) and IBM (Integrated Biodiversity Management) (Kiritani 2000).

References

Abe J, Sato S, Yukawa J (2010) Descriptions of two new endoparasitic cecidomyiids (Diptera: Cecidomyiidae) from Japan. Appl Entomol Zool 46:15–25

Barnes HF (1951) Gall midges of economic importance volume V. Trees. Crosby Lockwood & Son Ltd., London

Borkent A, Brown BV, Adler PH et al (2018) Remarkable fly (Diptera) diversity in a patch of Costa Rican cloud forest: why inventory is a vital science. Zootaxa 4402:53–90

Brown BV, Borkent A, Adler PH et al (2018) Comprehensive inventory of true flies (Diptera) at a tropical site. Communications Biology 1:21

Dorchin N, Harris KM, Stireman JO III (2019) Phylogeny of the gall midges (Diptera, Cecidomyiidae, Cecidomyiinae): systematics, evolution of feeding modes and diversification rates. Mol Phylogenet Evol 140:106602

Dregger-Jauffret F, Shorthouse JD (1992) Diversity of gall-inducing insects and their galls. In: Shorthouse JD, Rohfritsch O (eds) Biology of insect-induced galls. Oxford University Press, New York & Oxford, pp 8–33

Gagné RJ (1978) A systematic analysis of the pine pitch midges, *Cecidomyia* spp. (Diptera: Cecidomyiidae). Tech Bull USDA 1575:1–18

Gagné RJ (1989) The plant-feeding gall midges of North America. Cornell University Press, Ithaca

Gagné RJ (1994) The gall midges of the Neotropical region. Cornell University Press, Ithaca

Gagné RJ (1995) Revision of tetranychid (Acarina) mite predators of the genus *Feltiella* (Diptera: Cecidomyiidae). Ann Entomol Soc Am 88:16–30

Gagné RJ, Jaschhof M (2017) A catalog of the Cecidomyiidae (Diptera) of the world, 4th edn. Digital. [Cited 3 August 2020] Available from URL. https://www.ars.usda.gov/ARSUserFiles/80420580/Gagne_2017_World_Cat_4th_ed.pdf

Harris KM (1967) A systematic revision and biological review of the cecidomyiid predators (Diptera: Cecidomyiidae) on world Coccoidea (Hemiptera-Homoptera)

Harris KM (1973) Aphidophagous Cecidomyiidae (Diptera): taxonomy, biology and assessments of field populations. Bull Entomol Res 63:305–325

Hebert PDN, Ratnasingham S, Zakharov EV et al (2016) Counting animal species with DNA barcodes: Canadian insects. Philos Trans B 371:20150333

Jaschhof M (2000) Catotrichinae subfam. N.: a re-examination of higher classification in gall midges (Diptera: Cecidomyiidae). Entomol Sci 3:629–652

Jaschhof M, Jaschhof C (2009) The wood midges (Diptera: Cecidomyiidae: Lestremiinae) of Fennoscandia and Denmark. Stud Dipterol Suppl 18:1–333

Jaschhof M, Jaschhof C (2013) The Porricondylinae (Diptera: Cecidomyiidae) of Sweden, with notes on extralimital species. Stud Dipterol Suppl 20:1–392

Kiritani K (2000) Integrated biodiversity management in paddy fields: shift of paradigm from IPM toward IBM. Integr Pest Manag Rev 5:175–183

Mani MS (1964) Ecology of plant galls. Dr. W. Junk, The Hague

Möhn E (1955) Beiträge zur Systematik der Larven der Itonididae (= Cecidomyiidae, Diptera). 1. Teil: Porricondylinae und Itonidinae Mitteleuropas. Zoologica 105:1–247

Roskam JC (2005) Phylogeny of gall midges (Cecidomyiidae). In: Raman A, Schaefer CW, Withers TM (eds) Biology, Ecology, and Evolution of Gall–inducing Arthropods, vol 1. Science Publications, Enfield, pp 305–319

Sanui T, Yukawa J (1985) A new gall midge of the genus *Resseliella* (Diptera: Cecidomyiidae) inhabiting resin of the Japanese cedar, *Cryptomeria japonica* (Taxodiaceae). Appl Entomol Zool 20:27–33

Sikora T, Jaschhof M, Mantič M, Kaspřák D, Ševčík J (2019) Considerable congruence, enlightening conflict: molecular analysis largely supports morphology-based hypotheses on Cecidomyiidae (Diptera) phylogeny. Zool J Linnean Soc 185:98–110

Skuhravá M (1997) Family Cecidomyiidae. In: Papp L, Darvas B (eds) Contributions to a manual of Palaearctic Diptera (with special reference to flies of economic importance). Nematocera and lower Brachycera, vol 2. Science Herald, Budapest, pp 71–204

Stork NE, McBroom J, Gely C, Hamilton AJ (2015) New approaches narrow global species estimates for beetles, insects and terrestrial arthropods. Proc Natl Acad Sci USA 112:7519–7523

Yukawa J (1971) A revision of the Japanese gall midges (Diptera: Cecidomyiidae). Mem Fac Agric, Kagoshima Univ 8:1–203

Yukawa J, Haitsuka S (1994) A new cecidomyiid successor (Diptera) inhabiting empty midge galls. Jpn J Entomol 62:709–718

Yukawa J, Tokuda M, Uechi N, Ganaha-Kikumura T, Teshiba M (2018) Host and distribution ranges of some species of *Mycodiplosis* Rübsaamen, 1895 (Diptera: Cecidomyiidae) feeding on rust fungi, *Puccinia* spp. (ascomycetes) in Japan, Thailand and Vietnam. Acta Dipterol 29:1–7

Part II
Speciation and Adaptive Radiation

Chapter 2
Speciation

Junichi Yukawa, Makoto Tokuda, Shinsuke Sato, Tomoko Ganaha-Kikumura, and Nami Uechi

Abstract Most gall-inducing cecidomyiids are associated closely with particular plant taxa. Many examples of species radiation on a single or a few plant genera are known for mono- or oligophagous cecidomyiid genera. In the light of these examples, we consider that gall midges have diversified on a particular plant taxon under various sympatric conditions in addition to allopatric speciation. In this chapter, we show some of possible mechanisms leading to non-allopatric speciation. First, we demonstrate two case studies of mistaken oviposition. Then we refer to plant polyploidy, gall shape polymorphism including a geographic mosaic of coevolution, host range expansion, sexual isolation, ecological divergence, and host race formation. We also review two examples of allopatric speciation.

Keywords Ecological divergence · Gall shape polymorphism · Host race · Host range expansion · Mistaken oviposition · Plant polyploidy · Sexual isolation

J. Yukawa (✉)
Entomological Laboratory, Faculty of Agriculture, Kyushu University, Fukuoka, Japan
e-mail: galling-cecidomyiids@kyudai.jp

M. Tokuda
Laboratory of Systems Ecology, Faculty of Agriculture, Saga University, Saga, Japan
e-mail: tokudam@cc.saga-u.ac.jp

S. Sato
Horticultural Institute, Agricultural Center, Ibaraki, Japan
e-mail: shi_satou@pref.ibaraki.lg.jp

T. Ganaha-Kikumura
Okinawa Prefectural Central Wholesale Market, Okinawa, Japan
e-mail: kikumurt@pref.okinawa.lg.jp

N. Uechi
Institute of Plant Protection, National Agriculture and Food Research Organization, Ibaraki, Japan
e-mail: uechinmi@affrc.go.jp

J. Yukawa, M. Tokuda (eds.), *Biology of Gall Midges*, Entomology Monographs,
https://doi.org/10.1007/978-981-33-6534-6_2

2.1 Introduction

Among gall-inducing cecidomyiids listed in Gagné and Jaschhof (2017), most species are known from one plant species (tentatively defined as monophagous), some are from one plant genus or one plant family (oligophagous), and a few are from different plant families (polyphagous). Thus, most gall-inducing cecidomyiids are associated closely with particular plant taxa. Many examples of species radiation on a single or a few plant genera are known for mono- or oligophagous cecidomyiid genera such as *Caryomyia* on *Carya* (Jugulandaceae) (Gagné 2008), *Celticecis* on *Celtis* (Cannabaceae) (Gagné and Moser 2013), *Psectrosema* on *Tamarix* (Tamaricaceae), *Rabdophaga* on *Salix* (Salicaceae) and *Rhopalomyia* on *Artemisia* and related genera belonging chiefly to the tribe Anthemideae (Asteraceae) (e.g. Yukawa et al. 2005; Joy and Crespi 2007; Gagné and Jaschhof 2017). These data indicate that a cecidomyiid species radiated by altering gall shape, for example, as a result of galling-position shift after its arrival on a particular plant species. A good example has been demonstrated for North American *Asphondylia* species on creosote bush *Larrea tridentata* (de Candolle) Coville (Zygophyllaceae) (Gagné and Waring 1990; Joy and Crespi 2007). After radiation on a particular plant species, some gall midge species might expand their host range to congeneric plant species or to some related plant genera as has been noted for oligophagous *Rhopalomyia* species (e.g. Gagné and Jaschhof 2017).

Various factors have been proposed and discussed to explain such adaptive radiation of gall midge species. Besides allopatric speciation, we consider that gall midges have diversified under various sympatric conditions. This chapter demonstrates some of possible mechanisms leading to non-allopatric speciation such as mistaken oviposition, plant polyploidy, gall shape polymorphism including a geographic mosaic of coevolution, host range expansion, sexual isolation, ecological divergence, and host race formation. We also refer to two examples of allopatric speciation.

2.2 Mistaken Oviposition Leading Host Shift and Host Range Expansion

2.2.1 *Importance of Mistaken Oviposition*

Various arguments have been repeated about mechanisms of ecological speciation (e.g. Schluter 2000, 2001, 2009; Via 2001; Price 2003; Rundle and Nosil 2005; Nosil 2012), but factors that act as triggers for speciation have seldom been discussed. Mistaken oviposition on nonnatal host plant species must be one of the possible triggers. Mistaken oviposition may be a plausible case to open up a new adaptive zone for the species, in that shared leaf volatiles, rather than similar contact stimulants or visual cues, would play a role in triggering the mistakes (Berenbaum and Feeny 2008). Prior to species radiation on a particular plant, initial host shift to a novel plant by pioneer females is essential. They need to overcome various chemical barriers for oviposition and their progeny for feeding, particularly when the novel plant is genetically quite different from the natal host plant. Host range expansion to congeneric plant species or to related plant genera within the same family is also promoted by mistaken oviposition. In this case, the chemical barriers may be relatively low between related plant species or genera.

Mistaken oviposition has been observed for various taxa of herbivorous insects, such as lepidopteran (e.g. Yukawa et al. 2019b) and dipteran species (e.g. Larsson and Ekbom 1995). Larsson and Ekbom (1995) considered that gall-inducing cecidomyiids are likely to be victims of confusion in host selection, and thereby exhibit a non-discriminatory oviposition behavior.

The next question is what are triggers for promoting mistaken oviposition. There may be some sorts of trigger, but we need actual field data to support such prospects. However, evidences of mistaken oviposition by gall midges on genetically distant novel plants are hardly detected because the species identification of tiny ovipositing females is impossible in the field. In contrast, many instances of mistaken oviposition by butterflies on different plant families have been observed frequently in the field (e.g. Yukawa et al. 2019b). In this section, therefore, one of us, J. Yukawa provides two case studies of possible trigger for mistaken oviposition that promotes host range expansion to congeneric plant species by two gall-inducing cecidomyiids. The first example refers to a shortage of oviposition sites related to population density and abundance of host shoots, and the second to synchronization of adult emergence season with unexpected leaf-opening timing of a nonnatal congeneric plant species.

2.2.2 Exception to Mistaken Oviposition

Many studies have demonstrated preference-performance linkage exhibited by herbivorous insects but in some cases, females cannot discriminate susceptible and resistant plant genotypes despite the performance of their progeny is much better on the former (e.g. Hellqvist and Larsson 1998). Such indiscrimination is not included here in the definition of mistaken oviposition.

Females of *Masakimyia pustulae* Yukawa & Sunose (Diptera: Cecidomyiidae) lay their eggs on fresh leaves of *Euonymus japonicus* Thunberg (Celastraceae) for gall formation. They frequently lay some of their eggs also on 1- or 2-year old host leaves and on leaves of various understory plants including ferns but galls have never developed on these leaves (Yukawa and Sunose 1976). Nevertheless, such useless oviposition can be seen as a usual event every spring. They may not have an ability to select leaves of its proper host plant. Otherwise there may be an adaptive significance, such as a decoy against egg-larva parasitoids although not confirmed yet. This oviposition behavior cannot be included in the definition of mistaken oviposition.

2.2.3 *Case Study 1: Mistaken Oviposition by* **Pseudasphondylia neolitseae** *on* **Neolitsea** *spp.*

2.2.3.1 Life History and Daily Activity of *P. neolitseae*

Pseudasphondylia neolitseae Yukawa is a monophagous species forming leaf galls on *Neolitsea sericea* (Blume) Koidzumi (Lauraceae) (Fig. 2.1a) and is widely distributed in Japan (Honshu, Shikoku, Kyushu, and the Ryukyus) (Yukawa 1974). Each gall contains one gall midge larva. This gall midge is normally univoltine and passes the winter in the gall as a third instar. Pupation takes place in February or March, and adults emerge in late March or April in Kagoshima City, southern Kyushu, Japan. Females lay their eggs in the leaf buds (Yukawa 1974, 1983). A few individuals require 2 years to complete one generation (Takasu and Yukawa 1984).

Fig. 2.1 Photographs of gall midges, their galls, and host plants mentioned in Sect. 2.2. (**a**), galls induced by *Pseudasphondylia neolitseae* on a leaf of *Neolitsea sericea*; (**b**), females of *P. neolitseae* ovipositing in a leaf bud of *N. sericea*; (**c**), leaf buds of *Neolitsea aciculata*; (**d**), a female of *P. neolitseae* ovipositing in a leaf bud of *N. aciculata*; (**e**), eggs of *P. neolitseae* oviposited in a leaf blade of *N. aciculata*; (**f**), a female of *P. neolitseae* locating on a leaf bud of *Machilus thunbergii*; (**g**), normal galls of *Daphnephila machilicola* on under leaf-surface of *Machilus thunbergii*; (**h**), galls induced by *D. machilicola* on under-leaf surface of *Machilus japonica* (left: normal type, right: swollen type); (**i**), a gall induced by *D. machilicola* on under leaf-surface of *M. japonica*

According to Yukawa et al. (1976), both males and females emerge between 07:30 and 10:00 in late March or April. Males exhibit swarming around the host tree from which they emerged. After emergence, the females locate on the host leaves with galls from which they emerged and wait for the males to come. Some of swarming males fly towards females in response to pheromonal calling. Oviposition starts from 10:00 and lasts until 22:00, reaching a peak of oviposition at 15:00. They deposit their eggs in the terminal leaf buds of *N. sericea* (Fig. 2.1b). Leaf buds suitable for oviposition are those with a length ranging from 22 to 38 mm. The adult life span is several hours in males and about 12 h in females, and the female realizes 189 of 248 ovarian eggs on an average in their life.

2.2.3.2 Field Survey

From 1972 to 1996, J. Yukawa and his students at Kagoshima University studied the long-term population dynamics of *P. neolitseae* in evergreen broad-leaved forests in Kagoshima City (Yukawa and Akimoto 2006; Yukawa et al. 2018a, b, 2018a, b). During the 24 year-field surveys, we had an opportunity in May 1981 of finding several small galls on current leaves of *Neolitsea aciculata* (Blume) Koidzumi (Lauraceae) at Terayama Station for Education and Research on Nature, Faculty of Education, Kagoshima University (hereafter Terayama field station). We suspected that the galls were caused by mistaken oviposition by *P. neolitseae* on leaf buds of *N. aciculata* instead of *N. sericea*. The galls did not develop thereafter, and larvae seemed to die in the galls.

Sometimes *N. aciculata* grows nearby *N. sericea* in the forest. We can easily distinguish the two tree species because leaves are smaller and leaf buds are a little shorter and distinctly slenderer in *N. aciculata* (Fig. 2.1c) than in *N. sericea* (Fig. 2.1b, c). A question was what were triggers for *P. neolitseae* females make mistaken oviposition.

To answer this question, we observed the oviposition behavior of *P. neolitseae* females in April 1982 at Terayama field station and surveyed the maximum number of females locating on one host leaf bud at the same time to confirm their interference, if any, in relation to the number of leaf buds available for the females. We also attempted a preliminary test to examine their chemical reaction to leaf buds of non-host species, *N. aciculata* and *Machilus thunbergii* Siebold & Zuccarini (Lauraceae) by applying squeeze of leaf buds of *N. sericea* to those of non-hosts.

2.2.3.3 Mistaken Oviposition Observed

In the field, usually one female located on one host leaf bud and sometimes two were co-existing on the same bud (Fig. 2.1b). When the third female came to the bud, one of the three females exclusively left the bud because of interference among them. Such a female usually tried to locate again on a host leaf bud on another twig or tree if available near the original place. In addition to the interference, ovipositing females occasionally left the buds due to disturbance by other organisms such as ants and spiders walking the buds for hunting (Yukawa 1983).

During the daytime, no females of *P. neolitseae* landed on untreated leaf buds of *N. aciculata* and *M. thunbergii* when we approximated the buds to females flying around the host leaf buds. In contrast, the females landed without hesitation on non-host leaf buds that had been applied with squeeze of leaf buds of *N. sericea*. Immediately after landing, the females started and continued to lay eggs into the treated leaf buds of *N. aciculata* (Fig. 2.1d). Later, we confirmed existence of eggs and oviposition scars in the buds (Fig. 2.1e). Females landed on *M. thunbergii* walked the buds for a while as if they tried to measure the size of buds. A few females inserted their ovipositor into the buds (Fig. 2.1f) but soon they flew away probably because of thicker bud diameter in *M. thunbergii* than in the host buds. No eggs and oviposition scars were found in the buds.

After 17:30, however, we rarely observed that a few females that fall out of the host buds located and oviposited on leaf buds of *N. aciculata* growing near by the original place (Fig. 2.1d). By dissecting two of these *N. aciculata* buds, we found eggs and oviposition scars on the fresh leaves folded in the bud (Fig. 2.1e). Nevertheless, initial small galls did not appear thereafter on the fresh leaves opened from the remaining oviposited three buds. If initial leaf galls appeared, they did not develop normally.

2.2.3.4 Discussion

These results suggest that mistaken oviposition is caused by females that fell out of the host buds after interference among conspecific females. The interference is related to the relative abundance of females to the amount of host leaf buds available for oviposition, which depends on population dynamics of the gall midge and the annual changes in the number of leaf buds produced by the host tree. In addition, an availability of host buds is also influenced by the degree of synchronization between adult emergence season and the timing of host bud extension (buds suitable for oviposition with a length ranging from 22 to 38 mm) (Yukawa et al. 1976). The emergence season and degree of synchronization vary from year to year depending on climatic conditions (Yukawa and Akimoto 2006) (see also Sect. 10.2). Thus, the shortage of host buds is caused by biotic and abiotic factors and possibly promotes mistaken oviposition.

In our preliminary test, *P. neolitseae* females reacted to squeeze of leaf buds of *N. sericea* and landed on the leaf buds of *N. aciculata* and *M. thunbergii*, indicating that some chemical compounds included in the primary host buds act as oviposition stimulants for the females. Their mistaken oviposition did not seem to be exceedingly rare, but galls did not develop normally, and the first instars died in the galls. Therefore, we consider that *P. neolitseae* may not yet have a potential ability to overcome chemical barriers to develop on recently encountered novel plant species.

We also noted that the females that fell out of the host buds after interference never mistook in finding their proper host plant before evening. We observed mistaken oviposition in the evening when their ovarian eggs largely decreased. We speculate that females can search for the exact host plant at the beginning of oviposition, but they may lose delicate sensibility to distinguish oviposition stimulants between closely related congeneric plant species according to aging. Of course, we need further investigation to confirm this speculation.

2.2.4 *Case Study 2: Mistaken Oviposition by* Daphnephila *spp. on* Machilus *spp.*

2.2.4.1 *Daphnephila machilicola* and Its Allied Species

Daphnephila machilicola Yukawa is a monophagous species forming barrel-shaped galls on the under leaf-surface of *M. thunbergii* and widely distributed in Japan (Honshu, Shikoku, Kyushu, and the Ryukyus) (Yukawa 1974). Mean height of mature gall is 6.9 mm and mean maximum diameter is 5.0 mm (Fig. 2.1g). Each gall contains one gall midge larva. This gall midge is normally univoltine and passes the winter in the gall as second instars. Pupation takes place in April, and adults emerge in May and June in Kagoshima City. Females lay their eggs on under surface of fresh leaves and galls start growing in autumn, several months after larval hatching in spring (Yukawa 1974). A few individuals enter extended diapause at the first stadium, requiring 2 or 3 years to complete one generation (Maeda et al. 1982).

Daphnephila sp. is monophagous, forming angular, pear-shaped or club-shaped galls on the under leaf surface of *Machilus japonica* Siebold & Zuccarini ex Blume (Lauraceae) (Yukawa 1974) and widely distributed in Japan (Shikoku, Kyushu, Tanegashima, Yakushima, and the Southwest Islands) (Yukawa 1974; Yamauchi et al. 1982). Mean height of mature gall is 5.9 mm and mean maximum diameter is 2.5 mm (Fig. 2.1h). Each gall contains one gall midge larva. Adults emerge in April and May. When Yukawa (1974) described *D. machilicola*, this species was left unnamed to be *Daphnephila* sp. because they were morphologically remarkably similar to each other, but gall shapes and emergence seasons were distinctly different between them. Of course, more recently, genetic analysis revealed that they are distinct species (Tokuda and Yukawa unpublished data).

2.2.4.2 Host-Exchanging Experiments

In the 1970s, we did not have idea and tools of genetic analysis. Therefore, J. Yukawa and his students at Kagoshima University tried to examine the host range of *D. machilicola* by host-exchanging experiments. In late March 1977, we collected twigs of *M. thunbergii* with leaves of *D. machilicola* galls from Terayama field station. We kept them in an incubator at 20 °C and LD 13:11 (lighting conditions) for 3 weeks to make adults emerge earlier than usual to synchronize the emergence with the leaf-opening season of *M. japonica* that is about 1 month earlier than that of *M. thunbergii*. Then, we put the twigs into a breeding cage ($1 \times 1 \times 1$ m) and allowed the adults to emerge, swarm and mate. Some mated females were released in an insect net that covers shoots with half opened leaves of one of the *M. japonica* trees that were growing in a nursery of forest trees on the Kagoshima University campus.

In March 1978, we found two barrel-shaped galls on the under leaf-surface of the *M. japonica* tree (Fig. 2.1h). One was apparently a normal shaped gall that can be seen on *M. japonica*. The other gall was short but swollen and suggestive of *D. machilicola* influence on some extent. This sort of galls has never been observed previously on *M. japonica*. This result indicated that a female of *D. machilicola* might lay her eggs on the underside of a half-opened leaf of the *M. japonica* tree in the experiment. In May 1978, the third-stadium larva was found to be dead in the gall. This experiment may also indicate that *Machilus* plants have more influential power than *Daphnephila* gall midges in determining gall shape in this particular case.

2.2.4.3 Field Survey of Chronological Isolation Between the Two *Daphnephila* Species

In addition to the host-exchanging experiment, we surveyed the degree of chronological isolation in adult emergence season between *D. machilicola* on *M. thunbergii* from 1977 to 1980 and *Daphnephila* sp. on *M. japonica* from 1977 to 1995 in a mixed forest of evergreen broad-leaved trees and Moso bamboo, *Phyllostachys edulis* (Carrière) J. Houz (Poaceae), in Ono, Kagoshima City. We also surveyed, for each species, the degree of synchronization between the emergence season and leaf-opening season of the host plant.

For these purposes, we selected four census trees of *M. thunbergii* (sun tree; 130–195 cm in height) and five census trees of *M. japonica* (shade tree; 86–144 cm in height). The daily emergence of adult gall midges was counted every day or at intervals of 3–7 days during their emergence seasons by recording the number of exit holes opened by the gall midges on the tip of galls. The number of host plant shoots with half opened fresh leaves that are suitable for oviposition was recorded for *M. thunbergii* and *M. japonica* at intervals of 3–10 days.

In September 1982 at Terayama field station, a flurry of wind broke down some branches of canopy trees, such as *Castanopsis sieboldii* (Makino) Hatusima ex T. Yamazaki & Mashiba (Fagaceae) and *Cinnamomum camphora* (Linnaeus) J. Presl (Lauraceae). As a result, a small gap appeared in an evergreen broad-leaved forest and some *M. japonica* trees were exposed to direct sunshine for at least 3 h a day. To investigate the influence of direct sunshine on the leaf-opening phenology of the shade tree, we surveyed in 1983 and 1984 changes in the number of shoots with leaves opened for *M. japonica* trees in shade, those under sunshine, and trees of *M. thunbergii* under sunshine. In April 1984, we searched galls induced by larvae that had derived from eggs deposited on these trees during the emergence season in April to June 1993.

2.2.4.4 Results of Field Survey

Figure 2.2 shows selected emergence curves of the two *Daphnephila* species and seasons when half opened host leaves were available for oviposition. Emergence curves were not available for comparison in some years because the total number of

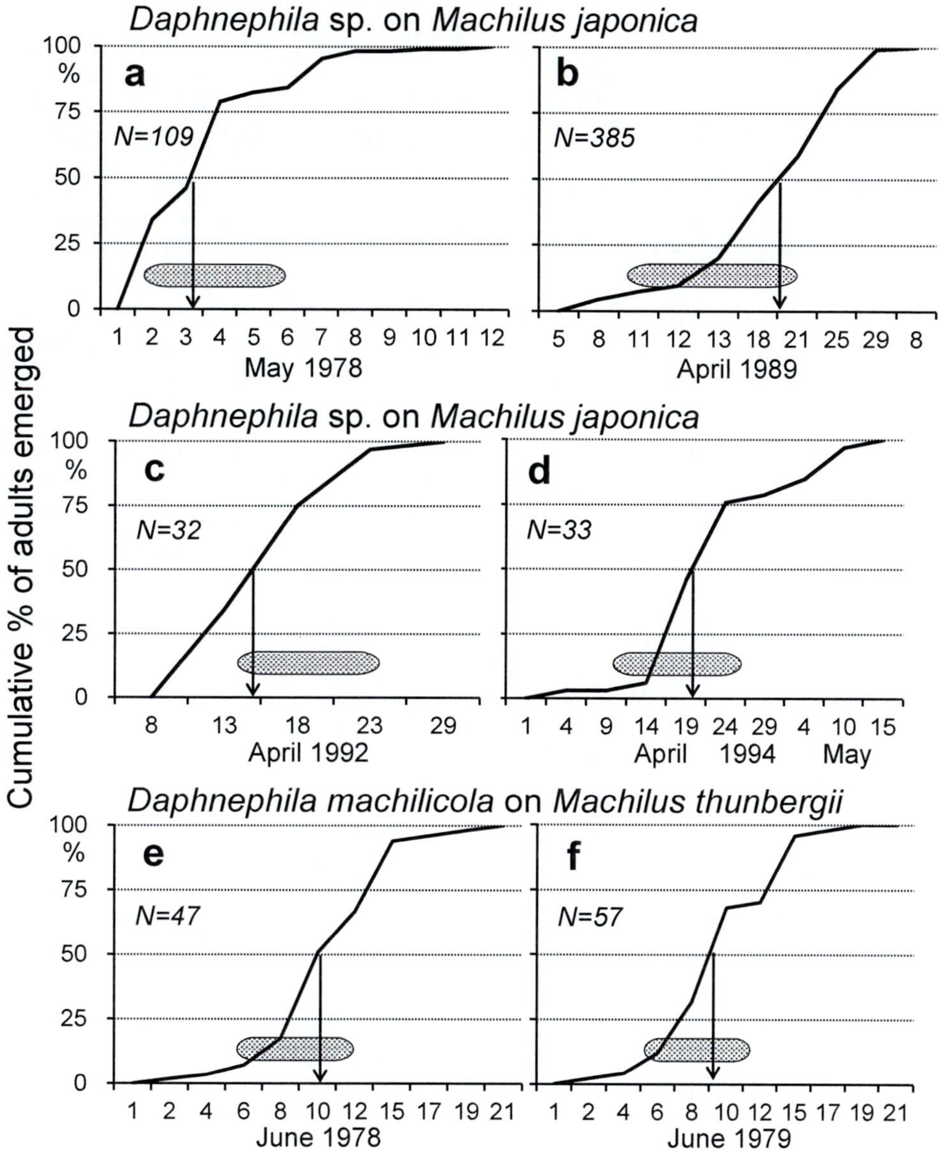

Fig. 2.2 Selected emergence curves of two *Daphnephila* species that indicate their 50% emergence date and synchronization with host plant phenology. (**a–d**), *Daphnephila* sp. on *Machilus japonica*; (**e**, **f**), *Daphnephila machilicola* on *Machilus thunbergii*. Arrows point toward 50% emergence date. Dotted areas schematically indicate the period in which young leaves were suitable for oviposition

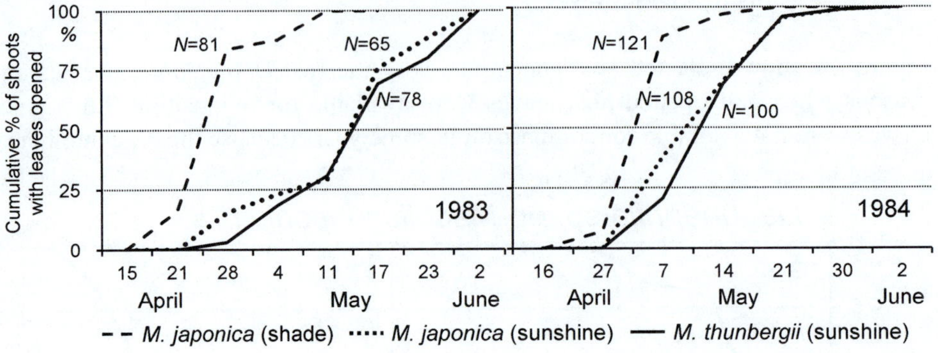

Fig. 2.3 Comparison in the leaf opening curves between *Machilus japonica* and *Machilus thunbergii* growing in shade and under sunshine at Terayama field station, Kagoshima City, Japan. *N* indicates the number of shoots examined

adults emerged was reduced to less than 30 mainly by the parasitism of *Platygaster* sp. (Platygastridae), *Gastrancistrus* spp. (Pteromalidae), *Bracon tamabae* Maeto, and *Simplicibracon curticaudis* Maeto (Braconidae). *Daphnephila* sp. emerged from the mid-April to early May (Fig. 2.2a–d) about 1 month earlier than *D. machilicola* that emerged in June (Fig. 2.2e, f). Because adult life span of gall-inducing cecidomyiids is short (1–3 days in the field) (e.g. Yukawa et al. 1976, 2012, 2013), the 1-month time lag is long enough for chronological sexual isolation between the two species. Emergence seasons of the two species more or less synchronized well with respective host available seasons, which means that the gall midges have adapted their life history traits to synchronize with host plant phenology.

At Terayama field station, we also confirmed in 1983 and 1984 that a distinct time lag existed in the leaf-opening season between *M. japonica* (growing in shade) and *M. thunbergii* (Fig. 2.3) as has been found for these trees in Ono (Fig. 2.2). However, the leaf-opening season of *M. japonica* trees that were growing under sunshine in the forest gap delayed 2–3 weeks and almost overlapped with that of *M. thunbergii* (Fig. 2.3). In 1983, we found many oviposition scars (green spots) on the under surface of fresh leaves of *M. japonica* (in shade) and *M. thunbergii* and in April 1984 a total of 170 and 56 mature galls appeared on 1-year old leaves of one of these trees, respectively. In contrast, a few oviposition scars were detected in 1983 on fresh leaves of *M. japonica* trees in the gap, indicating that *D. machilicola* females seldom laid eggs on *M. japonica* even though fresh leaves were available for oviposition. However, a small barrel-shaped gall was found in April 1984 on a 1-year old leaf of one of these trees (Fig. 2.1i).

2.2.4.5 Discussion

Our field surveys revealed that populations of the two *Daphnephila* species have been chronologically isolated and the emergence season of each species synchronized well with the season when its own host leaves were available for oviposition. Under such conditions, mistaken oviposition would not occur. However, a flurry of wind provided females of *D. machilicola* with unusual occasions to encounter a non-primary host plant, *M. japonica*. Such occasions seemed to lead some females to mistaken oviposition possibly because chemical oviposition stimulants of *M. japonica* fresh leaves are similar to those of *M. thunbergii* fresh leaves. Oviposition scars indicates that a few *D. machilicola* females oviposited on fresh leaves of *M. japonica* trees growing under the sunshine.

2.2.4.6 General Remarks

In this section, we provided two examples of possible trigger for mistaken oviposition that promotes host range expansion to congeneric plant species. The first example of trigger was caused by a biotic factor, shortage of oviposition sites that is primarily related to the population density of *P. neolitseae* and abundance of *N. sericea* shoots. The population density of *P. neolitseae* and its host abundance fluctuate from year to year, being influenced by various biotic and abiotic factors (Yukawa and Akimoto 2006; Yukawa et al. 2018a, b; Yukawa et al. 2019a). In the combination of high population density and low host availability, the shortage of oviposition sites occurs occasionally and promotes mistaken oviposition.

The second example of trigger was caused primarily by an abiotic factor, a flurry of wind that broke down some branches of canopy trees, resulted in the gap formation in the forest. The emergence season of *D. machilicola* synchronized unexpectedly with the delayed leaf-opening timing of a nonnatal congener, *M. japonica* that was growing under sunshine through the gap. In contrast to the relation between *P. neolitseae* and *N. aciculata*, females of *D. machilicola* lay their eggs of their own accord on the leaves of *M. japonica* without any external forces and some galls appeared although somewhat different in shape (Fig. 2.1i) from its own galls on the natal host, *M. thunbergii* (Fig. 2.1g). Opportunity would not be extremely limited for females of *D. machilicola* to encounter delayed leaf-opening timing of *M. japonica* because gap formation can be seen rather commonly in southern Japan where typhoons frequently attack (Yukawa et al. 2019a).

These examples indicate that mistaken oviposition is not rare and that chemical barriers for ovipositing, feeding, and inducing leaf galls on the nonnatal congeners are relatively lower. Actually *P. neolitseae* and *D. machilicola* oviposited on nonnatal congeneric plants and galls developed to some extent on *N. aciculata* and to maturity on *M. japonica* although the shape was not normal.

The 50% emergence date of gall midges can be predicted by the accumulation of temperature, while the 50% date of budburst cannot be predicted by temperature, because warm winter may delay the budburst or flowering season of host plants (Yukawa et al. 2013). Therefore, adult emergence does not always synchronize well with host plant phenology. Particularly under global-warming conditions, the emergence would become earlier and budburst or flowering season later than usual. Such an asynchrony enforces, for gall midges, inevitable host range expansion and host shift to a congeneric plant species, or even to plants across different families. No one knows when the progeny of a pioneer individual can establish on a novel plant species, but multiple trials may it be realized very long afterward. The species diversification of *Daphnephila* on *Machilus* species in Taiwan (Tokuda et al. 2008) indicates that such events actually occurred.

2.3 Plant Polyploidy That Possibly Leads Speciation of Gall Midges

2.3.1 Plant Polyploidy

Plant speciation has been known to initiate frequently by chromosomal changes (e.g. Wendel 2000; Levy and Feldman 2002; Liu and Wendel 2002; Ainouche et al. 2003; Osborn et al. 2003; Soltis and Soltis 2003). Mayrose et al. (2011) and Arrigo and Barker (2012) stated that ‘despite leaving a substantial legacy in plant genomes, only rare polyploids survive over the long term and most are evolutionary dead-ends’. Soltis et al. (2014) agreed with the long-espoused view that most polyploid entities likely go extinct shortly after formation, at the scale of small populations, before becoming established as evolutionarily significant entities or receiving taxonomic recognition. In that sense, polyploids are often ‘dead-ends’. However, Mayrose et al. (2011) and Arrigo and Barker (2012) addressed a timescale different from this initial phase of polyploidization. Then, Soltis et al. (2014) argued their statements by pointing out that there are important philosophical, statistical, analytical, sampling, and methodological issues with their approach to the problem. Wendel (2015) provided a figure showing the processes and patterns of polyploidy of plants and demonstrated that all modern flowering plant genomes derive from processes set in motion by a history of repeated, episodic whole-genome doubling, or polyploidy.

From the side of gall-inducing insects, a question arises whether or not they can discriminate host plant polyploidy at the time of oviposition. If not, they continuously use polyploid entities unconsciously and long afterward they will become an oligophagous species inducing a similar sort of gall on two or more congeneric plants if the entities would not go extinct before speciation. Life on different plant species may facilitate ecological diversification of the insects because environmental factors derived from different plant species will operate differently on insect populations, leading to speciation of gall-inducing insects. Up to the present, we have seen some gall-inducing cecidomyiids that use polyploid entities of the host plants for gall induction. It is an interesting study subject to investigate whether or not gall midges can discriminate polyploidy among their host plant individuals or populations in the course of host range expansion.

2.3.2 *Examples of Cecidomyiids That Induce Galls on Host Plants with Polyploidy*

Neolitsea sericea is a medium-sized evergreen tree up to 10 m tall and is widely distributed in Japan (Honshu, Shikoku, Kyushu and the Ryukyus), South Korea, Taiwan, and mainland China (e.g. Yukawa et al. 2019a). It is an only host plant of *P. neolitseae* (Yukawa 1974). All trees examined were diploid, except one triploid individual that grows in Takamori, Kumamoto Prefecture and seems to be a natural polyploid (Mishima and Yukawa 2007). Leaf galls of *P. neolitseae* were found on the triploid individual as well as on those diploids.

Aucuba japonica Thunberg (Garryaceae), an evergreen dioecious shrub, is widely distributed in the temperate forest zone from southwestern Hokkaido to Okinawa, Japan, in northernmost part of Taiwan and on the Ulleung-do and Jeju Island of South Korea (e.g. Ohi et al. 2003). By examining ploidy levels and cpDNA haplotypes throughout the whole geographic range of *A. japonica*, Ohi et al. (2003) revealed that there is a clear geographic differentiation between diploids in the southwestern part and tetraploids in the northeastern part of the Japanese Archipelago. *Asphondylia aucubae* Yukawa & Ohsaki has been known to induce

Fig. 2.4 Cecidomyiid galls induced on host plants with polyploidy. (**a**), fruit galls of *Asphondylia aucubae* on *Aucuba japonica* (reproduced with permission from Yukawa et al. 2016a); (**b**), a stem gall of *Lasioptera euphobiae* on *Eupatorium makinoi*; (**c**), leaf galls of *Cystiphora taraxaci* on *Taraxacum officinale*; (**d**), leaf galls of *Rhopalomyia cinerarius* on *Artemisia indica*; (**e**), leaf galls of *R. yomogicola* on *A. indica*; (**f**), axillary bud gall of *R. caterva* on *A. lancea*

fruit galls (Fig. 2.4a) both on diploid and on tetraploid individuals (Yukawa and Ohsaki 1988a; Uechi et al. 2002).

Eupatorium makinoi T. Kawahara & Yahara (Asteraceae) has well been known to exhibit polyploidy in Japan (e.g. Ito et al. 2000). On this plant, *Lasioptera euphobiae* Shinji had been known to form stem galls (Fig. 2.4b) in Hokkaido and northern Honshu, Japan until 1996 (Yukawa and Masuda 1996). Thereafter, it was detected in central and southwestern Honshu (Yukawa et al. 2014, 2018a, b, 2020), which means that *L. euphobiae* has been utilizing polyploid entities of *E. makinoi* for gall-induction in various localities within their distribution range.

European dandelion *Taraxacum officinale* Weber ex FH Wiggers (Asteraceae) has many hundred apomictic microspecies and has a great variety of chromosome numbers characterized by the occurrence of a polyploid series of 2× (2n = 16) to 6× (2n = 48) in addition to several aneuploids worldwide (e.g. Fedorov 1969; Richards 1973). A cecidomyiid, *Cystiphora taraxaci* (Kieffer) induces leaf galls on *T. officinale* (Fig. 2.4c) chiefly in Europe (Gagné and Jaschhof 2017) possibly without recognition of polyploid individuals.

2.3.3 Rhopalomyia *Species on Polyploid Entities of* Artemisia

There are about 270 species of the genus *Rhopalomyia* (Diptera: Cecidomyiidae) in the world and most species in this genus form galls on *Artemisia* and other genera (Asteraceae), chiefly of the tribe Anthemideae (Gagné and Jaschhof 2017). From Japan, at least 14 species of *Rhopalomyia* have been recorded, and 12 of them induce species-specific sorts of gall on five species of *Artemisia,* three of *Dendranthema,* and one of *Chrysanthemum* (Asteraceae) (Yukawa and Masuda 1996; Yukawa 2014). Among the five *Artemisia* species, *Artemisia indica* var. *maximowiczii* (Nakai) H. Hara (=*princeps*), *Artemisia montana* (Nakai) Pampanini, and *Artemisia lancea* Vaniot belong to the section Artemisia, *Artemisia capillaris* Thunberg and *Artemisia japonica* Thunberg to the section Dracunculus. The remaining two species, *Rhopalomyia callicarpae* Shinji on *Callicarpa* spp. (Lamiaceae) and *Rhopalomyia ilexifoliae* Shinji on *Ilex serrata* Thunberg ex Murray (Aquifoliaceae) should be excluded from *Rhopalomyia* because their host plants are not included in Asteraceae (Yukawa 2014).

Artemisia species frequently consist of individuals with a series of polyploidy in natural populations (e.g. Pellicer et al. 2011). To examine whether or not *Rhopalomyia* gall midges induce galls on the host plants regardless of polyploidy, we collected galled and ungalled plants of *A. indica* and *A. lancea* in November 2002 from Kitakyushu City, Fukuoka, Japan. Galls of *Rhopalomyia cinerarius* Monzen (Fig. 2.4d) and *Rhopalomyia yomogicola* (Matsumura) (Fig. 2.4e) were recognized on *A. indica* and those of *Rhopalomyia caterva* Monzen (Fig. 2.4f) on *A. lancea.* Both galled and ungalled leaves of *A. indica* and *A. lancea* was shredded, soaked into chemical substances, and subjected to a Ploidy Analyzer. Relative chromosomal number of

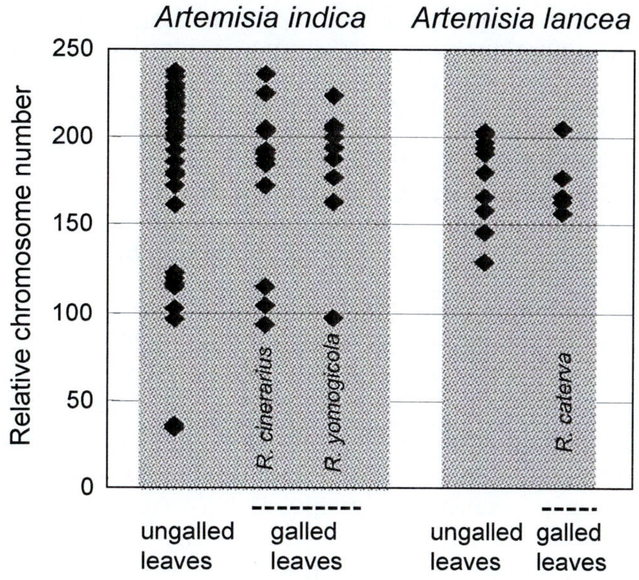

Fig. 2.5 Comparison in the relative DNA contents between galled and ungalled individuals of *Artemisia indica* and *A. lancea*. Each diamond indicates the relative mean chromosome number of an individual

each leaf was determined by reading the mean value of chromosomal numbers. There were no significant differences in the relative chromosome numbers between galled and ungalled leaves of *A. indica* and *A. lancea*. Galls of the three *Rhopalomyia* species were induced on the host plants with different chromosome numbers, which indicated that the three Japanese *Rhopalomyia* species examined could not discriminate differences in the polyploidy among their host plant individuals (Fig. 2.5).

2.3.4 *Species-Specific Relation Between* Asphondylia *Species and Polyploidy of Its Host Plant*

Gagné and Waring (1990) based on morphological features and Joy and Crespi (2007) based on genetic analysis demonstrated 15 *Asphondylia* species on creosote bush *L. tridentata* in North America are monophyletic and have diversified through gall-position shift, without host plant shift. Recently, O'Connor et al. (2019) found that one of the *Asphondylia* species distributed in Chihuahuan, Sonoran, and Mojave Deserts, North America was restricted to diploid entities of creosote bush while five were restricted to tetraploid and hexaploid host plants and demonstrated that polyploidy in plants can affect the biogeography of ecological communities. It is remarkable that females of some North American *Asphondylia* species can discriminates diploid, tetraploid and hexaplodid host plants at the time of oviposition. This may indicate that a particular relationship between gall midge and host polyploidy has been established long before the speciation of the *Asphondylia* species. We need to accumulate further such data to strengthen the hypothesis that plant polyploidy possibly leads speciation of gall midges to some extent.

2.4 Gall Shape Polymorphism

2.4.1 Various Types of Polymorphism

Polymorphism of insects can be seen not only between individuals within a population, but also between geographically separated populations (geographical polymorphism) and between generations at different seasons of the year (seasonal polymorphism). The phenomenon of mimicry and sexual dimorphism are included in polymorphism within a population. Polymorphism is also divided into two categories, genetic polymorphism that occurs when the morphs are a result of genetic determination (e.g. Zhang et al. 2019), and environmental polymorphism (often termed as polyphenism), where environmental factors play an important role rather than genetic heritability (e.g. Johnson 1966; Socha 2001; Olvido et al. 2003; Loskutova and Zhiltzova 2016; Kamioka and Iwasa 2017). Polymorphism has been studied in various aspects in relation to the aforementioned categories. Wing polymorphism has been most extensively and intensively studied for many insect groups, notably Hemiptera, Coleoptera, and Orthoptera (e.g. Harrison 1980; Roff 1986; Masaki and Shimizu 1995; Zera and Denno 1997; Sekimura and Nijhout 2017).

Drosopoulous et al. (2010) considered that the dorsal color polymorphism of spittlebugs *Philaenus* spp. (Hemiptera: Aphrophoridae) probably preceded speciation in the Mediterranean, and that the expression of polymorphism is developed separately in each species by visual selection (predator pressure). The southern green stink bug *Nezara viridula* Linnaeus (Hemiptera: Pentatomidae) exhibits several color polymorphs (e.g. Yukawa and Kiritani 1965), and a similar set of polymorphs (parallel polymorphism) can be seen also in a congener *Nezara antennata* Scott (Kiritani 1970), indicating that *Nezara* species has possibly diversified together with a set of polymorphs. In addition, type G of *N. viridula* is superior in reproductive ability but is inferior in the ability of surviving winters at least to types R and F (Kiritani 1970). Therefore, *N. viridula* needs to contain a set of polymorphs to exist in the distributional range by compensating each other, indicating that each of polymorphs would not independently differentiate into a separate species. In contrast, geographical polymorphism may have a chance to diversify into separate species.

2.4.2 Gall Shape Polymorphism

Gall shapes are generally species specific and have usually been regarded as extensions of the phenotypes of gall-inducing species (e.g. Fukatsu et al. 1994; Stern 1995; Stone and Cook 1998; Yukawa and Rohfritsch 2005). However, the shape and size of galls induced by a single species may vary, being frequently influenced by such factors as the presence or absence of parasitoids or inquilines, host plant species or varieties, and the number of gall inhabitants (e.g. Yukawa and Masuda 1996; Yukawa and Rohfritsch 2005). Variation in gall shape and size is divided into continuous and discontinuous patterns. Continuous variation is commonly seen and seldom causes misidentification of gall inducers when not accompanied by distinct shape variation (e.g. Yukawa and Masuda 1996). The discontinuous pattern has been regarded as gall polymorphism, which is exhibited by gall wasps (Hymenoptera: Cynipidae) in different seasons as unisexual and bisexual generation galls (e.g. Yukawa and Masuda 1996).

In gall-inducing cecidomyiids, gall polymorphism arises in the same generation and on the same host plant. For example, we have seen conical and hemispherical leaf galls of *Pseudasphondylia elaeocarpi* Tokuda & Yukawa on *Elaeocarpus sylvestris* (Loureiro) Poiret (Elaeocarpaceae) (Tokuda and Yukawa 2005) (Fig. 2.6a, b); upper and lower type leaf galls of *P. neolitseae* on *N. sericea* (Mishima and Yukawa 2007) (Fig. 2.6c, d); thick and thin leaf galls of *M. pustulae* on *E. japonicus* (Sunose 1985a; Fujii et al. 2014) (Fig. 2.6e, f); globular jar-shaped, jar-shaped and long jar-shaped types of axillary bud gall of *Rhopalomyia longitubifex* (Shinji) on *A. indica* and *A. montana* (Ganaha et al. 2007) (Fig. 2.6g–i); and cup-shaped and umbrella-shaped galls of *Bruggmanniella litseae* Lin, Yang & Tokuda on *Litsea acuminata* (Blume) S. Kurata (Lauraceae) (Lin et al. 2020).

2.4.3 Geographical Polymorphisms in Gall Midges

Among the aforementioned examples, leaf gall types of *P. neolitseae* and *M. pustulae* are known as geographical polymorphisms, and their distributional patterns are parapatric (Sunose 1985a; Mishima and Yukawa 2007). In northern Honshu, one of the regions where the two types of *M. pustulae* gall coexist, adults of *M. pustulae* from thin type galls emerge earlier than adults of a platygastrid (Hymenoptera: Platygastridae), while those from thick ones later than the platygastrid (Sunose 1985a). The platygastrid needs to emerge a little later than the host gall midge within a short period of host egg stage because it is an egg-larva endoparasitoid. This is a critical event for such short-lived insects as the gall midge and platygastrid (Sunose 1985b). Thus, the platygastrid is considered to have been acting as a selection pressure differently between the gall midge populations with different gall types. In addition, the larval color of *M. pustulae* is also divided into two distinct types, milky white and yellow, and the two-color types

Fig. 2.6 Polymorphism of cecidomyiid galls. (**a**), a leaf gall of *Pseudasphondylia elaeocarpi* on *Elaeocarpus sylvestris* (conical type); (**b**), ibid. (hemispherical type); (**c**), leaf galls of *Pseudasphondylia neolitseae* on *Neolitsea sericea* (upper type); (**d**), ibid. (lower type); (**e**), leaf galls of *Masakimyia pustulae* on *Euonymus japonicus* (thick type); (**f**), ibid. (thin type); (**g**), axillary

seem to fit simple Mendelian ratios by determining the intraphenotypic crosses (Sunose 1979). The distributional pattern of larval color types is incongruent with that of gall types (Sunose 1979, Sunose 1985a, b).

2.4.4 Sympatric Polymorphisms in Gall Midges

In contrast, distribution ranges of the three types of axillary bud gall of *R. longitubifex* on *Artemisia* (Fig. 2.6g–i) are entirely overlapped and sometimes coexist in the same area, exhibiting sympatric polymorphisms (Ganaha et al. 2007). Previously three different species of *Rhopalomyia* were considered to induce the three types of gall, respectively (Yukawa and Masuda 1996). However, molecular analysis revealed that these galls were caused by a single species, *R. longitubifex*, of which clade was supported by a 100% bootstrap value although the ratio of diameter to length of galls was significantly different between the types (Ganaha et al. 2007). In the clade of Japanese populations, seven haplotypes were found. The difference in mitochondrial DNA sequence was obscure among populations with different gall types, but was relatively clear among populations in different localities, namely South Korea, Hokkaido, northern Honshu (Aomori, Miyagi, and Ibaraki), and Kyushu (Fukuoka, Oita, and Miyazaki) (Ganaha et al. 2007). These results support the statement by Ganaha et al. (2004) that speciation in *R. longitubifex* would be initiated by geographic isolation rather than by host shifting.

2.4.5 Possible Relationship Between Polymorphism and Plant Polyploidy

In addition to the possibility of speciation by geographic isolation, we need to take an alternative possibility into consideration as mentioned below. The three gall types of *R. longitubifex* have never been found to occur on the same host individual of *Artemisia* even though they rarely coexist in the same area (Ganaha et al. 2007), contrasting with the polymorphism of *P. elaeocarpi* on the same leaf and with the situation of *R. cinerarius, R. yomogicola* and *R. caterva* on *Artemisia* individuals with polyploidy (see Sect. 2.3). Such a fact may suggest that the gall polymorphism of *R. longitubifex* is related to the ploidy level of *Artemisia* or some other intrinsic factors of the host plants. Although there are no actual data for *R. longitubifex* at the moment, larger sized galls may be expected to produce larger females that have more ovarian eggs. Differences among the polymorphs in gall size and gall wall thickness

Fig. 2.6 (continued) bud gall of *Rhopalomyia longitubifex* on *Artemisia indica* (globular jar-shaped type); (**h**), ibid. (jar-shaped type) (reproduced with permission from Ganaha et al. 2007); (**i**), ibid. (long jar-shaped type)

may affect the component of parasitoid community as has been demonstrated for *M. pustulae* (Sunose 1985a). In future, it is needed to accumulate actual data on the relationship between gall polymorphs and female fecundity, parasitoid component, or other biological factors, together with information on polyploidy of *Artemisia* individuals. Such ecological data would promote better understanding that gall polymorphism would lead to differentiation of selection pressure during the course of gall midge speciation.

2.4.6 Geographic Mosaic of Coevolution

Various researchers are now interested in a geographic mosaic of coevolution (hereafter, GMC) (e.g. Thompson 1999, 2005; Craig 2016). Recently, Craig et al. (2020) studied the GMC of a tritrophic interaction at small geographic scales in the gall-inducing fly *Eurosta solidaginis* (Cocquillet), (Diptera: Tephritidae) on tall goldenrod *Solidago altissima* Linnaeus (Asteraceae), and the galler's natural enemies in a region where geographically differentiated populations of all species come in contact with each other. They found that selection on gall diameter and length varies between prairie and forest habitats due to differences in host plants and natural-enemy communities. At the prairie–forest ecotone where prairie and forest habitats are intermixed, they found that geographic selection mosaics on gall diameter and length varied on a scale of a few kilometers. Gall diameter variation among sites correlated with selection on gall diameters, indicating local adaptation. In contrast, gall lengths did not correlate with selection, indicating that gene flow may have prevented local adaptation of this trait.

Geographical variation in cecidomyiid galls may also be very widespread because both plants and natural enemies vary geographically, and this could influence gall morphology. The studies of GMC motivate us to conduct ecological studies, even of a single gall midge species, not only in one habitat but also in two or more habitats, which possibly lead to the elucidation of diversification scenario in gall-inducing cecidomyiids and their associates.

2.5 Diversification Along with Host Range Expansion in *Salix* Gall Midges

2.5.1 Gall Midges on Salix

Willows (Salicaceae) are widely distributed in the Northern Hemisphere (e.g. Newsholme 1992). *Salix* is the largest genus in the family Salicaceae, consisting of at least 450 species (e.g. Newsholme 1992; Lauron-Moreau et al. 2015). So far, at least 103 cecidomyiid species have been known to occur on *Salix*

(Gagné and Jaschhof 2017). Among them, 75 species belong to the genus *Rabdophaga* (Gagné and Jaschhof 2017) and induce their galls on various plant organs, such as leaf buds, twigs, leaves, flower catkins, etc. (e.g. Gagné 1989; Nijveldt and Yukawa 1982; Bowser et al. 2018). They exhibit a close relationship with their host plants, like as other cecidomyiid genera, *Caryomyia* on *Carya* (Jugulandaceae), *Celticecis* on *Celtis* (Cannabaceae), and *Rhopalomyia* on *Artemisia* and some related genera (Asteraceae) (e.g. Gagné 2008; Gagné and Moser 2013; Gagné and Jaschhof 2017). Such examples imply possible sympatric speciation after their ancestor arrived at that host plant genus.

Within the aforementioned close relationship, many gall midge species are oligophagous, exhibiting a particular linkage of one gall midge species and several host species in a single plant genus. A connection of one *Rabdophaga* species with two or more *Salix* species has also been recorded for some *Rabdophaga* species. For example, *Rabdophaga heterobia* (Loew) and *Rabdophaga salicis* (Schrank) are oligophagous in the Palaearctic Region, and *Rabdophaga strobiloides* (Osten Sacken) in the Nearctic Region (e.g. Gagné and Jaschhof 2017). Accumulation of information on such linkages possibly leads us to roughly grasp the process towards speciation for one cecidomyiid species through its host range expansion.

This section demonstrates, based on Sato et al. (2020), an example of possible speciation scenario through host range expansion in the intraspecific populations of *Rabdophaga rosaria* (Loew) that induces rosette galls on the leaf buds of the various *Salix* species.

2.5.2 Rabdophaga rosaria

Rabdophaga rosaria is widely distributed in the Palaearctic Region and has been known to induce rosette galls on terminal buds (Fig. 2.7) of at least 26 *Salix* species (e.g. Stelter 1970; Nijveldt and Yukawa 1982; Yukawa and Masuda 1996; Skuhravá et al. 2014; Bowser et al. 2018) (see Table 1 in Sato et al. 2020). Each of these *Salix* species belongs to one of the three subgenera, *Salix*, *Vetrix* and *Chamaetia*, within the genus *Salix* (Newsholme 1992). This species is univoltine, overwinters in the galls on its host trees as final instars, and pupates in the galls in the following spring (Yukawa and Masuda 1996). Adults emerge from the galls during the period from late April to late May (Nijveldt and Yukawa 1982), and females lay their eggs on the under surface of newly opened leaves (Yukawa and Masuda 1996). In addition to its typical rosette galls, this species is distinguishable from the other congeners by having larval sternal spatula with a u-shaped emargination and larval pinkish-white body color (Nijveldt and Yukawa 1982).

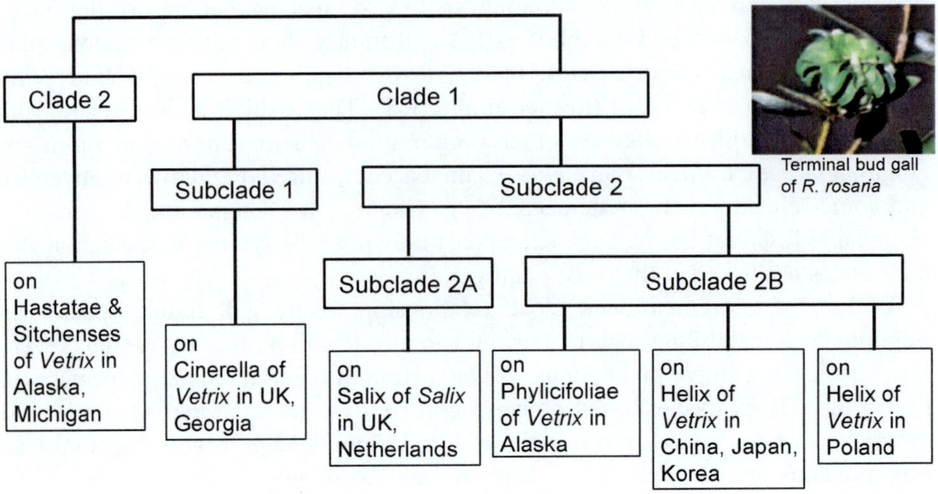

Fig. 2.7 Schematic representation of clade diversifications of *Rabdophaga rosaria* populations along with host range expansion on different sections of the subgenera *Vetrix* and *Salix* of the genus *Salix* (Salicaceae)

2.5.3 Molecular Phylogenetic Analysis of Salix *Species and* Rabdophaga *Gall Midges*

A total of 24 sequence datasets of the nuclear ribosomal DNA ITS region was obtained from Genbank for nine *Salix* species and an outgroup taxon, *Betula pendula* Roth (Betulaceae). The maximum likelihood tree for these plant species indicated a dichotomy between the subgenera *Salix* and *Vetrix* (Fig. 2 in Sato et al. 2020). In *Vetrix*, five sections were recognized but they were not clearly separated into clades (Sato et al. 2020).

A region of mitochondrial DNA COI was analyzed for samples of *R. rosaria*, *Rabdophaga strobiloides* (Osten Sacken) and their congeners from various *Salix* species from various localities in the Holarctic Region including Alaska, to confirm whether or not the clades of different populations are related to localities or sections of *Salix* (see Table 3 in Sato et al. 2020). A maximum likelihood tree indicated that *R. rosaria* and *R. strobiloides* populations were divided into Clades 1 and 2, respectively. Clade 1 was divided into Subclades 1 and 2, of which the latter was further divided into Subclades 2A and 2B (Fig. 2.7; see also Fig. 3 in Sato et al. 2020). Subclade 1 consisted of populations on several *Salix* species of the section Cinerella in the subgenus *Vetrix* in Georgia and the UK. Subclade 2A contains exclusively populations on *Salix alba* Linnaeus of the section Salix in the subgenus *Salix* in the Netherlands and the UK. Subclade 2B consisted of various populations on the section Helix in Poland, Phylicifoliae in Alaska, and *Salix* species in the Eastern Palaearctic Region.

Clade 2 consisted of Alaskan populations of *R. strobiloides* and *R.* cf. *strobiloides* (following Bowser et al. 2018, meaning tentatively identified species) on two *Salix* species, which belong to the sections Hastatae or Sitchenses of the subgenus *Vetrix* and are distributed in southwestern Alaska and along the Pacific coast.

2.5.4 *A Possible Diversification Scenario of* Rabdophaga rosaria

Based on the phylogenetic trees for *Rabdophaga* and *Salix*, Sato et al. (2020) proposed a possible diversification scenario of *R. rosaria* through the expansion of host plant ranges. An ancestor of *R. rosaria* seems to have started to use *Salix* species of the subgenus *Vetrix* as hosts somewhere in western Palaearctic Region. Thereafter, Clade 1 has been spreading over eastern Palaearctic Region using a wide range of *Salix* species about 1.50–2.66 Ma. Subclade 2A separated from 2B together with host range expansion to the section Salix (0.53–1.60 Ma). All populations in mainland China, Japan and South Korea were associated with *Salix* species of the section Helix and are included in Subclade 2B, but apparently separated from Alaskan, Polish and Russian Far East populations. Subclade 2B came from western to eastern Palaearctic Region and Alaska through at least two possible routes of expansion using *Salix* species of the subgenus *Vetrix* except the section Cinerella, one route being through northern Eurasia to Alaska, and another through central Asia to eastern Palaearctic Region.

The genetic differences among populations of Subclades 1, 2A and 2B ranged from 1.06 to 3.46% in minimum. Most percentages were distinctly higher than the 2% divergence criterion proposed by Hebert et al. (2003) as an acceptable distance to consider two closely related entities as distinct species. These data indicate that *R. rosaria* populations have diversified into the stage of sibling species through the expansion of host plant ranges although we do not know yet if these populations are reproductively isolated or not. As to the origin of Alaskan *R. strobiloides*, further molecular data are needed from its Holarctic congeners other than *R. rosaria*.

2.6 Sexual Isolation Between Intraspecific Populations of a *Fagus* Gall Midge

2.6.1 *Gall Midges on* Fagus

The genus *Fagus* (Fagaceae) contains at least 11 species in the world but future taxonomic studies will possibly add a few more species to the genus (Hara 1992). All known species are deciduous and distributed only in the temperate zone of the Northern Hemisphere (e.g. Hara 1992). Among the 11 species, the following four species, *Fagus crenata* Blume, *Fagus japonica* Maximowicz, *Fagus orientalis* Lipsky, and *Fagus sylvatica* Linnaeus are known to host 12 cecidomyiid species belonging to six genera (Table 2.1). In contrast, Sato and Yukawa (2001) could not find any cecidomyiid galls on *Fagus multinervis* Nakai in South Korea (Ulleung-do) and *Fagus grandifolia* Ehrhart and *Fagus mexicana* Martinez in North America despite their intensive field surveys.

In addition to known *Fagus* galls of identified cecidomyiid species (Fig. 2.8a–e), many sorts of gall induced by unidentified cecidomyiids have been found in eastern parts of Asia. At least 30 sorts of cecidomyiid gall have been known to occur on *F. crenata* (Fig. 2.8f–p) and *F. japonica* (Fig. 2.8q, r) in Japan (e.g. Yukawa and Masuda 1996; Sato et al. 2010). In Taiwan, unidentified gall midges are known to induce at least four sorts of leaf gall on *Fagus hayatae* Palibin ex Hayata (Yang et al. 2000). On various occasions of field survey from 2006 to 2016 in Hunan, Hubei, Yunnan, and Sichuan Provinces, mainland China, our Chinese and Japanese colleagues and we found a total of at least 16 sorts of cecidomyiid galls on *Fagus engleriana* Seemen, *Fagus longipetiolata* Seemen, and *Fagus lucida* Rehder et Wilson. Although these *Fagus* galls in the mainland China have not yet been

Table 2.1 List of gall midges that induce galls on the species of *Fagus* (Fagaceae) in the world[a]

Cecidomyiid species	*Fagus* spp.	Distribution
Polyphagous genera		
Contarinia fagi Rübsaamen	*F. sylvatica*	Widespread Europe
Janetiella infrafoli Monzen (Fig. 2.8e)	*F. crenata*	Japan
Macrolabis fagicola (Barnes)	*F. sylvatica*	Sweden, Germany, Czechoslovakia
Monophagous genera associated only with *Fagus*		
Hartigiola annulipes (Hartig)	*F. sylvatica*	Widespread Europe, Georgia, Turkey
Ibid.	*F. japonica*	Japan
Hartigiola faggalli (Monzen) (Fig. 2.8a, b)	*F. crenata*	Japan
Mikiola bicornis Sato & Yukawa (Fig. 2.8c)	*F. crenata*	Japan
Mikiola (?) *cristata* Kieffer[b]	*F. sylvatica*	France
Mikiola fagi (Hartig)	*F. sylvatica*[c], *F. orientalis*	Widespread Europe, W Asia
Mikiola glandaria Sato & Yukawa (Fig. 2.8d)	*F. crenata*	Japan
Mikiola orientalis Kieffer	*Fagus* sp.	Turkey, Georgia
Phegomyia fagicola (Kieffer)	*F. sylvatica*	Widespread Europe, W Asia.
Phegomyia tokunagai Sasakawa & Koyama	*F. crenata*	Japan

[a]Data based on Skuhravá (1986), Yukawa and Masuda (1996), Sato and Yukawa (2001, 2004, 2008), and Gagné and Jaschhof (2017)
[b]Generic position should be revised (Gagné and Jaschhof 2017)
[c]Host plant including *Fagus sylvatica pliocenica* (Saporta)

published, we recognized that some of them are similar in shape to those in Japan or Taiwan.

Gall shape indicates that cecidomyiid species responsible for most of the east Asian *Fagus* galls possibly belong either to *Hartigiola*, *Mikiola*, or *Phegomyia*. This means that species radiation occurred within each of the three genera on a *Fagus* species, at least three times in total, respectively, at different times, in different places, and on different *Fagus* species. Such multiple species radiations on a single plant genus are remarkably different from known examples of radiation within a single cecidomyiid genus on a single host plant genus, such as *Caryomyia* on *Carya* (Jugulandaceae) (Gagné 2008), *Celticecis* on *Celtis* (Cannabaceae) (Gagné and Moser 2013), *Psectrosema* on *Tamarix* (Tamaricaceae), *Rabdophaga* on *Salix* (Salicaceae) (Sect. 2.5) and *Rhopalomyia* on *Artemisia* and related genera belonging chiefly to the tribe Anthemideae (Asteraceae) (Sect. 2.3) (e.g. Yukawa et al. 2005; Joy and Crespi 2007; Gagné and Jaschhof 2017).

2.6.2 *Sibling Species*

In Cecidomyiidae, sibling species have been recognized with morphological similarity (e.g. Jaschhof and Jaschhof 2013), biochemical characterization (e.g. Makni et al. 2000), and molecular data (e.g. Grego et al. 1990; Windenfalk et al. 2002; Khemakhem et al. 2005). Cook et al. (2011) reported based on mating experiments that populations of *Dasineura oxycoccana* (Johnson) from cranberry *Vaccinium macrocarpon* Aiton and highbush blueberry *Vaccinium carymbosum* Linnaeus (Ericaceae) were reproductively isolated. In Cecidomyiidae, however, degree of sexual isolation between sibling species has never been investigated using I_{PSI} estimator (Rolán-Alvarez and Caballero 2000) together with field data of ecological and behavioral traits. I_{PSI} estimator is an effective index to estimate the degree of sexual isolation between two populations as has been used for various insects, such as *Rhagoletis* spp. (Diptera: Tephritidae) (Schwarz and McPheron 2007) and *Drosophila* spp. (Diptera: Drosophilidae) (e.g. Coyne et al. 2005). In this section, we refer to the intensive field study by Mishima et al. (2014) who demonstrated using I_{PSI} estimator that two intraspecific populations of a *Fagus* gall midge *Hartigiola faggalli* (Monzen) has been diversifying into sibling species through reproductive isolation.

2.6.3 *Gall Shape and Life History of* Hartigiola faggalli

Hartigiola faggalli induces leaf galls on the Japanese beech, *F. crenata*. The gall is single-chambered, containing one cecidomyiid larva (e.g. Yukawa and Masuda 1996; Sato and Yukawa 2004). The gall is a smooth, brown or blackish brown and bivalve-shaped swelling on upper- or lower surfaces of the leaf blade, with a height

Fig. 2.8 Some of *Fagus* galls in Japan. (**a**), galls of *Hartigiola faggalli* on *Fagus crenata* (upper type gall); (**b**), ibid. (lower type gall); (**c**), a gall of *Mikiola bicornis* on *F. crenata* (note that the gall shape is variable in height and thickness); (**d**), galls of *Mikiola glandaria* on *F. crenata*; (**e**), galls of *Janetiella infrafoli* on *F. crenata*; (**f–p**), galls induced by unidentified cecidomyiids on *F. crenata*; (**q, r**), galls induced by unidentified cecidomyiids on *Fagus japonica*

of 1.0–2.7 mm and a maximum width of 1.0–2.4 mm. Galls that are induced on the upper leaf-surface usually stand on lateral veins or rarely on the midrib (upper type gall, hereafter) (Fig. 2.8a), while those on the lower surface lie between the lateral veins (lower type gall, hereafter) (Fig. 2.8b). The two morphotypes frequently coexist on the same leaf blade (Sato and Yukawa 2004).

Hartigiola faggalli is univoltine and the adults emerge from late April to mid-May (Takizawa 1983; Sato and Yukawa 2004). The females alight on host leaves that are about to open and lay their eggs along lateral veins on the upper leaf-surface or between lateral veins on the under leaf-surface. Galls become conspicuous in June and larvae mature in the galls in October. Galls then drop to the ground while still attached to the leaf blade. Mature larvae overwinter in the galls on the ground and pupate in the galls from late March to mid-April.

2.6.4 *Sexual Isolation Between Intraspecific Populations of* Hartigiola faggalli

Upper and lower types of the leaf gall induced by *H. faggalli* on *F. crenata* have been regarded as two morphotypes (gall polymorphism: see Sect. 2.4), which frequently coexist on the same leaf blade (Sato and Yukawa 2004). Based on sequence data, Mishima et al. (2014) first confirmed that the differences between the two gall types were 6 bp (2.3%) among 438 bp examined and only one of 146 amino acid residues differed between the two intraspecific populations. The difference was also confirmed by analyzing 428 bp of ITS2 region for 24 individuals of *H. faggalli* collected from 14 localities. Based on these variations, the derivation of all adults that were caught during the field observations was determined.

Mishima et al. (2014) observed adult behavior of *H. faggalli* at an *F. crenata* stand in Iwate Prefecture, Japan during the adult emergence season in April and May from 2007 to 2009. Emergence, swarming, mating, and oviposition occurred sequentially each day and almost simultaneously in both populations (see also Chap. 7 for daily activity of gall midges). Thus, they were not isolated from each other in time or space. However, 85% of 134 swarming males flew to females of the same population when responding to female sex pheromone. About 92% of 251 mating pairs were homogenic, and I_{PSI} indicated a significantly homogenic mating. After mating, females of each population oviposited their eggs only on either the upper or lower surfaces of fresh leaves.

From the observation of mating and oviposition behavior, Mishima et al. (2014) considered that the chemical composition of the female sex pheromone, male sensitivity to the sex pheromone, and female sensitivity to volatiles from upper and lower leaf-surfaces seemed to differ between the two intraspecific populations. The strongly assortative mating combined with difference in pheromones, distinct oviposition-site selection, and gall morphology indicates that the two populations are almost completely reproductively isolated, and that they have diversified into the stage of sibling species. The slight difference in the sequence data indicates a possible recent diversification of the two intraspecific populations into the stage of sibling species. The presence of rare heterogenic pairs and progeny suggests that the post-mating isolation barrier has not been completed yet. The ratio and stereochemistry of pheromone compounds may be of importance in the sexual isolation between the two populations of *H. faggalli*, as has been noted for *Contarinia nasturtii* (Kieffer) (Hillbur et al. 2005).

Species diversification has been known to involve galling-site shift between different plant organs of the same host plant species (e.g. Condon and Steck 1997; Després et al. 2002; Joy and Crespi 2007). Therefore, oviposition-site selection by the two intraspecific populations is remarkable and will possibly facilitate their diversification. Oviposition-site selection by females may be related to different volatiles released from different plant parts and female sensitivity to the volatiles may diversify during the course of speciation.

We need further investigation as to behavioral diversification in response to chemical changes in female sex pheromone as well as a shift of galling position in relation to female sensitivity to plant volatiles.

2.7 Ecological Divergence Among Closely Related Congeners

2.7.1 Importance of Ecological Diversification in the Process of Speciation

Ecological divergence among populations evolves early in the process of speciation as the result of divergent natural selection and promotes the evolution of reproductive isolation (e.g. Rundle and Nosil 2005; Funk et al. 2006; Arnegard et al. 2014; Yukawa et al. 2019c). Studies of diverse taxa suggest that natural selection caused by shift in ecological interactions or by invasions of novel habitats plays an important role in adaptive divergence and speciation (e.g. Orra and Smith 1998). In the preceding sections of this chapter, we referred to triggers for invasions of novel habitats by a part of population. In the novel habitats, different sorts of natural selection would act on the newly invaded population, resulted in the promotion of ecological divergence. The genetic basis for the process of ecological diversification has become better understood in recent years (e.g. Fornaguera 2014; Pogson 2016; Huang et al. 2017; Lunds Universitet 2018; Schilling et al. 2018). However, further comparative ecological studies of closely related taxa are still needed in understanding the role of ecological divergence in speciation.

Uechi et al. (2005) and Uechi and Yukawa (2006a) found that the diversification of Japanese gall-inducing *Asphondylia* species has been associated with divergence in host plant use, shift in galled organs, and changes in voltinism. Dorchin et al. (2015) also found divergence of life history patterns, host ranges, galled organs, and gall shapes among morphologically and genetically similar *Asphondylia* species associated with North American goldenrods, *Solidago* and *Euthamia* (Asteraceae). All these studies indicate that ecological differentiation is important in speciation. In particular, differences in life history traits cause chronological reproductive isolation between species during the emergence season, particularly in short-lived insects such as gall midges (e.g. Yukawa 2000; see also Chap. 6).

2.7.2 **Asphondylia** ***Species in Japan***

In recent taxonomic studies, the following seven described species of the genus *Asphondylia* are known to occur in Japan: *A. aucubae*, *Asphondylia baca* Monzen, *Asphondylia itoi* Uechi & Yukawa, *Asphondylia morivorella* (Naito), *Asphondylia sphaera* Monzen, *Asphondylia tojoi* Elsayed & Tokuda, and *Asphondylia yushimai* Yukawa & Uechi (Uechi et al. 2012; Yukawa 2014; Elsayed et al. 2018a) (Table 4.4). *Asphondylia japonicus* (Monzen) is not included in these species because it should be synonymized with *A. baca*. Among the seven described species, *A. morivorella* is morphologically distinguishable from the other species by the shape of larval sternal spatula (Sunose 1983). The remaining six species are quite similar to each other in adult, larval and pupal morphological features (Yukawa 1971; Yukawa and Ohsaki 1988a; Yukawa et al. 2003; Uechi and Yukawa 2004; Elsayed et al. 2018a). In addition, at least 13 unidentified *Asphondylia* species have been found to induce galls on various plant species in Japan (e.g. Yukawa and Masuda 1996; Uechi et al. 2002, 2012; Uechi and Yukawa 2006a).

Life history traits of these described and undescribed species of Japanese *Asphondylia* have intensively been studied (Yukawa 1971; Yukawa and Miyamoto 1979; Sunose 1983; Yukawa and Ohsaki 1988b; Yukawa et al. 2003, 2016a, b, 2019c; Uechi and Yukawa 2004, 2006a, b; Uechi et al. 2004, 2005, 2018; Elsayed et al. 2018a). Such ecological information is crucial in understanding divergence. Ecological divergence of gall midges occurs together with host plant shift, host organ shift, gall shape polymorphism, or changes in life history traits such as lower developmental threshold temperature (hereafter LDT), thermal constant, voltinism, and diapausing season (e.g. Uechi et al. 2005; Uechi and Yukawa 2006a; Tokuda 2012; Dorchin et al. 2015; Yukawa et al. 2019c).

In addition, many *Asphondylia* species are known to induce ambrosia galls (Neger 1908, 1910). According to unpublished data by Uechi, for example, two species of *Botryosphaeria* (Ascomycota: Botryosphaeriaceae) were found in soybean pod galls of *A. yushimai*. Uechi has also collected various unidentified fungal symbiont from galls of other congeneric species. Such a combination of *Asphondylia* species and associated fungi should be included in the divergence of life history traits.

2.7.3 Actual Data of Ecological Diversification

Yukawa et al. (2019c) showed life history traits of four identified and two unidentified *Asphondylia* species in Japan. Their distribution ranges are almost overlapped in Japan. In addition to their morphological similarity, they are genetically similar to each other. For this section, one of us, N. Uechi, newly constructed a maximum likelihood tree for *Asphondylia* species in Japan (Fig. 2.9) and calculated genetic distances between the species. The minimum and maximum mean pairwise sequence divergences between the *Asphondylia* species were 0.2 and 16.9%, respectively, indicating that they separated from each other about 0.09–8.45 Ma. However, the clade including *A. itoi* and a morphologically different species *A. morivorella* is distinctly apart from the clade for the remaining species. When we exclude the two species, the minimum and maximum mean pairwise sequence divergences between the remaining six species were 0.2 and 8.0%, respectively.

In spite of similarities in morphological features and sequence divergences, important life history traits differed greatly among the species that are co-existing in Japan. For example, their host plants are different, and the LDT and thermal constant vary with the species. Such differences are reflected in their life history

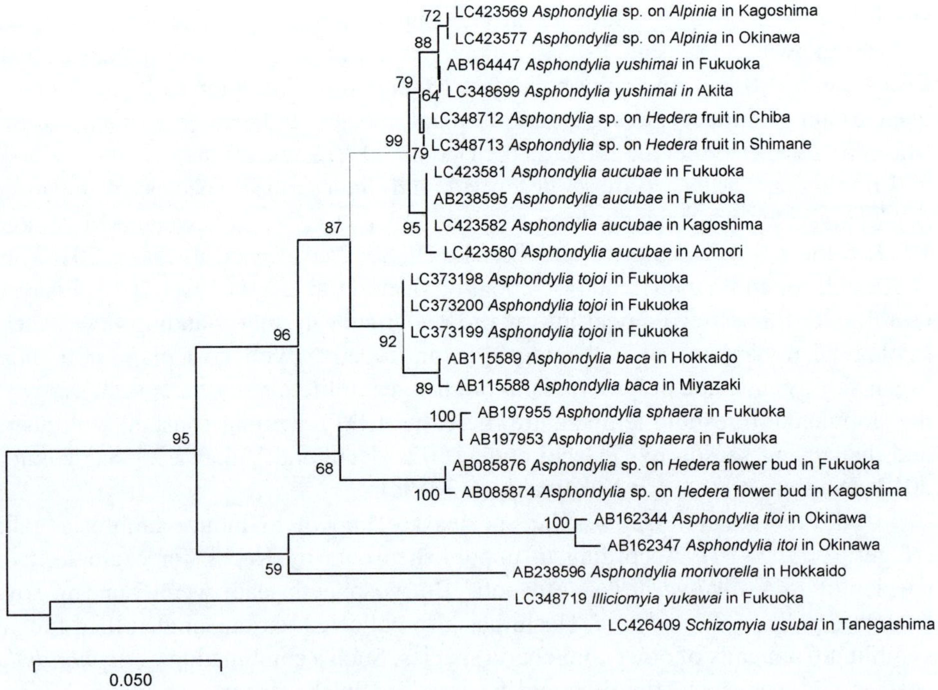

Fig. 2.9 Maximum likelihood tree for *Asphondylia* species in Japan based on 421 bp of the mitochondrial cytochrome oxidase subunit I gene. Bootstrap values are indicated for nodes gaining more than 58% support in 500 pseudoreplications. Evolutionary analyses were conducted with MEGA X. *Illiciomyia yukawai* and *Schizomyia usubai* were used as outgroup taxa

types that were categorized by Uechi and Yukawa (2006a). These data indicate that divergence in life history traits evolves before morphological divergence, and the ecological divergence may strengthen isolating barriers among the taxa (e.g. Schluter and Conte 2009; Matsubayashi et al. 2010; Nosil 2012; Arnegard et al. 2014; Funk et al. 2006). Changes in voltinism, LDT, thermal constant and diapausing season are important in allowing gall midges to establish themselves in new habitats because synchronization of adult emergence and host-plant phenology is a critical event for short-lived insects, such as gall-inducing cecidomyiids (Yukawa 2000; Yukawa and Akimoto 2006; see Chap. 10).

During the course of establishment on a newly colonized host plant, gall midges should gradually adjust their life history traits to the growth pattern of the new host, and this can lead to the evolution of reproductive isolation from the population on the ancestral host. This speciation scenario may be applicable to other gall midge genera such as *Contarinia*, *Dasineura*, and *Lasioptera*, which are speciose and polyphagous at the generic level.

2.8 Host Race Formation and Sympatric Speciation

2.8.1 Asphondylia borrichiae

Asphondylia borrichiae Rossi & Strong is distributed in southeastern parts of USA and Mexico, and induces multi-chambered galls on three species of halophytic Asteraceae, *Borrichia frutescens* (Linnaeus) de Candolle, *Iva frutescens* Linnaeus and *Iva imbricata* Walker (Stokes et al. 2012; Gagné and Jaschhof 2017). In this gall midge, deme formation was suggested in populations associated with *B. frutescens* (Stiling and Rossi 1997) and possible existence of *Borrichia*- and *Iva*-associated host races were reported (Stiling and Rossi 1995; Rossi et al. 1999). A recent molecular study examining mitochondrial haplotypes of approximately 250 gall midge individuals revealed that the genetic composition is different between populations associated with *B. frutescens* and *Iva* spp., confirming the host race formation in this species (Stokes et al. 2012).

2.8.2 Dasineura folliculi

Dasineura folliculi (Felt) is a multivoltine species distributed in northeastern parts of USA (Dorchin et al. 2009; Gagné and Jaschhof 2017). This species induces bud galls on goldenrods *Solidago gigantea* Aiton and *Solidago rugosa* Miller (Asteraceae), both are sympatrically distributed. According to Dorchin et al. (2009), adults emerged from *S. gigantea* and *S. rugosa* are a little different morphologically and the frequency of homogenic mating pairs from the same host species was significantly higher than heterogenetic pairs from different hosts in the laboratory experiment. In addition, mitochondrial haplotypes of the gall midge differ between individuals from respective host plants. Because genetic diversity of gall midges as well as parasitism rates of hymenopteran parasitoids were higher on *S. gigantea* than on *S. rugosa*, the former was considered to be the original and the latter to be a derived host (Dorchin et al. 2009). These results suggest that the gall midge is under the process from host races to sibling species.

2.9 Geographical Isolation and Allopatric Speciation

2.9.1 Pitydiplosis puerariae

Pueraria gall midge *Pitydiplosis puerariae* Yukawa, Ikenaga & Sato induces leaf galls on *Pueraria lobata* (Willdenow) Ohwi (Fig. 2.10a, b), and *Pueraria montana* (Loureiro) Merrill (Yukawa and Masuda 1996; Yukawa et al. 2012). Although *P. lobata* is frequently regarded as a variety of *P. montana*, the two species were suggested to have diverged 6.0–8.6 Ma (Sun et al. 2007). So, in this section, *P. lobata* is regarded as an independent species. *Pitydiplosis puerariae* is widely distributed in East Asia except for areas from Okinoerabu Island to Tarama Island in the Ryukyu Islands (Fig. 2.11) (Yukawa et al. 2012). Based on a molecular analysis, *P. puerariae* is phylogenetically subdivided into three clades: (1) 'YNT-*montana* group' associated with *P. montana* in the Yaeyama Islands and northern Taiwan, (2) 'CT-*montana* group' associated with *P. montana* in central Taiwan, and (3) 'JCK-*lobata* group' associated with *P. lobata* in Japan (from Hokkaido to Tokunoshima Island), South Korea, and mainland China (see Fig. 5 in Yukawa et al. 2012). Among them, the YNT-*montana* group is inferred to have diverged from the other two groups 1.37–1.77 Ma, and then the CT-*montana* and JCK-*lobata* groups have diverged 0.98–1.39 Ma (Yukawa et al. 2012). Because the divergence time between the YNT-*montana* group and others corresponds with the era when the Yaeyama Islands was separated from Chinese continent and Taiwan due to sea-level rise during the interglacial period, they were considered to have diverged by geographic isolation. After that, a part of population belonging to the YNT-*montana* group probably colonized northern Taiwan secondarily in a later glacial period when the Yaeyama Islands were connected to Taiwan. The divergence time between the CT-*montana* and JCK-*lobata* groups was much later than the divergence of their host plants, 6.0–8.6 Ma (Sun et al. 2007). Therefore, host shift should have occurred from *P. montana* to *P. lobata* in the gall midge, although the degree of geographical contributions (i.e. it happened allopatrically, parapatrically or sympatrically) is not known.

Fig. 2.10 Galls on *Pueraria* (Fabaceae) and *Ilex* (Aquifoliaceae). (**a**), lenticular leaf galls of *Pitydiplosis puerariae* on *Pueraria lobata* (a view of upper leaf-surface); (**b**), ibid. (a view of lower leaf-surface); (**c**), an axillary bug gall of *Schizomyia sasakii* on *Ilex crenata*; (**d**), axillary bud galls of *Schizomyia soyogo* on *Ilex integra*

2.9.2 Schizomyia sasakii *and* S. soyogo

Schizomyia sasakii (Monzen) and *Schizomyia soyogo* Kikuti are both endemic to Japan and induce axillary bud galls on *Ilex* species (Aquifoliaceae) (Tokuda et al.

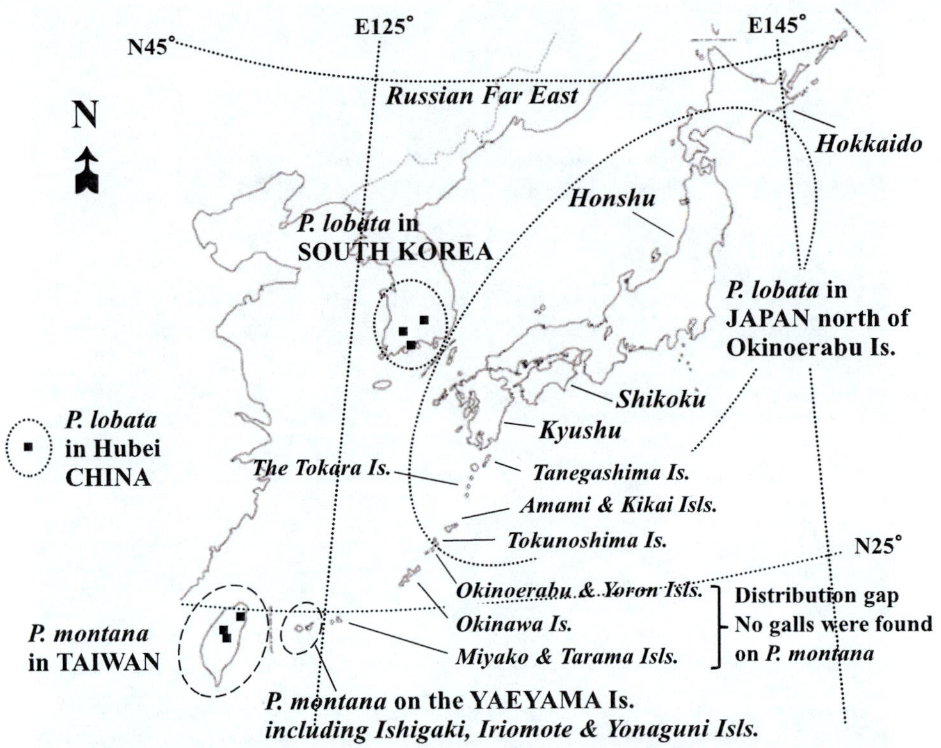

Fig. 2.11 Map showing areas surveyed in East Asia (modified from Yukawa et al. 2012). Circles with dotted line surround areas where thick lenticular galls on *Pueraria lobata* were found and circles with dashed line surround areas where those on *Pueraria montana* were found. A distribution gap is also indicated in the map. Black points in mainland China and South Korea indicate places where galls were found on *P. lobata*

2002, 2004; Tabuchi and Amano 2003a; Elsayed et al. 2018b). *Schizomyia sasakii* is distributed widely in Japan from Hokkaido to the Southwest Islands of Japan, but *S. soyogo* is known only from Honshu, Shikoku, and Kyushu. The host plant of *S. sasakii* is restricted only to *Ilex crenata* Thunberg (including two varieties *crenata* and *radicans*) (Fig. 2.10c) in wide areas from Hokkaido to Kyushu and to *Ilex maximowicziana* Loesener in the Southwest Islands of Japan, which means that *S. sasakii* is monophagous in respective distributional areas. In contrast, *S. soyogo* is oligophagous species associated with *Ilex chinensis* Sims, *Ilex integra* Thunberg (Fig. 2.10d), *Ilex leucoclada* (Maximowicz) Makino and *Ilex pedunculosa* Miquel through its distribution range. The two *Schizomyia* species are morphologically very similar and are regarded as sister species. As a result of molecular analysis, they are inferred to have diverged 3.2–4.3 Ma (Tokuda et al. 2004). In Honshu (Chiba) and Kyushu (Fukuoka), where the two species are coexisting, *S. sasakii* emerges slightly earlier in the season than *S. soyogo*, and males are attracted only to conspecific females (Tabuchi and Amano 2003b; Tokuda et al. 2004), suggesting that the component or composition of female sex pheromone is different between the two species. In comparisons of genetic diversity and molecular phylogenetic relationships, the two species were considered to have speciated allopatrically (Tokuda et al. 2004); *Schizomyia soyogo* had been distributed in Honshu, Shikoku, and Kyushu earlier than *S. sasakii*, which arrived later in these areas from the Southwest Islands of Japan (Tokuda et al. 2004). Further comparative studies of the *Schizomyia* species together with host plant phylogeny are needed to reveal evolutionary processes of their host usage patterns and to find the original host plant of the common ancestor.

References

Ainouche ML, Baumel A, Salmon A, Yannic G (2003) Hybridization, polyploidy, and speciation in *Spartina* (Poaceae). New Phytol 161:165–172

Arnegard ME, McGee MD, Matthews B, Marchinko KB, Conte GL, Kabir S, Bedford N, Bergek S, Chan YF, Jones FC, Kingsley DM, Peichel CL, Schluter D (2014) Genetics of ecological divergence during speciation. Nature 511:307–311

Arrigo N, Barker MS (2012) Rarely successful polyploids and their legacy in plant genomes. Curr Opin Plant Biol 15:140–146

Berenbaum MR, Feeny PP (2008) Chemical mediation of host-plant specialization: the Papilionid paradigm. In: Tilmon KJ (ed) Specialization, speciation, and radiation. University of California Press, Berkeley, Los Angeles, London, pp 3–19

Bowser M, Bowser AM, Bowser MD, Collet DM, Melvin T (2018) DNA barcoding Alaskan willow rosette gall makers (Diptera: Cecidomyiidae: *Rabdophaga*). Newsl Alaska Entomol Soc 11:8–14

Condon MA, Steck GJ (1997) Evolution of host use in fruit flies of the genus *Blepharoneura* (Diptera: Tephritidae): cryptic species on sexually dimorphic host plants. Biol J Linn Soc 60:443–466

Cook MA, Ozeroff SN, Fitzpatrick SM, Roitberg BD (2011) Host-associated differentiation in reproductive behaviour of cecidomyiid midges on cranberry and blueberry. Entomol Exp Appl 141:8–14

Coyne JA, Elwyn S, Rolán-Alvarez E (2005) Impact of experimental design on *Drosophila* sexual isolation studies: direct effects and comparison to field hybridization data. Evolution 59:2588–2601

Craig TP (2016) Geographic mosaic of coevolution. In: The encyclopedia of evolutionary biology, vol 2. Elsevier, Amsterdam, pp 201–207

Craig TP, Livingston-Anderson A, Itami JK (2020) A small-tiled geographic mosaic of coevolution between *Eurosta solidaginis* and its natural enemies and host plant. Ecosphere 11:e03182. https://doi.org/10.1002/ecs2.3182

Després L, Pettex E, Plaisance V, Pompanon F (2002) Speciation in the globeflower fly *Chiastocheta* spp. (Diptera: Anthomyiidae) in relation to host plant species, biogeography, and morphology. Mol Phylogenet Evol 22:258–268

Dorchin N, Joy JB, Hilke LK, Wise MJ, Abrahamson WG (2015) Taxonomy and phylogeny of the *Asphondylia* species (Diptera: Cecidomyiidae) of north American goldenrods: challenging morphology, complex host associations, and cryptic speciation. Zool J Linnean Soc 174:265–304

Dorchin N, Scott ER, Clarkin CE, Luongo MP, Jordan S, Abrahamson WG (2009) Behavioural, ecological and genetic evidence confirm the occurrence of host-associated differentiation in goldenrod gall-midges. J Evol Biol 22:729–739

Drosopoulous S, Maryanska-Nadachowska A, Kuznetsova VG (2010) The Mediterranean: area of origin of polymorphism and speciation in the spittlebug *Philaenus* (Hemiptera, Aphrophoridae). Zoosyst Evol 86:125–128

Elsayed AK, Matsuo K, Kim W, Uechi N, Yukawa J, Gyoutoku N, Tokuda M (2018a) A new *Asphondylia* species (Diptera: Cecidomyiidae) and a eulophid wasp (Hymenoptera) inducing similar galls on leaf buds of *Schoepfia jasminodora* (Schoepfiaceae), with reference to their ecological traits and a description of the new gall midge. Entomol Sci 21:324–339

Elsayed AK, Yukawa J, Tokuda M (2018b) A taxonomic revision and molecular phylogeny of the eastern Palaearctic species of the genera *Schizomyia* Kieffer and *Asteralobia* Kovalev (Diptera: Cecidomyiidae: Asphondyliini), with descriptions of five new species of *Schizomyia* from Japan. ZooKeys 808:123–160

Fedorov AA (1969) Chromosome numbers of flowering plants. Komarov Botanical Institute, Academy of Sciences, U.S.S.R., Leningrad. (In Russian)

Fornaguera FR (2014) The genomic basis of parallel ecological speciation. A thesis submitted for the degree of Doctor of Philosophy at the University of Queensland, Queensland

Fujii T, Matsuo K, Abe Y, Yukawa J, Tokuda M (2014) An endoparasitoid avoids hyperparasitism by manipulating immobile host herbivore to modify host plant morphology. PLoS One 9: e102508. https://doi.org/10.1371/journal.pone.0102508

Fukatsu T, Aoki S, Kurosu U, Ishikawa H (1994) Phylogeny of Cerataphidini aphids revealed their symbiotic microorganisms and basic structure of their galls: implications for host-symbiont coevolution and evolution of sterile soldier castes. Zool Sci 11:613–623

Funk DJ, Nosil P, Etges WJ (2006) Ecological divergence exhibits consistently positive associations with reproductive isolation across disparate taxa. Proc Natl Acad Sci USA 103:3209–3213

Gagné RJ (1989) The plant-feeding gall midges of North America. Cornell University Press, Ithaca

Gagné RJ (2008) The gall midges (Diptera: Cecidomyiidae) of hickories (Juglandaceae: *Carya*). Mem Am Entomol Soc 48:1–147

Gagné RJ, Jaschhof M (2017) A catalog of the Cecidomyiidae (Diptera) of the world, 4th edn. Digital. 762 pp. [Cited 10 May 2020] Available from URL https://www.ars.usda.gov/ARSUserFiles/80420580/Gagne_2017_World_Cat_4th_ed.pdf

Gagné RJ, Moser JC (2013) The North American gall midges (Diptera: Cecidomyiidae) of hackberries (Cannabaceae: *Celtis* spp.). Mem Am Entomol Soc 49:1–103

Gagné RJ, Waring GL (1990) The *Asphondylia* (Cecidomyiidae: Diptera) of creosote bush (*Larrea tridentata*) in North America. Proc Entomol Soc Wash 92:649–671

Ganaha T, Nohara M, Sato S, Uechi Yamagishi K, Yamauchi S, Yukawa J (2007) Polymorphism of axillary bud galls induced by *Rhopalomyia longitubifex* (Diptera: Cecidomyiidae) on *Artemisia princeps* and *A. montana* (Asteraceae) in Japan and Korea, with the designation of new synonyms. Entomol Sci 10:209–212

Ganaha T, Yukawa J, Uechi N, Nohara M, Paik JC (2004) Identifications of some species of the genus *Rhopalomyia* (Diptera: Cecidomyiidae) inducing galls on *Artemisia* (Asteraceae) in Korea. Esakia 44:45–55

Grego CL, Weis AE, Neil NO, Bretz CK (1990) Sympatric sibling species from three phenotypically distinct *Asteromyia* (Diptera: Cecidomyiidae) galls on the same host plant species. Ann Entomol Soc Am 83:149–154

Hara M (1992) The genus *Fagus* and *Fagus* stands in the world. In: Endo Y, Hara M, Ohba T (eds) Natural history of beech forests. Natural History Museum and Institute, Chiba, pp 43–48. (In Japanese)

Harrison RG (1980) Dispersal polymorphism in insects. Annu Rev Ecol Syst 11:95–118

Hebert PDN, Cywinska A, Ball SL, DeWaard JR (2003) Biological identifications through DNA barcodes. Proc R Soc B Biol Sci 270:313–321

Hellqvist S, Larsson S (1998) Host acceptance and larval development of the gall midge *Dasineura tetensi* (Diptera, Cecidomyiidae) on resistant and susceptible black currant. Entomol Fennica 9:95–102

Hillbur Y, Celander M, Baur R, Rauscher S, Haftmann J, Franke S, Francke W (2005) Identification of the sex pheromone of the swede midge, *Contarinia nasturtii*. J Chem Ecol 31:1807–1828
Huang BH, Huang CW, Huang CL, Liao PC (2017) Continuation of the genetic divergence of ecological speciation by spatial environmental heterogeneity in island endemic plants. Sci Rep 7:5465. https://doi.org/10.1038/s41598-017-05900-1
Ito M, Yahara T, King RM, Watanabe K, Oshita S, Yokoyama J, Crawford DJ (2000) Molecular phylogeny of Eupatorieae (Asteraceae) estimated from cpDNA RFLP and its implication for the polyploid origin hypothesis of the tribe. J Plant Res 113:91–96
Jaschhof M, Jaschhof C (2013) The Porricondylinae (Diptera: Cecidomyiidae) of Sweden with notes on extralimital species. Stud Dipterol Suppl 20:1–392
Johnson B (1966) Wing polymorphism in aphids iii. The influence of the host plant. Entomol Exp Appl 9:213–222
Joy JB, Crespi BJ (2007) Adaptive radiation of gall-inducing insects within a single host-plant species. Evolution 61:784–795
Kamioka T, Iwasa Y (2017) Evolution of density-dependent wing polymorphism in insects. Evol Ecol Res 8:335–348
Khemakhem MM, Marrakchi M, Makni H (2005) Genetic diversity of *Mayetiola destructor* and *Mayetiola hordei* (Diptera: Cecidomyiidae) by inter-simple sequence repeats (ISSRs). Afr J Biotechnol 4:601–606
Kiritani K (1970) Studies on the adult polymorphism in the southern green stink bug, *Nezara viridula* (Hemiptera: Pentatomidae). Res Popul Ecol 12:19–34
Larsson S, Ekbom B (1995) Oviposition mistakes in herbivorous insects: confusion or a step towards a new host plant. Oikos 72:155–160
Lauron-Moreau A, Pitre FE, Argus GW, Labrecque M, Brouillet L (2015) Phylogenetic relationships of American willows (*Salix* L., Salicaceae). PLoS One 10:e0121965. https://doi.org/10.1371/journal.pone.0121965
Levy A, Feldman M (2002) The impact of polyploidy on grass genome evolution. Plant Physiol 130:1587–1593
Lin SF, Tokuda M, Yang MM (2020) Leaf gall polymorphism and molecular phylogeny of a new *Bruggmanniella* species (Diptera: Cecidomyiidae: Asphondyliini) associated with *Litsea acuminate* (Lauraceae) in Taiwan, with ecological comparisons and a species description. Entomol Sci 23:10–22
Liu B, Wendel JF (2002) Non-Mendelian phenomena in allopolyploid genome evolution. Curr Genomics 3:1–17
Loskutova OA, Zhiltzova LA (2016) Wing and body size polymorphism in populations of the stonefly *Arcynopteryx dichroa* McL. (Plecoptera: Perlodidae) in the Ural Mountains, Russia. Polar Res. https://doi.org/10.3402/polar.v35.26596
Lunds Universitet (2018) Final report summary – genes and speciation (genetics of ecological speciation in birds). Sölvegatan, Lund, Sweden. [Cited 25 November 2018]. https://cordis.europa.eu/result/rcn/60107_en.html
Maeda N, Sato S, Yukawa J (1982) Polymodal emergence pattern of the machilus leaf gall midge, *Daphnephila machilicola* Yukawa (Diptera, Cecidomyiidae). Kontyû 50:44–50
Makni H, Marrakchi M, Pasteur N (2000) Biochemical characterization of sibling species in Tunisian *Mayetiola* (Diptera: Cecidomyiidae). Biochem Syst Ecol 28:101–109
Masaki S, Shimizu T (1995) Variability in wing form in crickets. Res Popul Ecol 37:119–128
Matsubayashi KW, Ohshima I, Nosil P (2010) Ecological speciation in phytophagous insects. Entomologia Exp Applicata 134:1–27
Mayrose I, Zhan SH, Rothfels CJ, Magnuson-Ford K, Barker MS, Rieseberg LH, Otto SP (2011) Recently formed polyploid plants diversify at lower rates. Science 333:1257
Mishima M, Sato S, Tsuda K, Yukawa J (2014) Sexual isolation between two known intraspecific populations of *Hartigiola* (Diptera: Cecidomyiidae) that induce leaf galls on upper and lower surfaces of *Fagus crenata* (Fagales: Fagaceae), indicating possible diversification into sibling species. Ann Entomol Soc Am 107:789–798

Mishima M, Yukawa J (2007) Dimorphism of leaf galls induced by *Pseudasphondylia neolitseae* (Diptera: Cecidomyiidae) on *Neolitsea sericea* (Lauraceae), representing geographic variations in Kyushu, Japan. Bull Kyushu Univ Museum 5:57–64

Neger FW (1908) Ambrosiapiltze. Berichte der Deutsche Botanische Gesellschaft 26:735–754

Neger FW (1910) Ambrosiapiltze – III Weitere Beobachtungen an Ambrosiagallen. Berichte der Deutsche Botanische Gesellschaft 28:455–480

Newsholme C (1992) Willows: the genus *Salix*. Timber Press, Inc, Portland

Nijveldt WC, Yukawa J (1982) A taxonomic study on *Salix*-inhabiting gall midges in Japan (Diptera, Cecidomyiidae). Bull Kitakyushu Museum Nat Hist 4:23–56

Nosil P (2012) Ecological Speciation. Oxford University Press, Oxford

O'Connor TK, Laport RG, Whiteman NK (2019) Polyploidy in creosote bush (*Larrea tridentata*) shapes the biogeography of specialist herbivores. J Biogeogr 2019:1–14

Ohi T, Kajita T, Murata J (2003) Distinct geographic structure as evidenced by chloroplast DNA haplotypes and ploidy level in Japanese *Aucuba* (Aucubaceae). Am J Bot 90:1645–1652

Olvido AE, Elvington ES, Mousseau TA (2003) Relative effects of climate and crowding on wing polymorphism in the southern ground cricket, *Allonemobius socius* (Orthoptera: Gryllidae). Fla Entomol 86:158–164

Orra MR, Smith TB (1998) Ecology and speciation. Trends Ecol Evol 13:502–506

Osborn TC, Pires JC, Birchler JA, Auger DL, Chen ZJ, Lee HS, Comai L, Madlung A, Doerge RW, Colot V, Martienssen RA (2003) Understanding mechanisms of novel gene expression in polyploids. Trends Genet 19:141–147

Pellicer J, Garnatje T, Vallés J (2011) *Artemisia* (Asteraceae): understanding its evolution using cytogenetic and molecular systematic tools, with emphasis on subgenus *Dracunculus*. Rec Adv Pharm Sci 2011:199–222

Pogson GH (2016) Studying the genetic basis of speciation in high gene flow marine invertebrates. Curr Zool 62:643–653

Price PW (2003) Macroevolutionary theory on macroecological patterns. Cambridge University Press, Cambridge

Richards AJ (1973) The origin of *Taraxacum* agamospecies. Bot J Linn Soc 66:189–211

Roff D (1986) Evolution of wing polymorphism and its impact on life cycle adaptation in insects. In: Taylor F, Karban R (eds) The evolution of insect life cycles. Springer-Verlag, New York, pp 204–221

Rolán-Alvarez E, Caballero A (2000) Estimating sexual selection and sexual isolation effects from mating frequencies. Evolution 54:30–36

Rossi AM, Stiling P, Cattell MV, Bowdish TI (1999) Evidence for host-associated races in a gall-forming midge: trade-offs in potential fecundity. Ecol Entomol 24:95–102

Rundle HD, Nosil P (2005) Ecological speciation. Ecol Lett 8:336–352

Sato S, Harris KM, Collet DM, Kim W, Yukawa J (2020) Genetic variation in intraspecific populations of *Rabdophaga rosaria* (Diptera: Cecidomyiidae) indicating possible diversification scenarios into sibling species along with host range expansion on willows (Salicaceae: *Salix*). Zool J Linnean Soc 189:1426–1437

Sato S, Tsuda K, Yukawa J (2010) Decline in spring emergents of gall midge (Diptera: Cecidomyiidae) inducing leaf galls on *Fagus crenata* (Fagaceae) in Kyushu. Jpn J Environ Entomol Zool 21:7–13. (In Japanese with English summary)

Sato S, Yukawa J (2001) Absence record of *Fagus* gall midges (Diptera: Cecidomyiidae) on Ulleung Island, Korea and in North America. Esakia 41:17–25

Sato S, Yukawa J (2004) Redescription of *Hartigiola faggalli* (Monzen) comb. n. (Diptera: Cecidomyiidae) inducing leaf galls on *Fagus crenata* (Fagaceae) in Japan. Esakia 44:13–26

Sato S, Yukawa J (2008) Descriptions of two new *Mikiola* species (Diptera: Cecidomyiidae) that induce leaf galls on *Fagus crenata* (Fagaceae) in Japan. Stud Dipterol 15:151–164

Schilling MP, Mullen SP, Kronforst M, Safran RJ, Nosil P, Feder JF, Gompert Z, Flaxman SM (2018) Transitions from single- to multi-locus processes during speciation with gene flow. Genes 9:274. https://doi.org/10.3390/genes9060274

Schluter D (2000) The ecology of adaptive radiation. Oxford University Press, Oxford
Schluter D (2001) Ecology and the origin of species. Trends Ecol Evol 16:372–380
Schluter D (2009) Evidence for ecological speciation and its alternative. Science 323:737–741
Schluter D, Conte GL (2009) Genetics and ecological speciation. Proc Nat Acad Sci USA 106 (suppl. 1):9955–9962
Schwarz D, McPheron BA (2007) When ecological isolation breaks down: sexual isolation is an incomplete barrier to hybridization between *Rhagoletis* species. Evol Ecol Res 9:829–841
Sekimura T, Nijhout HF (2017) Diversity and evolution of butterfly wing patterns: an integrative approach. Springer Open, Singapore
Skuhravá M (1986) Family Cecidomyiidae. In: Soós A, Papp L (eds) Catalogue of Palaearctic Diptera: Sciaridae and Cecidomyiidae. Akadémiai Kiadó, Budapest. Vol. 4, pp 72–297
Skuhravá M, Karimpour Y, Sadeghi H, Gol A, Joghataie M (2014) Gall midges (Diptera: Cecidomyiidae) of Iran: annotated list and zoogeographical analysis. Acta Societatis Zoologicae Bohemoslovaca 78:269–301
Socha R (2001) Latitudinal gradient in response of wing polymorphism to photoperiod in a flightless bug, *Pyrrhocoris apterus* (Heteroptera: Pyrrhocoridae). Eur J Entomol 98:167–169
Soltis DE, Segovia-Salcedo MC, Jordon-Thaden I, Majure L, Miles NM, Mavrodiev EV, Mei W, Cortez MB, Soltis PS, Gitzendanner MA (2014) Are polyploids really evolutionary dead-ends (again)? A critical reappraisal of Mayrose *et al.* (2011). New Phytol 202:1105–1117
Soltis DE, Soltis PS (2003) Tate JA (2003) advances in the study of polyploidy since plant speciation. New Phytol 161:173–191
Stelter H (1970) Untersuchungen über Gallmücken XIV. *Rhabdophaga rosaria* (H. Loew, 1850) Kieffer, 1913 syn.: *Cecidomyia cinerearum* Hardy, 1850. Deutsche Entomologische Zeitschrift N.F 17:215–225
Stern DL (1995) Phylogenetic evidence that aphids, rather than plants, determine gall morphology. Proc R Soc Lond Ser B 260:85–89
Stiling P, Rossi AM (1995) Coastal insect herbivore communities are affected more by local environmental conditions than by plant genotype. Ecol Entomol 20:184–190
Stiling P, Rossi AM (1997) Deme formation in a dispersive gall-forming midge. In: Mopper S, Strauss SY (eds) Genetic structure and local adaptation in natural insect populations. Chapman & Hall, New York, pp 22–36
Stokes K, Stiling P, Gilg MR, Rossi AM (2012) The gall midge *Asphondylia borrichiae* (Diptera: Cecidomyiidae): an indigenous example of host-associated genetic divergence in sympatry. Environ Entomol 45:1246–1254
Stone GN, Cook JM (1998) The structure of cynipid oak galls: patterns in the evolution of an extended phenotype. Proc R Soc Lond Ser B 265:979–988
Sun Y, Shaw PC, Fung KP (2007) Molecular authentication of Radix Puerariae lobatae and Radix Puerariae thomsonii by ITS and 5S rRNA spacer sequencing. Biol Pharm Bull 30:173–175
Sunose T (1979) Two colour types of the euonymus gall midge, *Masakimyia pustulae* Yukawa and Sunose (Diptera: Cecidomyiidae). Kontyû 47:352–357
Sunose T (1983) Redescription of *Asphondylia morivorella* (Naito), comb. n. (Diptera: Cecidomyiidae), with notes on its bionomics. Appl Entomol Zool 18:22–29
Sunose T (1985a) Geographical distribution of two gall types of *Masakimyia pustulae* Yukawa & Sunose (Diptera, Cecidomyiidae) and reproductive isolation between them by a parasitoid. Kontyû 53:677–689
Sunose T (1985b) Population regulation of the euonymus gall midge *Masakimyia pustulae* Yukawa and Sunose (Diptera: Cecidomyiidae) by hymenopterous parasitoids. Res Popul Ecol 27:287–300
Tabuchi K, Amano H (2003a) Polymodal emergence pattern and parasitoid composition of *Asteralobia sasakii* (Monzen) (Diptera: Cecidomyiidae) on *Ilex crenata and I. integra* (Aquifoliaceae). Appl Entomol Zool 38:493–500

Tabuchi K, Amano H (2003b) Host-associated differences in emergence pattern, reproductive behavior and life history of *Asteralobia sasakii* (Monzen) (Diptera: Cecidomyiidae) between populations on *Ilex crenata* and *I. integra* (Aquifoliaceae). Appl Entomol Zool 38:501–508

Takasu K, Yukawa J (1984) Two-year life history of the neolitsea leaf gall midge, *Pseudasphondylia neolitseae* Yukawa (Diptera, Cecidomyiidae). Kontyû 52:596–604

Takizawa Y (1983) Life history of the beech leaf gall-midge, *Oligotrophus faggalli* (Diptera: Cecidomyiidae) in the Tohoku district, Japan. Nichirin Kantou Shiron 44:125–126. (In Japanese)

Thompson JN (1999) Specific hypotheses on the geographic mosaic of coevolution. Am Nat 153S: S1–S14

Thompson JN (2005) The geographic mosaic of coevolution. University of Chicago Press, Chicago, IL

Tokuda M (2012) Biology of Asphondyliini (Diptera: Cecidomyiidae). Entomol Sci 15:361–383

Tokuda M, Tabuchi K, Yukawa J, Amano H (2004) Inter- and intraspecific comparisons between *Asteralobia* gall midges (Diptera: Cecidomyiidae) causing axillary bud galls on *Ilex* species (Aquifoliaceae): species identification, host range, and mode of speciation. Ann Entomol Soc Am 97:954–970

Tokuda M, Uechi N, Yukawa J (2002) Distribution of *Asteralobia* gall midges (Diptera: Cecidomyiidae) causing axillary bud galls on Ilex species (Aquifoliaceae) in Japan. Esakia 42:19–31

Tokuda M, Yang MM, Yukawa J (2008) Taxonomy and molecular phylogeny of *Daphnephila* gall midges (Diptera: Cecidomyiidae) inducing complex leaf galls on Lauraceae, with descriptions of five new species associated with *Machilus thunbergii* in Taiwan. Zool Sci 25:533–545

Tokuda M, Yukawa J (2005) Two new and three known Japanese species of genus *Pseudasphondylia* Monzen (Diptera: Cecidomyiidae: Asphondyliini) and their life history strategies. Ann Entomol Soc Am 98:259–272

Uechi N, Kim WG, Tokuda M, Fujii T, Kikuchi H, Kakizaki M, Iwasaki A, Paik JC, Yukawa J (2018) Genetic and ecological differences between *Asphondylia yushimai* and the ivy gall midge, *Asphondylia* sp. (Diptera: Cecidomyiidae) with a new distribution record of the former from Hokkaido and South Korea. Appl Entomol Zool 53:363–371

Uechi N, Tokuda M, Yukawa J (2002) Distribution of *Asphondylia* gall midges (Diptera; Cecidomyiidae) in Japan. Esakia 42:1–10

Uechi N, Yukawa J (2004) Description of *Asphondylia itoi* sp. n. (Diptera: Cecidomyiidae) inducing fruit galls on *Distylium racemosum* (Hamamelidaceae) in Japan. Esakia 44:27–43

Uechi N, Yukawa J (2006a) Life history patterns and host ranges of the genus *Asphondylia* (Diptera: Cecidomyiidae). In: Ozaki K, Yukawa J, Ohgushi T, Price PW (eds) Galling arthropods and their associates: ecology and evolution. Springer-Verlag, Tokyo, pp 275–285

Uechi N, Yukawa J (2006b) Host range and life history of *Asphondylia sphaera* Monzen (Diptera: Cecidomyiidae): use of short-term alternate hosts. Ann Entomol Soc Am 99:1165–1171

Uechi N, Yukawa J, Usuba S (2005) Discovery of an additional winter host of the soybean pod gall midge, *Asphondylia yushimai* (Diptera: Cecidomyiidae) in Japan. Appl Entomol Zool 40:597–607

Uechi N, Yukawa J, Usuba S, Gyoutoku N, Mitamura T (2012) Findings of new cecidomyiid galls induced by *Asphondylia* segregates (Diptera: Cecidomyiidae) in Japan. Esakia 52:51–57

Uechi N, Yukawa J, Yamaguchi D (2004) Host alternation by gall midges of the genus *Asphondylia* (Diptera: Cecidomyiidae). In: Evenhuis NL, Kaneshiro KY (eds) D. Elmo Hardy memorial volume. Contributions to the systematics and evolution of diptera. Bishop Museum Press, Honolulu, pp 53–66. Bishop Museum Bulletin in Entomology 12

Via S (2001) Sympatric speciation in animals: the ugly duckling grows up. Trends Ecol Evol 16:381–390

Wendel JF (2000) Genome evolution in polyploids. Plant Mol Biol 42:225–249

Wendel JF (2015) The wondrous cycle of polyploidy in plants. Am J Bot 102:1753–1756

Windenfalk O, Gyllenstrand N, Sylvén E, Solbreck C (2002) Identity and phylogenetic status of two sibling gall midge species (Diptera: Cecidomyiidae: *Contarinia*) on the perennial herb *Vincetoxicum hitundinaria*. Syst Entomol 27:519–528

Yamauchi S, Ikenaga H, Yukawa J (1982) Midge galls collected from the south-west islands of Japan. Satsuma 31:1–23. (In Japanese with English summary)

Yang SY, Chen MY, Yang JT (2000) Plant galls in Taiwan. Taiwan Council of Agriculture, Taipei. (In Chinese)

Yukawa J (1971) A revision of the Japanese gall midges (Diptera: Cecidomyiidae). Mem Fac Agric Kagoshima Univ 8:1–203

Yukawa J (1974) Descriptions of new Japanese gall midge (Diptera, Cecidomyiidae, Asphondyliidi) causing leaf galls on Lauraceae. Kontyû 42:293–304

Yukawa J (1983) Arthropod community centred upon the neolitsea leaf gall midge, *Pseudasphondylia neolitseae* Yukawa (Diptera, Cecidomyiidae) and its host plant, *Neolitsea sericea* (Blume) Koidz. (Lauraceae). Mem Fac Agric Kagoshima Univ 19:89–96

Yukawa J (2000) Synchronization of gallers with host plant phenology. Popul Ecol 42:105–113

Yukawa J (2014) Family Cecidomyiidae. In: Nakamura T, Saigusa T, Suwa M (eds) Catalogue of the insects of Japan. Entomological Society of Japan, Fukuoka, pp 126–160. (In Japanese)

Yukawa J, Akimoto K (2006) Influence of synchronization between adult emergence and host plant phenology on the population density of *Pseudasphondylia neolitseae* (Diptera: Cecidomyiidae) inducing leaf galls on *Neolitsea sericea* (Lauraceae). Popul Ecol 48:13–21

Yukawa J, Fujimoto K, Ganaha-Kikumura T, Wanggyu K (2019a) Effects of typhoons on an arthropod community centered upon a leaf gall midge, *Pseudasphondylia neolitseae* (Diptera: Cecidomyiidae) and its host plant, *Neolitsea sericea* (Lauraceae) through leaf fall and late-season shoot production. Ecol Res 35:252–264

Yukawa J, Harris KM, Kim W (2020) Descriptions of two new species of the genus *Lasioptera* (Diptera: Cecidomyiidae) that infest tomato in the Mediterranean region and Hokkaido, Japan. Appl Entomol Zool 55:129–140

Yukawa J, Ichinose M, Wanggyu K, Uechi N, Gyotoku N, Fujii T (2016a) Lower development threshold temperatures and thermal constants for four species of *Asphondylia* (Diptera: Cecidomyiidae) in Japan and their larval developmental delay caused by heat stress. Appl Entomol Zool 57:71–80

Yukawa J, Ikenaga H, Sato S, Tokuda M, Ganaha-Kikumura T, Uechi N, Matsuo K, Mishima M, Tung GH, Paik JC, Ren BQ, Don XO (2012) Description and ecological traits of a new species of *Pitydiplosis* (Diptera: Cecidomyiidae) that induces leaf galls on *Pueraria* (Fabaceae) in East Asia, with a possible diversification scenario of intraspecific groups. Entomol Sci 15:81–98

Yukawa J, Kiritani K (1965) Polymorphism in the southern green stink bug. Pacific Insects 7:639–642

Yukawa J, Masuda H (1996) Insect and mite galls of Japan in colors. Zenkoku Nôson Kyôiku Kyôkai, Tokyo. (In Japanese with English explanations for color plates)

Yukawa J, Matoba I, Matoba M, Takasu H (2018a) Cecidomyiid galls found in Wakayama prefecture. Nanki Seibutsu 60:1–15

Yukawa J, Miyamoto K (1979) Redescription of *Asphondylia sphaera* Monzen (Diptera, Cecidomyiidae), with notes on its bionomics. Mem Fac Agric Kagoshima Univ 15:99–106

Yukawa J, Miyamoto K, Yamaguchi T, Takesaki K, Uechi N, Matsuo K (2016b) Key factor/key stage analysis of long-term life table data for *Asphondylia sphaera* (Diptera: Cecidomyiidae) that induces fruit galls on *Ligustrum japonicum* (Oleaceae). Ecol Entomol 41:516–526

Yukawa J, Nakagawa K, Saigou T, Awa T, Fukuda T, Higashi M (2013) Adult behavior of an ambrosia gall midge *Illiciomyia yukawai* (Diptera: Cecidomyiidae) and synchronization between its emergence and host plant phenology. Entomol Sci 16:400–412

Yukawa J, Nishida R, Fukuda H, Inoue R (2019b) Aristolochiaceae- and Asteraceae-feeding by larvae of *Papilio xuthus* L. (Lepidoptera: Papilionidae) in Japan: a review. Entomol Sci 22:355–364

Yukawa J, Ohsaki N (1988a) Separation of the aucuba fruit midge, *Asphondylia aucubae* sp. nov. from the ampelopsis fruit midge, *Asphondylia baca* Monzen (Diptera, Cecidomyiidae). Kontyû 56:365–376

Yukawa J, Ohsaki N (1988b) Adult behaviour of the aucuba fruit midge, *Asphondylia aucubae* Yukawa & Ohsaki (Diptera: Cecidomyiidae). Kontyû 56:645–652

Yukawa J, Rohfritsch O (2005) Biology and ecology of gall-inducing Cecidomyiidae (Diptera). In: Raman A, Schaefer CW, Withers TM (eds) Biology, ecology, and evolution of gall-inducing arthropods, vol 1. Science Publications, Enfield, pp 273–304

Yukawa J, Sunose T (1976) Description of a new gall midges (Diptera, Cecidomyiidae) on *Euonymus*, with notes on its bionomics. Kontyû 44:159–168

Yukawa J, Takahashi K, Ohsaki N (1976) Population behaviour of the neolitsea leaf gall midge, *Pseudasphondylia neolitseae* Yukawa (Diptera, Cecidomyiidae). Kontyû 44:358–365

Yukawa J, Tokuda M, Uechi N, Yasuda K, Ganaha-Kikumura T. Matsuo K, Shimizu Y, Yamaguchi D (2019c) Ecological divergence among morphologically and genetically related *Asphondylia* species (Diptera: Cecidomyiidae), with new life history data for three congeners including the *Alpinia* fruit gall midge. Entomol Sci 22:437–449

Yukawa J, Tokuda M, Yamagishi K (2014) Host plant ranges and distribution records of identified and unidentified species of the genus *Lasioptera* (Diptera: Cecidomyiidae) in Japan. Esakia 54:1–15

Yukawa J, Uechi N, Horikiri M, Tuda M (2003) Description of the soybean pod gall midge, *Asphondylia yushimai* sp. n. (Diptera: Cecidomyiidae), a major pest of soybean and findings of host alternation. Bull Entomol Res 93:73–86

Yukawa J, Uechi N, Tokuda M, Sato S (2005) Radiation of gall midges (Diptera: Cecidomyiidae) in Japan. Basic Appl Ecol 5:453–461

Yukawa J, Yamamura K, Fujimoto K, Tokuda M, Muroya S, Maki T, Sadoshima T, Fukuda T (2018b) Key-factors and key-stages that determine the leaf longevity of an evergreen broad-leaved tree, *Neolitsea sericea* (Lauraceae) at different growing sites in southern Japan. Ecol Res 33:175–190

Zera AJ, Denno RF (1997) Physiology and ecology of dispersal polymorphism in insects. Annu Rev Entomol 42:207–230

Zhang CX, Brisson JA, Xu HJ (2019) Molecular mechanisms of wing polymorphism in insects. Annu Rev Entomol 64:297–314

Chapter 3
Adaptive Radiation

Makoto Tokuda

Abstract Adaptive radiation process in herbivorous insects is a key subject to understand the diversity of extant insects. Some gall midges have been radiating on certain host taxa. In this chapter, I introduce examples of adaptive radiation in gall midges: *Asphondylia* on creosote bush, *Daphnephila* on *Machilus* (Lauraceae), and East Asian *Bruggmanniella* on Lauraceae.

Keywords Adaptive radiation · *Asphondylia* · *Daphnephila* · *Bruggmanniella*

3.1 Introduction

Herbivorous insects account for nearly a half of extant insect species (Price et al. 2011). So, their radiation mechanisms are important for understanding the diversification of insects. As mentioned in Chap. 1, the family Cecidomyiidae is presumed as the most speciose family in Diptera and its species richness is largely owing to herbivorous species inducing galls on their host plants. Almost all species of gall-inducing cecidomyiids are monophagous or oligophagous and their host ranges are restricted to a single plant species or genus (Yukawa et al. 2005; Yukawa and Rohfritsch 2005). Then, at the generic level, some large genera including *Asphondylia*, *Contarinia*, *Dasineura*, *Lasioptera*, and *Neolasioptera* are associated with various plant families, while others are restricted to certain plant taxa. Examples are shown in *Caryomyia* on *Carya* (Juglandaceae), *Procontarinia* on *Mangifera* (Anacardiaceae), *Rabdophaga* on *Salix* (Salicaceae), *Semudobia* on *Betula* (Betulaceae), and *Thecodiplosis* on *Pinus* (Pinaceae) and so on (Yukawa et al. 2005).

In this chapter, I introduce some examples of gall midges radiated on certain plant taxa, i.e., *Asphondylia* species on creosote bush, *Daphnephila* species on *Machilus* (Lauraceae), and East Asian *Bruggmanniella* species on Lauraceae.

M. Tokuda (✉)
Laboratory of Systems Ecology, Faculty of Agriculture, Saga University, Saga, Japan
e-mail: tokudam@cc.saga-u.ac.jp

J. Yukawa, M. Tokuda (eds.), *Biology of Gall Midges*, Entomology Monographs,
https://doi.org/10.1007/978-981-33-6534-6_3

3.2 Radiation of *Asphondylia* Species on Creosote Bush

Fifteen species of *Asphondylia* are associated with creosote bush *Larrea tridentata* (Augustin Pyramus de Candolle) Coville (Zygophyllaceae) in North America and each induces galls on host buds, stems or leaves. Joy and Crespi (2007) analyzed their phylogenetic relationships, including allied congeners, based on approximately 2000 bp of mitochondrial and nuclear DNA sequences. Because the *Larrea*-associated *Asphondylia* became a monophyletic group, they concluded that the *Asphondylia* species have diversified on *L. tridentata*.

O'Connor et al. (2019) surveyed polyploidy level of creosote bush in various localities and examined its association with *Asphondylia* species. As a result, one *Asphondylia* species was restricted to diploid hosts and five to tetraploid or hexaploid hosts. In areas where different plant cytotypes are parapatrically distributed, the contact zones acted as dispersal barriers for the gall midges. In addition, in localities where two cytotypes are sympatrically distributed, gall midges induced galls on their typical host cytotypes more frequently. These results suggest that plant polyploidy levels facilitate the diversification of associated herbivores on the host (Sect. 2.3).

Importance of plant polyploidy was reported also for *Dasineura cardaminis* Winnertz that induces flower bud galls on *Cardamine pratensis* Linnaeus (Brassicaceae). It selected only octoploid host populations for galling in Sweden (Arvanitis et al. 2010).

3.3 *Daphnephila* Species Associated with *Machilus*

Daphnephila is distributed in the eastern Palaearctic and Oriental Regions and comprises ten nominal species. One of them *Daphnephila linderae* Kieffer is associated with *Lindera pulcherrima* Bentham in India, but all the remaining species induce diverse shaped leaf or stem galls on *Machilus* (Lauraceae) (Fig. 3.1) (Kieffer 1905; Yukawa 1974; Tokuda et al. 2008; Pan et al. 2015; Gagné and Jaschhof 2017). There are some more undescribed species associated with *Machilus* or *Persea* in Southeast Asia, Taiwan and Japan (Yukawa 1974; Tokuda and Yukawa 2007; Tung and Yang 2018). *Persea* is very close to *Machilus*, both of them belonging to the '*Persea* group' sensu Chanderbali et al. (2001) in Lauraceae.

Based on phylogenetic analyses of the Oriental and eastern Palaearctic *Daphnephila*, a stem galler situated at the most basal clade in the genus (Tokuda et al. 2008; Pan et al. 2015). So, gall induction on stems is considered as an ancestral state and leaf-galling habit is descendent. Further studies are needed to clarify the radiation process of leaf gallers in relation to geographic distribution and host plants.

Fig. 3.1 Selected examples of cecidomyiid galls induced by identified or unidentified species of *Daphnephila* on *Machilus japonica* in Japan (**a**), *M. thunbergii* in Japan (**b**–**c**) and in Taiwan (**d**–**h**), and *M. zuihoensis* in Taiwan (**i**). (**a**), a club-shaped gall of *Daphnephila* sp.; (**b**), barrel-shaped galls of *Daphnephila machilicola*; (**c**), a crown-shaped gall of *Daphnephila* sp.; (**d**), banana-shaped galls of *D. stenocalia*; (**e**), ovoid galls of *D. taiwanensis*; (**f**), swan-shaped galls of *D. ornithocephala*; (**g**), obovate galls of *D. sueyenae*; (**h**), spindle-shaped galls of *D. truncicola*; (**i**), a urn-shaped gall of *Daphnephila urnicola*

3.4 Asian *Bruggmanniella* Species Associated with Lauraceae

Bruggmanniella was previously reported only from the Neotropical and southern Nearctic Regions (e.g. Möhn 1961; Gagné 2004), but recently several species have been described from Japan and Taiwan (Tokuda and Yukawa 2006; Gagné and Jaschhof 2017; Lin et al. 2019, 2020). In the New World, *Bruggmanniella* is associated with various plant families such as Anacardiaceae, Annonaceae, Celastraceae, Dilleniaceae, Lauraceae, Malpighiaceae, Moraceae and Sapotaceae (Gagné and Jaschhof 2017). In contrast, in the eastern Palearctic Region, the genus has radiated exclusively on Lauraceae, including *Cinnamomum*, *Litsea* (treated also as *Actiondaphne*, an independent genus from *Litsea*), and *Neolitsea* (Tokuda and Yukawa 2006; Gagné and Jaschhof 2017; Lin et al. 2019, 2020).

The tribe Asphondyliini is divided into two subtribes Schizomyiina and Asphondyliina, and the former is regarded as a paraphyletic group (Tokuda and

Yukawa 2007; Tokuda 2012). In Asphondyliini, *Asphondylia* is the most speciose, widespread, and radiating genus (Yukawa et al. 2005). *Bruggmanniella* is also distributed both in the Old and New Worlds, but the genus is phylogenetically considered as an old taxon of Asphondyliina (Tokuda and Yukawa 2007), which shares some morphological features with several other Old World genera such as *Illiciomyia*, *Probruggmanniella*, and *Pseudasphondylia*. In contrast, Neotropical genera of Asphondyliina other than *Bruggmanniella* are morphologically closer to *Asphondylia*, suggesting that they have later origins than *Bruggmaniella*. The wide distribution range and its phylogenetic position imply that *Bruggmanniella* previously radiated worldwide, as seen in the present situation of *Asphondylia*, but now the members remain only in the Neotropical Region and East Asia. Further studies dealing with the divergent time between Asian and Neotropical *Bruggmanniella* will reveal the possible origin and diversification scenario of this genus.

References

Arvanitis L, Wiklund C, Münzbergova Z, Dahlgren JP, Ehrlén J (2010) Novel antagonistic interactions associated with plant polyploidization influence trait selection and habitat preference. Ecol Lett 13:330–337

Chanderbali AS, Van Der Werff H, Renner SS (2001) Phylogeny and historical biogeography of Lauraceae: evidence from the chloroplast and nuclear genomes. Ann Mo Bot Gard 88:104–134

Gagné RJ (2004) A catalog of the Cecidomyiidae (Diptera) of the world. Mem Entomol Soc Washington 25:1–408

Gagné RJ, Jaschhof M (2017) A catalog of the Cecidomyiidae (Diptera) of the world, 4th edn. Digital. [Cited 10 June 2020] Available from URL https://www.ars.usda.gov/ARSUserFiles/80420580/Gagne_2017_World_Cat_4th_ed.pdf

Joy JB, Crespi BJ (2007) Adaptive radiation of gall-inducing insects within a single host-plant species. Evolution 61:784–795

Kieffer JJ (1905) Étude sur de nouveaux insectes et phytoptides gallicoles du Bengale. Annales de la SociÉtÉ Scientifique de Bruxelles 29:143–200

Lin SF, Tokuda M, Yang MM (2019) The first record of genus *Bruggmanniella* (Diptera: Cecidomyiidae) from Taiwan with description of a new species inducing bud galls on *Neolitsea parvigemma* (Lauraceae). J Asia Pac Entomol 22:203–207

Lin SF, Tokuda M, Yang MM (2020) Leaf gall polymorphism and molecular phylogeny of a new *Bruggmanniella* species (Diptera: Cecidomyiidae: Asphondyliini) associated with *Litsea acuminata* (Lauraceae) in Taiwan, with ecological comparisons and a species description. Entomol Sci 23:10–22

Möhn E (1961) Gallmücken (Diptera, Itonididae) aus El Salvador. 4. Zur Phylogenie der Asphondyliidi der neotropischen und holarktischen Region. Senckenberg Biol 42:131–330

O'Connor TK, Laport RG, Whiteman NK (2019) Polyploidy in creosote bush (*Larrea tridentata*) shapes the biogeography of specialist herbivores. J Biogeogr 2019:1–14

Pan LY, Chiang TC, Weng YC, Chen WN, Hsiao SC, Tokuda M, Tsai CL, Yang MM (2015) Taxonomy and biology of a new ambrosia gall midge *Daphnephila urnicola* sp. nov. (Diptera: Cecidomyiidae) inducing urn-shaped leaf galls on two species of *Machilus* (Lauraceae) in Taiwan. Zootaxa 3955:371–388

Price PW, Denno RF, Eubanks MD, Finke DL, Kaplan I (2011) Insect ecology: behavior, populations and communities. Cambridge University Press, Cambridge

Tokuda M (2012) Biology of Asphondyliini (Diptera: Cecidomyiidae). Entomol Sci 15:361–383
Tokuda M, Yang MM, Yukawa J (2008) Taxonomy and molecular phylogeny of *Daphnephila* gall midges (Diptera: Cecidomyiidae) inducing complex leaf galls on Lauraceae, with descriptions of five new species associated with *Machilus thunbergii* in Taiwan. Zool Sci 25:533–545
Tokuda M, Yukawa J (2006) First records of genus *Bruggmanniella* (Diptera: Cecidomyiidae: Asphondyliini) from Palaearctic and oriental regions, with descriptions of two new species that induce stem galls on Lauraceae in Japan. Ann Entomol Soc Am 99:629–637
Tokuda M, Yukawa J (2007) Biogeography and evolution of gall midges (Diptera: Cecidomyiidae) inhabiting broad-leaved evergreen forests in oriental and eastern Palaearctic regions. Orient Insects 41:121–139
Tung GS, Yang MM (2018) Let's gall Taiwan – A guidebook on insect galls. Council of Agriculture Executive Yuan. Forestry Bureau, Taipei
Yukawa J (1974) Descriptions of new Japanese gall midges (Diptera, Cecidomyiidae, Asphondyliidi) causing leaf galls on Lauraceae. Kontyû 42:293–304
Yukawa J, Rohfritsch O (2005) Biology and ecology of gall–inducing Cecidomyiidae (Diptera). In: Raman A, Schaefer CW, Withers TM (eds) Biology, ecology, and evolution of gall–inducing arthropods, vol 1. Science Publications, Enfield, pp 273–304
Yukawa J, Uechi N, Tokuda M, Sato S (2005) Radiation of gall midges (Diptera: Cecidomyiidae) in Japan. Basic Appl Ecol 5:453–461

Part III
Diversity and Distribution

Chapter 4
Diversity of Gall-Inducing Cecidomyiid in Japan

Junichi Yukawa

Abstract First, each gall sort is defined based on gall structure and host plant information. Then, the number of sorts of gall was compared between different gall-inducing arthropod taxa, and cecidomyiid galls were confirmed to be most numerous in Japan. Up to present, 122 identified species of gall-inducing cecidomyiid have been known to occur in Japan. Faunistic comparison between China and Japan indicated no significant difference between the two areas. Some of Japanese cecidomyiids expanded their distribution range northward by overcoming big thermal differences and have established themselves in cold areas. Information on gall sorts, host plants, and distribution in this Chapter would be useful for young students to find target cecidomyiid galls in the field.

Keywords Distribution range · Faunistic comparison · Gall-inducing cecidomyiid · Northern limit · Sorts of gall · Tohoku district

4.1 Definition of Gall Sorts

In this book, each gall sort is defined based on galled organ and position, gall shape, size, hardness, inner structure (the number and size of larval chambers), presence or absence of protuberances on gall surface, solitary or gregarious, and host plant information on taxonomy and phenology. Usually, similar sorts of gall induced on congeneric host plants are gathered into one sort. Such a way of gall sort definition has been supported by morphological and molecular studies of gall midges (e.g. Nijveldt and Yukawa 1982; Tokuda et al. 2004; Sato and Yukawa 2006; Uechi and Yukawa 2006a; Elsayed et al. 2019a, b, c; Kim et al. 2019; Yukawa et al. 2020). Therefore, the number of gall sorts roughly represents the potential number of gall-inducing cecidomyiid species, although we need to adjust both the

J. Yukawa (✉)
Entomological Laboratory, Faculty of Agriculture, Kyushu University, Fukuoka, Japan
e-mail: galling-cecidomyiids@kyudai.jp

J. Yukawa, M. Tokuda (eds.), *Biology of Gall Midges*, Entomology Monographs,
https://doi.org/10.1007/978-981-33-6534-6_4

numbers in cases when the species exhibit gall shape polymorphism (e.g. Ganaha et al. 2007; Sect.2.3), sister species relation (e.g. Tokuda et al. 2004; Sect. 2.8), host plant alternation (Uechi and Yukawa 2006b; Sect. 6.4), and polyphagous host range (e.g. Uechi et al. 2003; Sect. 12.1).

4.2 Cecidomyiid Galls in Japan

Galls are induced by various species of arthropod taxa, such as Agromyzidae, Cecidomyiidae, Chloropidae, Tephritidae (Diptera), Cynipidae, Chalcidoidea, Tenthredinidae (Hymenoptera), Adelgidae, Aleyrodidae, Aphidoidea, Coccoidea, Phylloxeridae, Psylloidea, Tingidae (Hemiptera), Pyralidae, Sesiidae, Tortricidae (Lepidoptera), Apionidae, Buprestidae, Curculionidae (Coleoptera), Phlaeothripidae, Thripidae (Thysanoptera), and Eriophyoidea (Acarina). According to Yukawa and Masuda (1996), galls induced by Cecidomyiidae were most numerous in Japan in terms of the number of gall sorts, followed by those of Cynipidae, Aphidoidea, and Eriophyoidea (Table 4.1). This tendency is common in the Holarctic Region (e.g. Felt 1965; Roskam 2019) as will be mentioned in Sect. 5.3. Unlike cynipids and tenthredinids that use a few restricted plant families for galling, cecidomyiids induce galls on various plant families. Therefore, their galls are most diverse among arthropod galls in shape and structure.

After Yukawa and Masuda (1996), many amateur naturalists, my colleagues, and students, and I have found various sorts of cecidomyiid gall in many parts of Japan. At present, they approximately amount to 734 sorts on 88 families of 38 vascular plant orders (Table 4.2). Sorts of gall are dominant in such large families as Asteraceae (90), Fagaceae (59), Rosaceae (40), Fabaceae (35), Lamiaceae (33), Lauraceae (28), Poaceae (22), and Hydorangeaceae (21). In Fagaceae, in contrast to cynipids that exclusively use *Quercus* for galling, cecidomyiids use both *Quercus* and *Fagus*, of which the latter hosts more diverse sorts of gall than the former (Sect. 2.5). Lauraceae consists of many evergreen broad-leaved trees that are distributed widely from southern parts of Japan to southeastern Asia. On these trees, various sorts of gall are induced by species of Asphondyliini (Chap. 3).

4.3 Identified Gall-Inducing Cecidomyiids in Japan

In the Catalogue of the Insects of Japan, Yukawa (2014) listed 139 identified species of the subfamily Cecidomyiinae together with 14 questionable or erroneously recorded species of Cecidomyiinae from Japan. In this Chapter, the following 35 non-gall inducing species of Cecidomyiinae are excluded from the 139 species. They are 10 predacious species belonging to *Aphidoletes*, *Diadiplosis*, *Endaphis*, *Feltiella, and Tricontarinia*, three resin-feeding species each belonging to *Cecidomyia*, *Lestodiplosis*, and *Resseliella*, four phytophagous but non-gall inducing species of *Resseliella*, and 18 species of which generic position and feeding habit

Table 4.1 The number of sorts of arthropod gall found in Japan until 1996 and the number of plant families that were used for gall induction

Gall-inducing taxon	No. of sorts of gall	%	No. of plant families used	Name of host plant family
Cecidomyiidae	628	44.1	65	Many families
Cynipidae	180	12.6	3	Fagaceae, Rosaceae, Asteraceae
Aphidoidea	178	12.5	38	Many families
Eriophyoidea	171	12.0	45	Many families
Psylloidea	73	5.1	24	Many families
Lepidoptera other than Sesiidae	41	2.9	12	Many families
Tenthredinidae	30	2.1	2	Salicaceae, Caprifoliaceae
Coccoidea	20	1.4	5	Pinaceae, Fagaceae, Lauraceae, Schisandraceae, Apocynaceae
Coleoptera	20	1.4	9	Many families
Sesiidae	17	1.2	5	Fagaceae, Rosaceae, Vitaceae, Cucurbitaceae, Rubiaceae
Chalcidoidea	13	0.9	6	Several families
Tephritidae	12	0.8	1	Asteraceae
Thysanoptera	10	0.7	9	Many families
Tingidae	8	0.6	3	Salicaceae, Lamiaceae, Asteraceae
Adelgidae	7	0.5	1	Pinaceae
Chloropidae	6	0.4	1	Poaceae
Phylloxeridae	4	0.3	3	Fagaceae, Ulmaceae, Vitaceae
Aleyrodidae	3	0.2	1	Theaceae
Agromyzidae	2	0.1	1	Fabaceae
Total	1423	100	92	

Rearranged the data in Yukawa and Masuda 1996 and added information on host plant families

are uncertain because of insufficiency in the original descriptions. Then, 17 new (Elsayed et al. 2017, 2018a, b, 2019a, b, c, 2020; Kim et al. 2014, 2019; Yukawa et al. 2020) and two described species (Yukawa et al. 2017; Kim et al. 2019) that were newly found in Japan from 2015 to 2020 are added to the list. As a result, 123 identified species of gall-inducing cecidomyiids are now known to occur in Japan (Table 4.3). They are responsible for at least 125 sorts of gall including those caused by host-alternating species.

Gall-inducing cecidomyiids for about 600 remaining sorts of gall are still waiting for species identification. Some of them have been identified to the generic level based on larval morphological characters, gall shape, galled organ, and host plant information. They are species of *Asphondylia* (Uechi et al. 2002), *Contarinia* (Tokuda and Yukawa 2004; Uechi et al. 2011), *Daphnephila* (Yukawa 1974), *Lasioptera* (Yukawa et al. 2014), and *Rhopalomyia* (Yukawa and Masuda 1996). However, most of them remain unidentified because mature larvae quit the galls before winter and adults are hardly reared from the soil or from galls on the ground.

Table 4.2 The number of sorts of cecidomyiid galls that have been found in Japan until 2020

Order and family of host plant[a]	Gall sorts	Order and family of host plant[a]	Gall sorts	Order and family of host plant[a]	Gall sorts
Polypodiales		Vitales		Cornales	
Dennstaedtiaceae	1	Vitaceae	17	Cornaceae + Helwingiaceae	5
Pinales		Crossosomatales		Hydorangeaceae	21
Pinaceae	8	Staphyleaceae	2	Ericales	
Taxaceae	1	Stachyuraceae	1	Balsaminaceae	1
Cupressaceae	6	Myrtales		Primulaceae	2
Austrobaileyales		Lythraceae	1	Theaceae	7
Schisandraceae	4	Myrtaceae	1	Pentaphylacaceae	5
Piperales		Celastrales		Symplocaceae	9
Piperaceae	1	Celastraceae	6	Styracaceae	18
Magnoliares		Malpighiales		Actinidiaceae	2
Magnoliaceae	6	Salicaceae	17	Clethraceae	1
Laurales		Violaceae	2	Ericaceae	11
Lauraceae	28	Euphorbiaceae	7	Order not determined	
Dioscoreales		Clusiaceae	2	Boraginaceae	1
Dioscoreaceae	2	Oxalidales		Garryales	
Liliales		Oxalidaceae	1	Garryaceae	2
Smilacaceae	4	Elaeocarpaceae	1	Gentianales	
Liliaceae	2	Fabales		Rubiaceae	17
Asparagales		Fabaceae	35	Gentianaceae	1
Orchidaceae	1	Rosales		Apocynaceae	7
Poales		Rosaceae	40	Solanales	
Cypraceae	2	Elaeagnaceae	3	Solanaceae	5
Poaceae	22	Rhamnaceae	2	Lamiales	
Zingiberales		Ulmaceae	5	Oleaceae	11
Zingiberacaea	1	Cannabaceae	10	Acanthaceae	1
Ranunculales		Moraceae	15	Lamiaceae	33
Lardizabalaceae	5	Urticaceae	12	Phrymaceae	3
Menispermaceae	2	Cucurbitales		Plantaginaceae	2
Berberidaceae	1	Cucurbitaceae	1	Aquifoliales	
Ranunculaceae	10	Fagales		Helwingiaceae	2
Buxales		Fagaceae	59	Aquifoliaceae	9
Buxaceae	2	Myricaceae	1	Apiales	
Caryophyllales		Betulaceae	12	Araliaceae	16
Polygonaceae	5	Juglandaceae	3	Apiaceae	8
Caryophyllaceae	4	Malvares		Dipsacales	
Amaranthaceae	2	Malvaceae	7	Adoxaceae	17
Phytolaccaceae	1	Sapindales		Caprifoliaceae	16
Santalales		Sapindaceae	11	Asterales	
Schoepfiaceae	1	Anacardiaceae	1	Campanulaceae	7
Loranthaceae	1	Simaroubaceae	1	Asteraceae	90

(continued)

Table 4.2 (continued)

Order and family of host plant[a]	Gall sorts	Order and family of host plant[a]	Gall sorts	Order and family of host plant[a]	Gall sorts
Saxifragales		Rutaceae	3		
Hamamelidaceae	2		**Total no. of plant families**		**89**
Saxifragaceae	1		**Total sorts of gall**		**734**

[a]Names of orders and families of host plants are arranged following syllabus of the vascular plants of Japan (Ohba 2009)

In future, young dipterists will be able to find many new species of gall-inducing cecidomyiids in Japan.

4.4 Faunistic Comparison between Japan and China

Jiao and Bu (2014) provided a checklist of the Chinese species of Cecidomyiidae, which includes 177 species belonging to 77 genera recorded from 1900 to 2012. Among 177 species, 114 species belonging to 53 genera were included in the subfamily Cecidomyiinae. From the 114 species, 42 species, such as predacious species, resin-feeding species, phytophagous but non-gall inducing species, and a few species of which feeding habit are uncertain, were excluded for faunistic comparison between Japan and China (including Taiwan) in the number of gall-inducing species belonging to different genera of Cecidomyiinae (Table 4.4). Although there are many genera without common species, statistical analysis indicates no significant difference between Japan and China in the aforementioned number of species (Chi-square test for independence, $\chi^2 = 80.545$, $df = 52$, $P = 0.068 > P = 0.01$, *Cramer's* $V = 0.6427$).

Because both Japan and China are included in the East Palaearctic Region and partly in the Oriental Region, faunistic similarity is natural but to some extent as shown with the *Cramer's V* that is nearer to 0.5 (intermediate) than to 1.0 (coincidence). Many undescribed species of cecidomyiids await discovery both in Japan and China. We need further faunistic comparison when cecidomyiid fauna of Japan and China becomes clearer in future.

4.5 Similarity in the Gall Midge Fauna between Taiwan and Southwestern Japan

Tung and Yang (2018) provided many photographs of insect galls found in Taiwan. Among them galls induced by seven identified and 38 unidentified species of cecidomyiids are included. Most of the galls occur on evergreen broad-leaved trees of Aquifoliaceae, Elaeocarpaceae, Fagaceae, Schisandraceae, Lauraceae, Rutaceae, Symplocaceae, Theaceae, etc. The gall midges belong to one of the

Table 4.3 Identified species of gall-inducing cecidomyiids in Japan with their host plants

Higher taxonomic position	Cecidomyiid species[a]	Host plant species (family) in Japan	Note
Cecidomyiidi			
Asphondyliini			
Asphondyliina			
Asphondylia	*aucubae* Yukawa & Ohsaki	*Aucuba japonica* (Garryaceae)	
	baca Monzen	*Ampelopsis glandulosa* var. *heterophylla* (Vitaceae), *Weigela floribunda, W. coraeensis, W. japonica, W.* spp. (Caprifoliaceae)	HPA[b]
	itoi Uechi & Yukawa	*Distylium racemosum* (Hamamelidaceae)	
	japonicus (Monzen)	*Weigela hortensis* (Caprifoliaceae)	Need ID[c]
	morivorella (Naito)	*Morus alba* (Moraceae)	
	sphaera Monzen	*Ligustrum japonicum, L. lucidum, L. obtusifolium, L. ovalifolium* (Oleaceae), *Rhus succedanea, R. sylvestris* (Anacardiaceae)	HPA[b]
	tojoi Elsayed & Tokuda	*Schoepfia jasminodora* (Schoepfiaceae)	
	yushimai Yukawa & Uechi	*Glycine max* (Fabaceae), many other fabaceous plants, *Laurocerasus zippeliana* (Rosaceae), *Osmanthus heterophyllus* (Oleaceae)	HPA[b]
Bruggmanniella	*actinodaphnes* Tokuda & Yukawa	*Actinodaphne lancifolia* (Lauraceae)	
	cinnamomi Tokuda & Yukawa	*Cinnamomum japonicum* (Lauraceae)	
Daphnephila	*machilicola* Yukawa	*Machilus thunbergii* (Lauraceae)	
Illiciomyia	*yukawai* Tokuda	*Illicium anisatum* (Schisandraceae)	
Pseudasphondylia	*elaeocarpini* Tokuda & Yukawa	*Elaeocarpus sylvestris* (Elaeocarpaceae)	
	kiritanii Tokuda & Yukawa	*Cornus controversa* (Cornaceae)	HPA[b]
	matatabi (Yuasa & Kumazawa)	*Actinidia polygama* (Actinidiaceae)	HPA[b]
	neolitseae Yukawa	*Neolitsea sericea* (Lauraceae)	
	rokuharensis Monzen	*Viburnum dilatatum* (Caprifoliaceae)	

(continued)

Table 4.3 (continued)

Higher taxonomic position	Cecidomyiid species[a]	Host plant species (family) in Japan	Note
	tominagai Elsayed & Tokuda	*Eleutherococcus spinosus* (Araliaceae)	HPA[b]
Kiefferia	*ezoensis* Kim & Yukawa	*Angelica ursina* (Apiaceae)	
	olla Kim & Yukawa	*Oenanthe javanica* (Apiaceae)	
	pericarpiicola (Bremi)	*Angelica edulis, A. sachalinensis, Spuriopimpinella calycina, Torilis japonica* (Apiaceae)	
Schizomyiina			
Ampelomyia	*conicocoricis* Elsayed & Tokuda	*Vitis coignetiae, V. ficifolia, V. flexuosa* (Vitaceae)	
Oxycephalomyia	*styraci* (Shinji)	*Styrax japonicus* (Styracaceae)	
Schizomyia	*achyranthesae* Elsayed & Tokuda	*Achyranthes bidentata* (Amaranthaceae)	
	broussonetiae Elsayed & Tokuda	*Broussonetia papyrifera* (Moraceae)	
	castanopsisae Elsayed & Tokuda	*Castanopsis sieboldii* (Fagaceae)	
	diplocyclosae Elsayed & Tokuda	*Diplocyclos palmatus* (Cucurbitaceae)	
	doellingeriae Kovalev	*Aster scaber (Asteraceae)*	
	humuli (Shinji)	*Humulus japonicus* (Cannabaceae)	
	paederiae Elsayed & Tokuda	*Paederia foetida* (Rubiaceae)	
	patriniae (Shinji)	*Patrinia villosa* (Valerianaceae)	
	sasakii (Monzen)	*Ilex crenata, I. maximowicziana* (Aquifoliaceae)	
	soyogo Kikuti	*Ilex chinensis, I. integra, I. leucoclada, I. pedunculosa* (Aquifoliaceae)	
	uechiae Elsayed & Tokuda	*Ampelopsis glandulosa* var. *brevipedunculata* (Vitaceae)	
	usubai Elsayed & Tokuda	*Trachelospermum asiaticum* (Apocynaceae)	
Cecidomyiini			

(continued)

Table 4.3 (continued)

Higher taxonomic position	Cecidomyiid species[a]	Host plant species (family) in Japan	Note
Contarinia	*hydorangeae* Shinji	*Hydrangea paniculata* (Hydrangeaceae)	
	inouyei Mani	*Cryptomeria japonica* (Cupressaceae)	
	maculipennis Felt	*Dendrobium phalaenopsis* (Orchidaceae), *Jasminum sambac* (Oleaceae), *Momordica charantia* (Cucurbitaceae), *Pseuderanthemum laxiflorum* (Amaranthaceae)	Polyphagy Invasive
	mali Barnes	*Marus pumila* (Rosaceae)	
	matusintome Haraguti & Monzen	*Pinus densiflora, P. thunbergii* (Pinaceae)	
	okadai (Miyoshi)	*Citrus unshiu, C. hassaku, C. iyo, C. natsudaidai, C.* spp. (Rutaceae)	
	salicola Shinji	*Salix* sp. (Salicaceae)	
	tiliarum (Kieffer)	*Tilia japonica, T. maximowicziana* (Malvaceae)	Need ID[c]
Macrodiplosis	*selenis* Kim & Yukawa	*Quercus serrata, Q. dentata, Q. mongolica* (Fagaceae)	
Paradiplosis	*manii* (Inouye)	*Abies sachalinensis* var. *sachalinensis, A. firma, A. homolepis* (Pinaceae)	
Procontarinia	*mangicola* (Shi)	*Mangifera indica* (Anacardiaceae)	Invasive
Stenodiplosis	*sorghicola* (Coquillett)	*Sorghum bicolor* (Poaceae)	Invasive
Thecodiplosis	*japonensis* Uchida & Inouye	*Pinus densiflora, P. thunbergii* (Pinaceae)	
Clinodiplosini			
Ametrodiplosis	*acutissima* (Monzen)	*Quercus acutissima* (Fagaceae)	Need ID[d]
Sitodiplosis	*mosellana* (Géhin)	*Triticum aestivum, Hordeum vulgare, Agropyron ciliare* var. *minus, Elymus tsukushiensis, E. racemifer* var. *japonensis* (Poaceae)	
Lopesiini			
Obolodiplosis	*robiniae* (Haldeman)	*Robinia pseudoacacia* (Fabaceae)	Invasive
Unplaced to tribe			
Aschistonyx	*eppoi* Inouye	*Juniperus chinensis* (Cupressaceae)	
Etsuhoa	*okayamana* Inouye	*Juniperus rigida* (Cupressaceae)	
Geromyia	*nawai* (Monzen)	*Pleioblastus simonii, P. chino, P. c.* var. *viridis* (Poaceae)	
Orseolia	*miscanthi* (Shinji)	*Miscanthus sinensis, M. condensatus* (Poaceae)	

(continued)

Table 4.3 (continued)

Higher taxonomic position	Cecidomyiid species[a]	Host plant species (family) in Japan	Note
Pitydiplosis	*puerariae* Yukawa, Ikenaga & Sato	*Pueraria lobata*, *P. montana* (Fabaceae)	
Tokiwadiplosis	*matecola* Simbolon & Yukawa	*Lithocarpus edulis* (Fagaceae)	
Lasiopteridi			
Dasineurini			
Dasineura	*asteriae* (Shinji)	*Aster scaber* (Asteraceae)	
	ezomatsue Uchida & Inouye	*Picea jezoensis, P. glehnii* (Pinaceae)	
	fulvicola Shinji	Use vacated galls of *Rhopalomyia* sp. on *Artemisia indica* or *A. japonica*	Successor
	nipponica Inouye	*Larix kaempferi* (Pinaceae)	
	oxycoccana (Johnson)	*Vaccinium* spp. (Ericaceae)	Invasive
	procera Rübsaamen	*Aster yomena* (Asyeraceae)	Need ID[c]
	pteridis (Müller)	*Pteridium aquilinum* (Dennstaedtiaceae)	Need ID[c]
	shinjii Skuhravá	*Vicia pseudo-orobus, V. unijuga* (Fabaceae)	
	symplocos Elsayed & Tokuda	*Symplocos cochinchinensis* (Symplocaceae)	
	viciae (Kieffer)	*Vicia cracca* (Fabaceae)	Need ID[c]
	viticola (Osten Sacken)	*Vitis ficifolia* var. *lobata*, *V. saccharifera, V.* spp., *Ampelopsis glandulosa* var. *brevipedunculata* (Vitaceae)	Need ID[c]
	wistariae Mani	*Wisteria floribunda* (Fabaceae)	
Hartigiola	*annulipes* (Hartig)	*Fagus japonica* (Fagaceae)	Need ID[c]
	faggalli (Monzen)	*Fagus crenata* (Fagaceae)	
Janetiella	*infrafoli* Monzen	*Fagus crenata* (Fagaceae)	
	kimurai Inouye	*Pinus parviflora* (Pinaceae)	
Lygocecis	*yanagi* (Shinji)	*Salix babylonica, S. eriocarpa, S. gracilistyla, S.* spp. (Salicaceae)	
Rabdophaga	*clavifex* (Kieffer)	*Salix integra, S. jessoensis, S. sachalinensis, S.* sp. (Salicaceae)	
	rosaeformis Kovalev	*Salix jessoensis, S. subfragilis* (Salicaceae)	

(continued)

Table 4.3 (continued)

Higher taxonomic position	Cecidomyiid species[a]	Host plant species (family) in Japan	Note
	rosaria (Loew)	*Salix gilgiana, S. integra* (Salicaceae)	
	salicifoliae Shinji	*Salix* sp. (Salicaceae)	
	saliciperda (Dufour)	*Salix koriyanagi, S.* sp. (Salicaceae)	
	salicis (Schrank)	*Salix babylonica, S. caprea, S. gracilistyla, S.* spp. (Salicaceae)	
	salicivola Shinji	*Salix babylonica, S. bakko, S. gilgiana, S. jessoensis, S.* spp. (Salicaceae)	
Lasiopterini			
Lasioptera	*achyranthii* Shinji	*Achyranthes bidentata* (Amaranthaceae)	
	artemisifoliae Shinji	*Artemisia japonica* (Asteraceae)	
	astericola Shinji	*Aster scaber* (Asteraceae)	
	azami Shinji	*Cirsium nipponicum* var. *incomptum, Hemistepta lyrata* (Asteraceae)	
	callicarpae (Shinji)	*Callicarpa japonica* (Lamiaceae)	
	camelliae Ohno & Yukawa	*Camellia japonica* (Theaceae)	
	euphobiae Shinji	*Eupatorium chinense, E. euphobiae, E. makinoi* (Asteraceae)	
	gibaushi Shinji	*Aster scaber* (Asteraceae)	
	lespedezae Shinji	*Lespedeza bicolor, L. cyrtobotrya, L. thunbergii* var. *patens* (Fabaceae)	
	paederiae (Shinji)	*Paederia foetida* (Rubiaceae)	
	puerariae (Shinji)	*Pueraria lobata* (Fabaceae)	
	rubi (Schrank)	*Rubus parvifolius, R. phoenicolasius* (Rosaceae)	
	ukogi Shinji	*Acanthopanax spinosus* (Araliaceae)	
	yadokariae Yukawa & Haitsuka	Use vacated galls induced by other gall midges (see Sect. 10.3)	Successor
	yoichiensis Yukawa & Kim	*Solanum lycopersicum* (Solanaceae)	Invasive
Oligotrophini			
Oligotrophus	*betheli* Felt	*Juniperus horizontalis* (Cupressaceae)	Invasive
	nezu Kikuti	*Juniperus rigida* (Cupressaceae)	
	uetsukii (Inouye)	*Juniperus rigida* (Cupressaceae)	

(continued)

Table 4.3 (continued)

Higher taxonomic position	Cecidomyiid species[a]	Host plant species (family) in Japan	Note
Rhopalomyiini			
Rhopalomyia	*abdominalis* Shinji	*Artemisia montana* (Asteraceae)	
	artemisiae (bouche)	*Artemisia capillaris, A. japonica, A. lancea* (Asteraceae)	
	caterva Monzen	*Artemisia capillaris, A. japonica, A. lancea* (Asteraceae)	
	chrysanthemum Monzen	*Chrysanthemum morifolium, C. pacificum* (Asteraceae)	
	cinerarius Monzen	*Artemisia indica, A. montana, A. capillaris, A. lancea* (Asteraceae)	
	foliorum (Loew)	*Artemisia indica, A. montana, A. japonica* (Asteraceae)	
	giraldii Kieffer & Trotter	*Artemisia indica, A. montana, A. capillaris, A. lancea* (Asteraceae)	
	iwatensis Shinji	*Artemisia indica, A. montana, A. capillaris* (Asteraceae)	
	japonica Monzen	*Artemisia indica, A. montana, A. japonica* (Asteraceae)	
	longitubifex (Shinji)	*Artemisia indica, A. montana, A. capillaris* (Asteraceae)	
	struma Monzen	*Artemisia indica, A. montana, A. japonica* (Asteraceae)	
	yomogicola (Matsumura)	*Artemisia indica, A. montana, A. japonica, A. lancea* (Asteraceae)	
Unplaced to tribe			
Celticecis	*japonica* Yukawa & Tsuda	*Celtis sinensis, C. jessoensis* (Cannabaceae)	
Hasegawaia	*sasacola* Monzen	*Sasa veitchii, S. kurilensis, S. palmate, Sasaella ramose* (Poaceae)	
Iteomyia	*capreae* (Winnertz)	*Salix caprea, S. hultenii, S. rorida, S.* spp. (Salicaceae)	
Masakimyia	*pustulae* Yukawa & Sunose	*Euonymus japonicus, E. fortunei* (Celastraceae)	
Massalongia	*nakamuratetsui* Elsayed & Tokuda	*Betula grossa* (Betulaceae)	
Mikiola	*bicornis* Sato & Yukawa	*Fagus crenata* (Fagaceae)	
	glandaria Sato & Yukawa	*Fagus crenata* (Fagaceae)	

(continued)

Table 4.3 (continued)

Higher taxonomic position	Cecidomyiid species[a]	Host plant species (family) in Japan	Note
Phegomyia	*tokunagai* Sasakawa & Koyama	*Fagus crenata* (Fagaceae)	
Procystiphora	*uedai* Sato & Yukawa	*Sasa nipponica* (Poaceae)	
Semudobia	*betulae* (Winnertz)	*Betula platyphylla* (Betulaceae)	
	skuhravae Roskam	*Betula ermanii* (Betulaceae)	
	tarda Roskam	*Betula platyphylla* (Betulaceae)	

[a]Species of the genera *Cecidomyia* and *Resseliella* are excluded from the list because they are phytophagous but not gall inducers. Some of the species that were previously described or identified in Japan are also not included in this list because their generic positions and feeding habits are doubtful as mentioned in Yukawa (2014)
[b]Host plant-alternating species
[c]Species identification is needed based on more precise morphological comparison
[d]Generic position should be reexamined

following nine genera, *Asphondylia*, *Bruggmanniella*, *Daphnephila*, *Illiciomyia*, *Pseudasphondylia*, *Schizomyia* (= *Asteralobia*), *Dasineura*, *Lasioptera*, and *Pitydiplosis*. All these genera are distributed commonly in warm southwestern Japan, except *Asphondylia* on various host plants and *Pitydiplosis* on *Pueraria* (Fabaceae) that are distributed widely in Japan (see a later paragraph of this Chapter). Gall shape, position and host plant information indicate that at least 12 of the 38 unidentified Taiwanese species seem to be identical to or closely related to Japanese species. These data suggest that gall midges associated with evergreen broad-leaved trees have been expanding their distribution ranges northward from the Oriental Region through Taiwan.

4.6 Distribution

The distribution range of gall-inducing cecidomyiids is fundamentally determined by the presence or absence of their host plants in addition to climatic conditions of the habitat. In southwestern parts of Japan, there are many species of evergreen broad-leaved trees that bear various sorts of cecidomyiid galls. Most of them belong to the family Lauraceae (Table 4.2). Because these tree species are considered to have originated in the Oriental Region, the distribution range of most cecidomyiid species on these trees is naturally restricted to warm southwestern parts of Japan.

However, two evergreen broad-leaved species, *Machilus thunbergii* Siebold & Zuccarini and *Neolitsea sericea* (Blume) Koidzumi (Lauraceae) have expanded their range northward to the northern Tohoku District despite the other evergreen

Table 4.4 Faunistic comparison between China and Japan in the number of gall-inducing species belonging to different genera of Cecidomyiinae

Genera of Cecidomyiinae	No. of species			Genera of Cecidomyiinae	No. of species		
	China[a]	Japan[b]	Common		China[a]	Japan[b]	Common
Ametrodiplosis	0	1	0	*Macrodiplosis*	1	1	0
Amperomyia	0	1	0	*Masakimyia*	0	1	0
Artemisiomyia	1	0	0	*Massalongia*	0	1	0
Aschistonyx	0	1	0	*Mayetiola*	1	0	0
Asphondylia	3	8	2	*Mikiola*	0	2	0
Baldratia	1	0	0	*Obolodiplosis*	1	1	1
Bruggmanniella	0	2	0	*Oligotrophus*	0	3	0
Careopalpis	1	0	0	*Orseolia*	1	1	0
Celticecis	1	1	1	*Oxycephalomyia*	0	1	0
Contarinia	4	8	0	*Paradiplosis*	0	1	0
Daphnephila	5	1	0	*Phegomyia*	0	1	0
Dasineura	8	12	1	*Pitydiplosis*	1	1	1
Etsuhoa	0	1	0	*Procontarinia*	3	1	1
Eumerosema	1	0	0	*Procystiphora*	1	1	0
Geromyia	0	1	0	*Psectrosema*	6	0	0
Giraudiella	1	0	0	*Pseudasphondylia*	2	6	0
Halodiplosis	1	0	0	*Rabdophaga*	6	7	3
Hartigiola	0	2	0	*Rhopalomyia*	4	12	1
Hasegawaia	0	1	0	*Schizomyia*	1	12	1
Horidiplosis	1	0	0	*Semudobia*	0	3	0
Hybolasioptera	1	0	0	*Sitodiplosis*	2	1	1
Illiciomyia	0	1	0	*Stefaniola*	6	0	0
Iteomyia	1	1	1	*Stenodiplosis*	2	1	1
Janetiella	0	2	0	*Thecodiplosis*	1	1	1

(continued)

Table 4.4 (continued)

Genera of Cecidomyiinae	No. of species			Genera of Cecidomyiinae	No. of species		
	China[a]	Japan[b]	Common		China[a]	Japan[b]	Common
Kiefferia	0	3	0	*Tokiwadiplosis*	0	1	0
Lasioptera	1	15	1	Total (species)	72	123	18
Litchiomyia	1	0	0	Total (genera)	33	41	15
Lygocecis	1	1	1				

[a]Data from Jiao and Bu (2014)

[b]Data from Table 4.4. No significant difference between the data of China and Japan (Chi-square test for independence, $\chi^2 = 80.545$, $df = 52$, $P = 0.068 > P = 0.01$, *Cramer's* $V = 0.6427$)

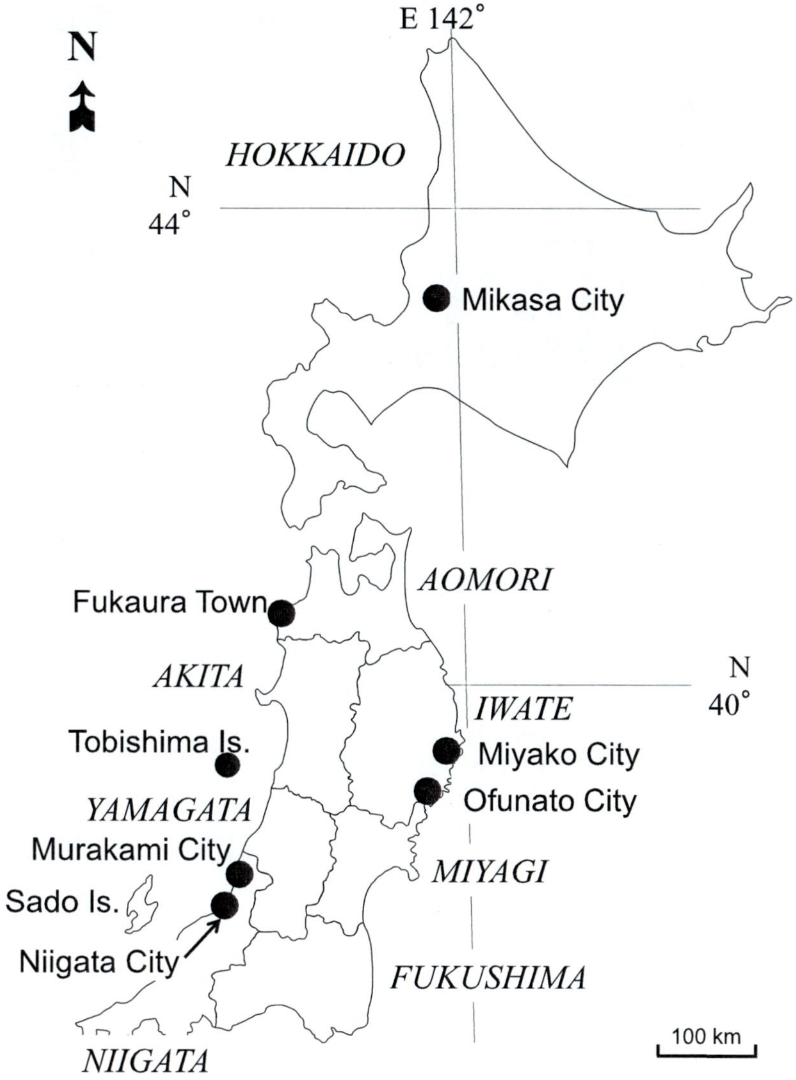

Fig. 4.1 Map of Hokkaido and the Tohoku District, northern Japan

lauraceous trees, such as *Machilus japonica* Siebold & Zuccarini ex Blume, *Neolitsea aciculata* (Blume) Koidzumi, and *Cinnamomum yabunikkei* H. Ohba, remain in southwestern Japan (Horikawa 1972). The northern limit of *M. thunbergii* is Fukaura Town, Aomori Prefecture (west side of the Tohoku District) and Miyako City, Iwate Prefecture (east side), and that of *N. sericea* is Sado Island, Niigata Prefecture (west side) and Ofunato City, Iwate Prefecture (east side) (Fig. 4.1).

Leaf galls of *Daphnephila machilicola* Yukawa induced on *M. thunbergii* (Figs. 4.1g and 3.1b) were found in Murakami City and its vicinity, Niigata Prefecture (Yukawa and Sunose 1988). Later, one of my colleagues, H. Ikenaga

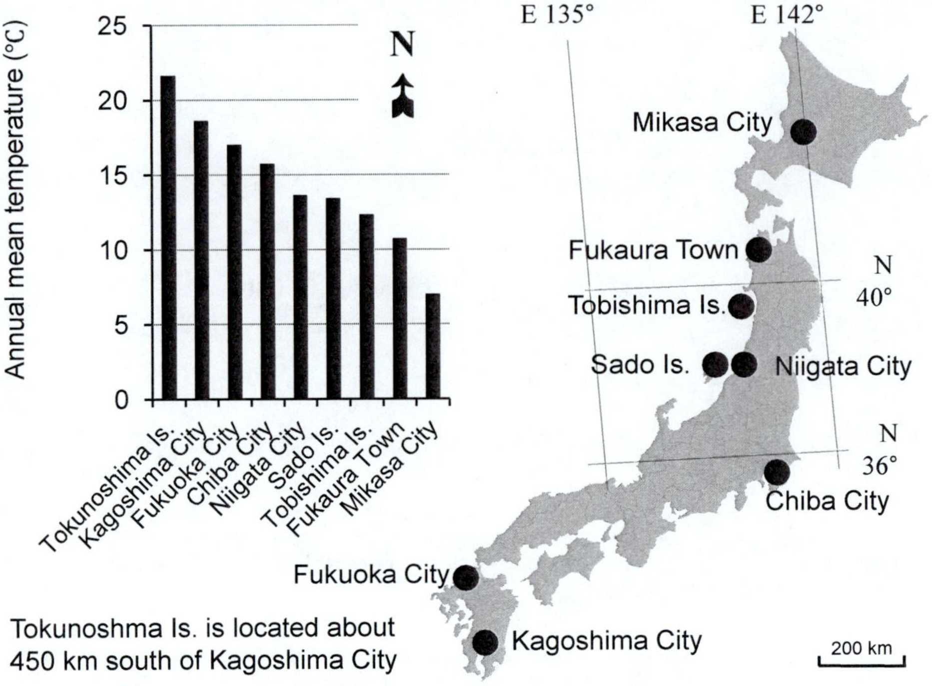

Fig. 4.2 Annual mean temperatures (1981–2010) at different localities shown in the map attached. Data obtained from Japan Meteorological Agency (2020)

informed me that thousands of galls of *D. machilicola* were found in May 2005 on Tobishima Island, Yamagata Prefecture, which is about 160 km south of Fukaura Town, the northern limit of *M. thunbergii* (Fig. 4.1). Leaf galls of *Pseudasphondylia neolitseae* Yukawa induced on *N. sericea* (Fig. 2.1a) were collected in 1981 from Sado Island, Niigata Prefecture, the northern limit of *N. sericea* (Sunose 1982). In 1987, I examined at least 1000 *N. sericea* trees (70–300 cm in height) in Niigata City, Murakami City, and several towns between the two cities, but no galls were found (Yukawa and Sunose 1988). Unfortunately, the distribution range of *D. machilicola* and *P. neolitseae* have never been intensively surveyed in the eastern coast of the Tohoku District.

Both *D. machilicola* and *P. neolitseae* are univoltine and widely distributed in southwestern parts of Japan (Yukawa and Masuda 1996) as well as the other Lauraceae-associated cecidomyiids. However, *D. machilicola* has expanded its range northward to Tobishima Island, near the northern limit of *M. thunbergii*, and *P. neolitseae* to Sado Island, the northern limit of *N. sericea* (Fig. 4.1). It should be contrasted to the distribution range of *Daphnephila* sp. that induces crown-shaped galls on the leaves of *M. thunbergii* only in Okinawa, the most southern prefecture of Japan (Fig.3.1c).

The annual mean temperatures (1981–2010) are distinctly lower in northern areas than in southern areas (Fig. 4.2) (Japan Meteorological Agency 2020). However,

D. machilicola and *P. neolitseae* might have overcome the big thermal differences, adjusted their emergence season with host plant phenology, and established themselves in such cold areas.

Here is another example in which the distribution range of a gall midge almost completely overlaps with the range of its host plant. *Pitydiplosis puerariae* Yukawa, Ikenaga & Sato induces leaf galls on *Pueraria lobata* (Willdenow) Ohwi (Fig. 2.10a, b) in mainland China, South Korea, and Japan, and on *Pueraria montana* (Loureiro) Merrill in Taiwan and on the Yaeyama Islands, Japan (Yukawa and Masuda 1996; Yukawa et al. 2012) (Sect. 2.8). In Japan, *P. puerariae* is univoltine and widely distributed from Mikasa City, Hokkaido, the northern limit of *P. lobata*, to Tokunoshima Island, Kagoshima Prefecture, the southern limit (Fig. 2.11). Thus, *P. puerariae* also has established itself in cold Hokkaido by overcoming the big thermal differences.

These examples indicate that if gall midges have tolerance to cold temperature, they can expand their distribution range northward wherever their host plants exist.

References

Elsayed AK, Ogata K, Kaburagi K, Yukawa J, Tokuda M (2017) A new *Dasineura* species (Diptera: Cecidomyiidae) associated with *Symplocos cochinchinensis* (Loureiro) (Symplocaceae) in Japan. Jpn J Syst Entomol 23:81–86

Elsayed AK, Matsuo K, Kim W, Uechi N, Yukawa J, Gyoutoku N, Tokuda M (2018a) A new *Asphondylia* species (Diptera: Cecidomyiidae) and a eulophid wasp (hymenoptera) inducing similar galls on leaf buds of *Schoepfia jasminodora* (Schoepfiaceae), with reference to their ecological traits and a description of the new gall midge. Entomol Sci 21:324–339

Elsayed AK, Yukawa J, Tokuda M (2018b) A taxonomic revision and molecular phylogeny of the eastern Palaearctic species of the genera *Schizomyia* Kieffer and *Asteralobia* Kovalev (Diptera: Cecidomyiidae: Asphondyliini), with descriptions of five new species of *Schizomyia* from Japan. ZooKeys 808:123–160

Elsayed AK, Uechi N, Yukawa J, Tokuda M (2019a) *Ampelomyia*, a new genus of Schizomyiina (Diptera: Cecidomyiidae) associated with *Vitis* (Vitaceae) in Palearctic and Nearctic regions, with description of a new species from Japan. Canadian Entomol 151:149–162

Elsayed AK, Yukawa J, Tokuda M (2019b) Two new species of *Schizomyia* (Diptera: Cecidomyiidae) from Japan, with an updated key to larval, pupal and adult *Schizomyia* in Japan. Zootaxa 4688:348–360

Elsayed AK, Yukawa J, Tokuda T (2019c) *Pseudasphondylia tominagai,* a new gall midge species (Diptera: Cecidomyiidae) inducing flower bud galls on *Eleutherococcus spinosus* (Araliaceae) in Japan. Biodiver Data J 7:e35673. https://doi.org/10.3897/BDJ.7.e35673

Elsayed AK, Skuhravá M, Ohta K, Yoshida S, Tokuda M (2020) Revision of the birch-associated genus *Massalongia* (Diptera, Cecidomyiidae), with description of a new species from Japan and a taxonomic key to worldwide species. ZooKeys 958:1–27

Felt EP (1965) Plant galls and gall makers. Hafner Publishing Company, New York and London

Ganaha T, Nohara M, Sato S, Uechi Yamagishi K, Yamauchi S, Yukawa J (2007) Polymorphism of axillary bud galls induced by *Rhopalomyia longitubifex* (Diptera: Cecidomyiidae) on *Artemisia princeps* and *A. montana* (Asteraceae) in Japan and Korea, with the designation of new synonyms. Entomol Sci 10:209–212

Horikawa Y (1972) Atlas of the Japanese Flora II, an introduction to plant sociology of East Asia. Gakken, Tokyo

Japan Meteorological Agency (2020) Information on meteorological statistics: meteorological observation. http://www.data.jma.go.jp. (In Japanese)

Jiao K, Bu W (2014) Gall midges (Diptera: Cecidomyiidae) recorded from China during the period from 1900 to 2012, with faunistic comparison between China and Japan. Jpn J Syst Entomol 20:201–215

Kim W, Yukawa J, Harris KM, Minami T, Matsuo K, Skrzypczyeńska M (2014) Description, host range and distribution of a new *Macrodiplosis* species (Diptera: Cecidomyiidae) that induces leaf-margin fold galls on deciduous *Quercus* (Fagaceae) with comparative notes on Palaearctic congeners. Zootaxa 3821:222–238

Kim W, Minami T, Tokuda M, Matsuo K, Harris KM, Yukawa J (2019) Detection of two new cryptic species of *Kiefferia* (Diptera: Cecidomyiidae) by means of morphological, molecular and ecological studies. Entomol Sci 22:450–462

Nijveldt WC, Yukawa J (1982) A taxonomic study on *Salix*-inhabiting gall midges in Japan (Diptera, Cecidomyiidae). Bull Kitakyushu Museum Nat Hist 4:23–56

Ohba H (2009) Syllabus of the Vascular Plants of Japan. Aboc-sha, Kamakura, Japan

Roskam JC (2019) Plant galls of Europe. KNNV Publishers, Zeist

Sato S, Yukawa J (2006) Resurrection of *Rabdophaga salicivora* Shinji, a Japanese gall midge formerly misidentified as a north American species, *Rabdophaga rigidae* (Osten Sacken), (Diptera: Cecidomyiidae), with observations on the phylogenetic relationships of its populations in Japan and the Russian Far East. Entomol Sci 9:423–434

Sunose S (1982) Midge galls of Sado and Okushiri Islands (Diptera: Cecidomyiidae). Akitu, New Series 43:1–6

Tokuda M, Yukawa J (2004) Occurrence of *Contarinia* sp. (Diptera: Cecidomyiidae) causing leaf-fold galls on cultivated roses in Japan. Proc Assoc Plant Prot Kyushu 50:77–81. (In Japanese with English summary)

Tokuda M, Tabuchi K, Yukawa J, Amano H (2004) Inter- and intraspecific comparisons between *Asteralobia* gall midges (Diptera: Cecidomyiidae) causing axillary bud galls on *Ilex* species (Aquifoliaceae): species identification, host range and mode of speciation. Ann Entomol Soc Am 97:954–970

Tung GS, Yang MM (2018) Let's gall Taiwan - a guidebook of insect galls. Council of Agriculture Executive Yuan/Forestry Bureau, Taipei

Uechi N, Yukawa J (2006a) Life history patterns and host ranges of the genus *Asphondylia* (Diptera: Cecidomyiidae). In: Ozaki K, Yukawa J, Ohgushi T, Price PW (eds) Galling arthropods and their associates: ecology and evolution. Springer-Verlag, Tokyo, pp 275–285

Uechi N, Yukawa J (2006b) Host range and life history of *Asphondylia sphaera* Monzen (Diptera: Cecidomyiidae): use of short-term alternate hosts. Ann Entomol Soc Am 99:1165–1171

Uechi N, Tokuda M, Yukawa J (2002) Distribution of *Asphondylia* gall midges (Diptera; Cecidomyiidae) in Japan. Esakia 42:1–10

Uechi N, Tokuda M, Yukawa J, Kawamura F, Teramoto KK, Harris KM (2003) Confirmation by DNA analysis that *Contarinia maculipennis* (Diptera: Cecidomyiidae) is a polyphagous pest of orchids and other unrelated cultivated plants. Bull Entomol Res 93:545–551

Uechi N, Yukawa J, Tokuda M, Ganaha-Kikumura T, Taniguchi M (2011) New information on host plants and distribution ranges of an invasive gall midge, *Contarinia maculipennis* (Diptera: Cecidomyiidae), and its congeners in Japan. Appl Entomol Zool 46:383–389

Yukawa J (1974) Descriptions of new Japanese gall midges (Diptera, Cecidomyiidae, Asphondyliidi) causing leaf galls on Lauraceae. Kontyû 42:293–304

Yukawa J (2014) Family Cecidomyiidae. In: Nakamura T, Saigusa T, Suwa M (eds) Catalogue of the insects of Japan, vol 8. Entomological Society of Japan, Fukuoka, pp 126–160. (In Japanese)

Yukawa J, Masuda H (1996) Insect and mite galls of Japan in colors. Zenkoku Nôson Kyôiku Kyôkai, Tokyo. (In Japanese with English explanation for color plates)

Yukawa J, Sunose T (1988) Midge galls of Niigata prefecture (Diptera: Cecidomyiidae). Trans Essa Entomol Soc, Niigata 66:45–58. (In Japanese with English abstract)

Yukawa J, Ikenaga H, Sato S, Tokuda M, Ganaha-Kikumura T, Uechi N, Matsuo K, Mishima M, Tung GH, Paik JC, Ren BQ, Don XO (2012) Description and ecological traits of a new species of *Pitydiplosis* (Diptera: Cecidomyiidae) that induces leaf galls on *Pueraria* (Fabaceae) in East Asia, with a possible diversification scenario of intraspecific groups. Entomol Sci 15:81–98

Yukawa J, Tokuda M, Yamagishi K (2014) Host plant ranges and distribution records of identified and unidentified species of the genus *Lasioptera* (Diptera: Cecidomyiidae) in Japan. Esakia 54:1–15

Yukawa J, Yoshimura H, Matsuo K, Kim W (2017) New record of an alien gall midge, *Oligotrophus betheli* (Diptera: Cecidomyiidae) on a north American *Juniperus horizontalis* (Cupressaceae) in Japan, with reference to its ecological traits and possibility of further dispersal and host range expansion. Appl Entomol Zool 52:417–427

Yukawa J, Harris KM, Kim W (2020) Descriptions of two new species of the genus *Lasioptera* (Diptera: Cecidomyiidae) that infest tomato in the Mediterranean region and Hokkaido, Japan. Appl Entomol Zool 55:129–140

Chapter 5
Island Biogeography

Junichi Yukawa, Makoto Tokuda, and Wanggyu Kim

Abstract Arthropod galls are appropriate materials for comparative faunistic studies because galls are conspicuous in shape and enable us to obtain faunistic data easily. Particularly they have been used for island biogeography. In this chapter, we refer to the data obtained from the Izu Islands, the Tokara Islands, the Tsushima Islands between the Korean Peninsula and Kyushu, and the Krakatau Islands, Indonesia. On the Izu Islands, relationship between land area and the number of gall midge species was discussed in relation to the influence of volcanic activity and the distance from the mainland to respective islands. Data from the Tokara Islands suggested that the Tokara Strait (= Watase Line), which is the border between the Palearctic and Oriental Regions, has been acting as a geographical barrier for gall midges to expand the distribution range from north to south but not from south to north. Tsushima is considered to play an important role as a stepping stone island in the northward expansion of Japanese insects than the southward expansion of Korean insects. The Krakatau Islands have been recolonized by gall midges together with vegetation succession since the world famous big volcanic eruption in 1883. They came to the islands mainly from Java directly rather than from Sumatra through stepping stone islands, Sebesi and Sebuku.

Keywords Colonization · Izu · Krakatau · Stepping stone · Tsushima · Tokara

J. Yukawa (✉)
Entomological Laboratory, Faculty of Agriculture, Kyushu University, Fukuoka, Japan
e-mail: galling-cecidomyiids@kyudai.jp

M. Tokuda
Laboratory of Systems Ecology, Faculty of Agriculture, Saga University, Saga, Japan
e-mail: tokudam@cc.saga-u.ac.jp

W. Kim
Animal Resources Division, National Institute of Biological Resources, Incheon, South Korea
e-mail: entowk@korea.kr

J. Yukawa, M. Tokuda (eds.), *Biology of Gall Midges*, Entomology Monographs,
https://doi.org/10.1007/978-981-33-6534-6_5

5.1 Introduction

Arthropod galls are appropriate materials for comparative faunistic studies. Galls are conspicuous in shape and enable us to obtain faunistic data easily because their inhabitants are sedentary, and evidence of gall induction exists on the host plants long after the event. In addition, many gall-inducing arthropods are monophagous or oligophagous, and the appearance, structure and position of galls are specific to gall-inducing species in many instances (e.g. Yukawa and Masuda 1996; Yukawa and Rohfritsch 2005). Therefore, gall-inducing arthropods can be frequently identified to the species level based on host plant information and the appearance and structure of galls. In the tropics, however, species identification is limited at least to the family level because most gall inducers have not been described yet. Nevertheless, we can assume that different galls on different plant species or genera are induced by different species in most cases. By enumerating arthropod galls found on the target areas, we can easily compare arthropod faunas between localities and different survey periods. Thus, arthropod galls are ideal materials for faunistic studies, particularly those related to island biogeography. In this chapter, we refer to the data of arthropod galls obtained from the Izu Islands, Tokyo, Japan (Tokuda et al. 2012a, b, 2013, 2015; Tokuda and Kawauchi 2013), the Tokara Islands, Kagoshima, Japan (Tokuda 2017, 2018), the Tsushima Islands, Nagasaki, Japan between the Korean Peninsula and Kyushu, Japan (Kim et al. 2015) and the Krakatau Islands and adjacent area, Indonesia (Partomihardjo et al. 2011).

5.2 The Izu Islands and the Tokara Islands

5.2.1 Locations and Geographic Traits

The Izu islands contain nine inhabited islands, Ohshima, Toshima, Niijima, Shikinejima, Kouzushima, Miyakejima, Mikurajima, Hachijojima, and Aogashima, which are located south of the Izu Peninsula, Honshu, Japan (Fig. 5.1). They are originally volcanic and oceanic islands and extend for about 230 km from north to south. The Islands are characterized by unique biota of insects (e.g. Kurosawa 1978; Inoue and Amano 1986; Takaoka and Saito 2005), reptiles (e.g. Hasegawa 2003), and plants (e.g. Inoue 1988; Oiki et al. 2001; Miyake and Inoue 2003).

The Tokara Islands are located between Yakushima Island and Amami-ohshima Island, south of Kyushu, Japan, and comprise seven inhabited and five uninhabited islands extending 160 km from northeast to southwest. They are mostly considered as volcanic origins but, unlike the Izu Islands, all of them are continental islands connected previously with the Eurasian Continent (Osozawa et al. 2012). These islands are situated on the boundary of Biogeographic Regions. Among the seven inhabited islands, northern five (Kuchinoshima, Nakanoshima, Tairajima, Suwanosejima, and Akusekijima) and southern two (Kodakarajima and Takarajima)

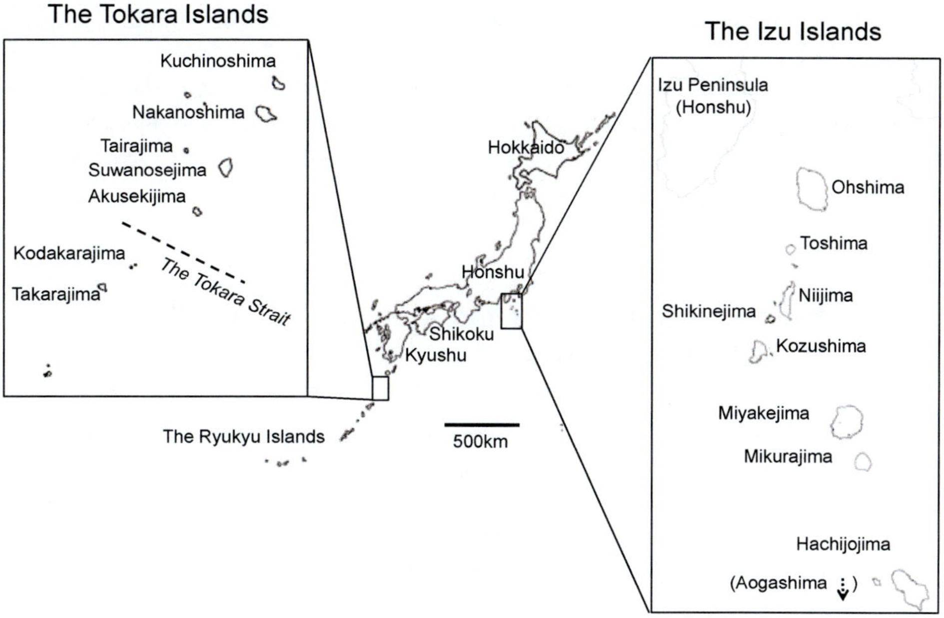

Fig. 5.1 Map of the Izu and the Tokara Islands

were separated from each other by the Watase Line along the Tokara Strait (Okada 1927), and the former belongs to the Palearctic and the latter to the Oriental Region.

The Izu Islands and the Tokara Islands are approximately 1000 km apart from each other, but some studies referred to similarities in insect fauna between them (e.g. Kurosawa 1990) possibly because both Islands are situated on the flow path of Kuroshio Current.

5.2.2 Study History of Gall Midges

Gall midges distributed on the Izu Islands have been recorded in some reports until the 2000s (Ishizawa 1942; Yukawa 1971; Sunose 1981b; Hachijojima Interpretation Association 2007). Sunose (1981b) remarked that *Masakimyia pustulae* Yukawa & Sunose did not exist on the Izu Islands and the distribution of *Pseudasphondylia neolitseae* Yukawa was restricted to areas at an elevation of 250–500 m asl. on Miyakejima and 300–500 m asl. on Hachijojima. In 2010–2014, gall midge fauna of the Izu Islands was intensively surveyed by M. Tokuda and his colleagues (Tokuda et al. 2012a, b, 2013, 2015; Tokuda and Kawauchi 2013). Through the field surveys, we also noticed several interesting phenomena including the population decline of *P. neolitseae* on Miyakejima (Tokuda 2014), outbreaks of *Schizomyia castanopsisae* Elsayed & Tokuda in the southern parts of the Izu Islands (Tokuda 2014), and predation by birds on mature larvae of *S. castanopsisae* on the ground (Naito and Tokuda 2018).

Gall midge fauna of the Tokara Islands was almost unknown until recently except Yamauchi et al. (1982) and Yukawa (1988). M. Tokuda visited all inhabited islands between 2015 and 2017 to survey gall-inducing cecidomyiid fauna (Tokuda 2017, 2018).

5.2.3 Relationship between Land Area and the Number of Gall Midge Species

As summarized in Table 5.1, the land area of each island is generally larger on the Izu Islands than on the Tokara Islands, and the number of gall midge species ranges from 4 to 24 on the Izu Islands while from 1 to 13 on the Tokara Islands. The log land areas and the log number of gall midge species were not significantly correlated when respectively analyzed either for the Izu ($y = 0.218x + 0.861$; $R^2 = 0.150$; $p = 0.164$) or for the Tokara Islands ($y = 0.446x + 0.322$; $R^2 = 0.308$; $p = 0.114$).

Table 5.1 Land areas and the number of gall midge species recorded on each island of the Izu and the Tokara Islands

Name of island	Land area (km^2)	Gall midge species[a]
The Izu Islands		
Ohshima	91.1	13
Hachijojima	69.5	24
Miyakejima	55.5	19
Niijima	23.9	18
Mikurajima	20.6	16
Kouzushima	18.5	14
Aogashima	6.0	4
Toshima	4.1	17
Shikinejima	3.9	11
The Tokara Islands		
Nakanoshima	34.5	13
Suwanosejima	27.8	4
Kuchinoshima	13.3	8
Akusekijima	7.5	10
Takarajima	7.1	8
Tairajima	2.1	1
Kodakarajima	1.0	3

[a]The number of gall midge species recorded on respective islands were calculated based on the following literatures: Ishizawa (1942), Yukawa (1971), Sunose (1981a, b), Hachijojima Interpretation Association (2007), Tokuda et al. (2012a, b, 2013, 2015), Tokuda and Kawauchi (2013) for the Izu Islands, and Yamauchi et al. (1982), Yukawa (1988), Tokuda (2017, 2018) for the Tokara Islands

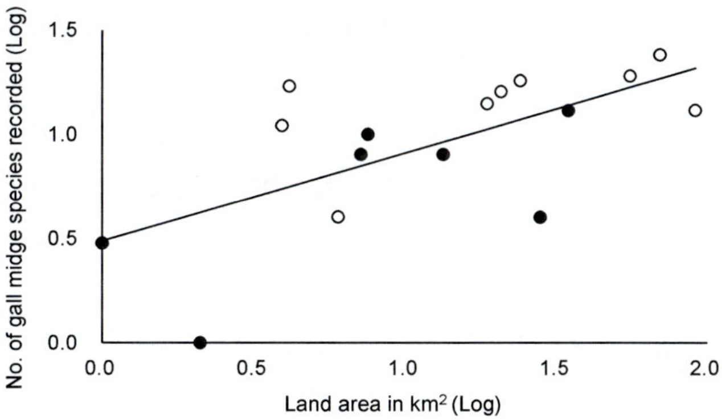

Fig. 5.2 Relationship between the land area of each island on the Izu (open circles) and the Tokara (solid circles) Islands and the number of gall midge species recorded

However, a significant positive correlation was detected when the data of both Islands were combined ($y = 0.423x + 0.487$; $R^2 = 0.377$; $p = 0.007$) (using R ver. 3.5.1; R Core Team 2018) (Fig. 5.2). As has been well noted in island biogeography (MacArthur and Wilson 1963, 1967), the land area of these islands was confirmed to be one of the factors determining the species richness for galling-cecidomyiids. Relatively low coefficient of determination in the relationship between the land area and the gall midge species may be attributed to data on some islands where only a few gall midge species were recorded (e.g. Aogashima on the Izu Islands, and Suwanosejima and Tairajima on the Tokara Islands). The low gall midge diversity on Aogashima is probably because of the long distance from the mainland (Tokuda et al. 2012a). On Suwanosejima, frequent volcanic activities have been preventing field surveys and establishments of gall midge species on the island. Vegetation on Tairajima is very poor, being almost fully covered with pure communities of *Pleioblastus linearis* (Hackel) Nakai (Poaceae) possibly as a result of anthropogenic disturbance (Shiuchi and Hotta 2015). Such conditions are likely to result in extremely low gall midge diversity (only one species) on this island.

5.2.4 Species Composition of Gall Midges

Distributional information on some gall midge species recorded on the Izu and the Tokara Islands were summarized in Table 5.2. *Rhopalomyia* sp. inducing bud galls on *C. pacificum* was recorded on the Izu Islands as well as in Honshu and Kyushu but not on the Tokara Islands. *Asphondylia baca* Monzen exhibits typical Palearctic distribution pattern, which has been recorded only from Honshu, Kyushu, the Izu Islands and Palearctic parts of the Tokara Islands (Table 5.2). Similarly, *M. pustulae* also shows Palearctic distribution pattern, but it is not reported on the Izu Islands (Sunose 1981b; Tokuda 2014). The dissimilarity between the two species is possibly due to differences in dispersal and host recognition abilities of adults. *Asphondylia* species are known to have relatively strong flight abilities for long distance (e.g. Yukawa et al. 2019; see also Chap. 7) but flight distance by other species is

Table 5.2 Distribution of gall midges in Honshu, the Izu Islands, Kyushu, the Tokara Islands, and the Ryukyu Islands (including the Amami Islands)

	Palearctic Region				Oriental Region		
Gall midge[a]	Honshu	Izu	Kyushu	N Tokara[b]	S Tokara[c]	Ryukyus	Remarks
Asphondylia baca	○	○	○	○	–	×	Distributed only in Palearctic (to N Tokara)
Asphondylia sp.[d]	–	×	○	○	○	○	Oriental but expanding to Kyushu
Asphondylia sphaera	○	○	○	○	○	○	Distributed in Palearctic and Oriental
Daphnephila machilicola	○	○	○	○	×	○	Distributed in Palearctic and Oriental
Illiciomyia yukawai	○	×	○	○	–	○	Distributed in Palearctic (not on Izu Isls) and Oriental
Lasioptera camelliae	○	○	○	○	–	○	Distributed in Palearctic and Oriental
Lasioptera sp. 6[e]	○	×	○	○	○	×	Palearctic but expanding to S Tokara
Masakimyia pustulae	○	×	○	○	×	×	Distributed only in Palearctic (not on Izu Isls)
Pseudasphondylia elaeocarpi	○	×	○	○	–	○	Distributed in Palearctic (not on Izu Isls) and Oriental
Pseudasphondylia neolitseae	○	○	○	○	×	○	Distributed in Palearctic and Oriental
Rhopalomyia chrysanthemum	○	○	○	×	○	×	Palearctic but expanding to S Tokara
Rhopalomyia sp.[f]	○	○	○	×	×	×	Distributed only in Palearctic (not on Tokara Isls)

(continued)

Table 5.2 (continued)

Gall midge[a]	Palearctic Region				Oriental Region		Remarks
	Honshu	Izu	Kyushu	N Tokara[b]	S Tokara[c]	Ryukyus	
Schizomyia castanopsisae	×	○	○	×	–	○	Oriental but expanding to Kyushu and Izu Isls
Schizomyia soyogo	○	○	○	○	○	×	Palearctic but expanding to S Tokara
Tokiwadiplosis matecola	×*	×*	○	○	○	○	Oriental but expanding to Kyushu
Unidentified gen. sp. 1[g]	○	○	○	○	×	○	Distributed in Palearctic and Oriental
Unidentified gen. sp. 2[h]	–	–	○	○	○	○	Oriental but expanding to Kyushu
Unidentified gen. sp. 3[i]	–	–	–	○	○	○	Oriental but expanding to N Tokara

[a]The distributional information on each island was fundamentally based on the following literatures: Tokuda et al. (2012a, b, 2013, 2015), Tokuda and Kawauchi (2013), Tokuda (2017, 2018)
[b]N Tokara includes Kuchinoshima, Nakanoshima, Tairajima, Suwanosejima and Akusekijima Islands (Palaearctic Region)
[c]S Tokara includes Kodakarajima and Takarajima Islands (Oriental Region)
[d]This species induces fruit galls on *Alpinia intermedia* Gagnep. (Zingiberaceae)
[e]This species induces stem galls on *Trichosanthes cucumeroides* (Ser.) Maxim. ex Franch. et Sav. (Cucurbitaceae)
[f]This species induces bud galls on *Chrysanthemum pacificum* Nakai (Asteraceae)
[g]This species induces flat leaf galls on *Machilus thunbergii* Siebold & Zucc. (Lauraceae)
[h]This species induces leaf galls on *Ficus superba* (Miq.) Miq. var. *japonica* Miq. (Ficaceae)
[i]This species induces leaf galls on *Ficus microcarpa* L.f. (Ficaceae)
○: Gall midges were distributed; ×; host plants were distributed but gall midges were not found (*not naturally distributed but anthropogenically introduced); –: host plants were not distributed or not yet intensively surveyed

considered to be relatively short (e.g. Tokuda 2012). In addition, *M. pustulae* seems to lack exact host recognition ability and lay their eggs even on non-host plants (Yukawa and Sunose 1976; Sunose & Yukawa 1979) (see also Sect. 2.2). Such behavioral traits of *M. pustulae* probably prevent range expansions to oceanic islands including the Izu Islands (Sunose 1981b). Some other gall midges such as *Illiciomyia yukawai* Tokuda and *Pseudasphondylia elaeocarpi* Tokuda & Yukawa also seem to have not yet colonized the Izu Islands (Tokuda 2014). *Lasioptera* sp. 6 associated with *Trichosanthes cucumeroides* (Seringe) Maximowicz ex Franchet & Savatier (Cucurbitaceae) (Yukawa et al. 2014), *Schizomyia soyogo*

Kikuti associated with *Ilex* species (Aquifoliaceae) (Tokuda et al. 2002; Tokuda et al. 2004; Elsayed et al. 2018b), and *Rhopalomyia chrysanthemum* Monzen (Diptera: Cecidomyiidae) associated with *Chrysanthemum pacificum* Nakai (Asteraceae) are also Palaearctic species. However, they were confirmed to have expanded their ranges to Oriental parts of the Tokara Islands (Tokuda 2017). *Tokiwadiplosis matecola* Simbolon & Yukawa and unidentified cecidomyiids inducing galls on *Ficus superba* (Miquel) Miquel var. *japonica* Miquel and *Ficus microcarpa* Linnaeus (Ficaceae) are considered to be Oriental origins because their hosts are mainly distributed in the Oriental Region. These species are recorded not only from the Oriental Region but also from Palearctic parts of the Tokara Islands and in Kyushu. These results suggest that the Tokara Strait (= Watase Line), which is the border between the Palearctic and Oriental Regions, has been acting as a geographical barrier for gall midges to expand the distribution range from north to south but not from south to north. Because the movement of most gall midges is largely dependent on wind directions (Yukawa and Rohfritsch 2005; Tokuda 2012; Kim et al. 2015), a westerly wind may have been playing an important role for Oriental gall midges to expand their range northward but not for Palearctic species to expand southward.

5.3 Tsushima Island in Relation to the Korean Peninsula

5.3.1 Tsushima Island

The area around the Japanese Archipelago was a continuous coastal margin of the ancient Asian continent until the end of the Carboniferous Period (about 289 Ma), and thereafter the archipelago separated from the continent during the Cretaceous and Paleogene Periods (25–130 Ma) and was alternately connected and separated from the continent by the repetition of glacial periods until the Pleistocene Era (0.01–2.0 Ma) (Taira and Tashiro 1987; Taira 1990). The Japanese Archipelago, including the Tsushima Islands, is now completely separated from the continent, and the distance between the Korean Peninsula and Kyushu (the nearest part of the mainland of Japan) is about 200 km (Fig. 5.3) (Kim et al. 2015).

The Tsushima Islands consist of a major island and more than 100 small adjacent islets. The major island was artificially separated into three islands by the construction of canals first in 1672 and later in 1900. Because the three islands are now connected with bridges, they are regarded as one major island 'Tsushima'. The island is long and narrow, being 696 km^2 in area, and the highest point is 648.5 m in altitude (the northern and southern most latitude: N 34°42′ and N 34°05′, respectively, and the eastern and western most longitude: E 129°30′ and E 129°10′, respectively) (Fig. 5.3). About 89% of the Tsushima Islands is covered with secondary forests of broad-leaved evergreen and deciduous trees and plantations of conifers, mainly *Cryptomeria japonica* D. Don (Cupressaceae) (Forestry Administration Division and Forest Department Office, Agriculture and Forestry

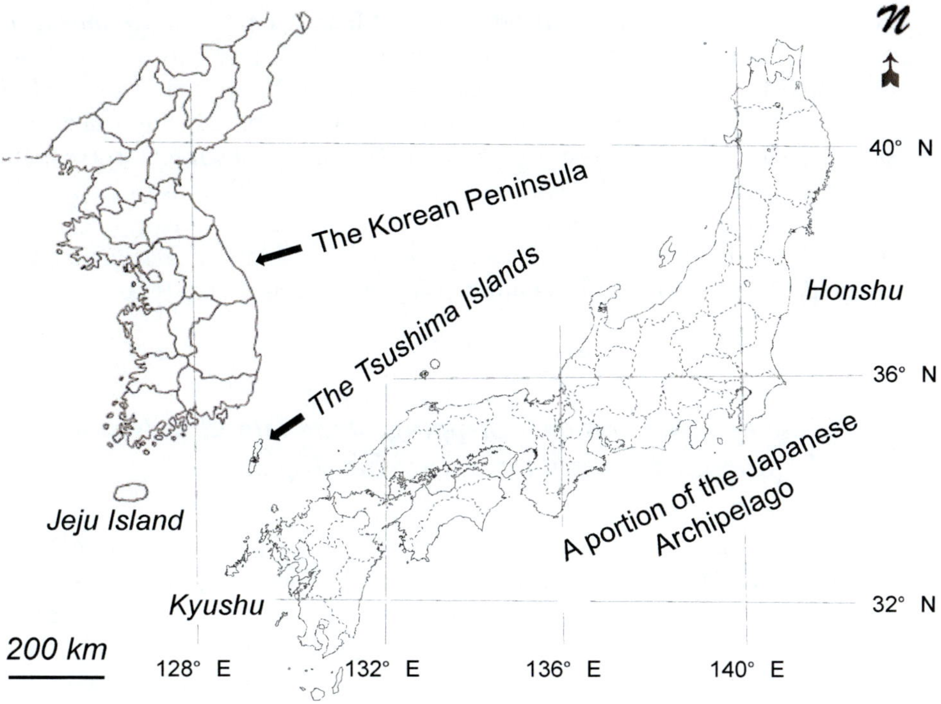

Fig. 5.3 Map of the Korean Peninsula, the Tsushima Islands, and a portion of the Japanese Archipelago (modified from Kim et al. 2015)

Department, Nagasaki Prefecture 2015). The climate of the islands is relatively mild due to the influence of a warm current called "Tsushima Current" and the average of the annual temperature being 15.8 °C and the average annual precipitation being 2132.6 mm (Meteorological data in Izuhara, Tsushima, Japan Meteorological Agency).

Tsushima is about 50 km south of the Korean Peninsula and about 100 km north of Kyushu, Japan. Because Tsushima has been regarded as one of the important stepping stone islands between the peninsula and the mainland of Japan, the entomofauna of the island has been surveyed for various insect taxa to compare with those of the Korean Peninsula and of the mainland of Japan (e.g. for general biota by Nagasaki Biological Society 1976; for Trichoptera by Kobayashi 1985; for Chironomidae, Diptera by Sasa and Suzuki 1999; for Cecidomyiidae, Diptera by Kim et al. 2015).

5.3.2 Cecidomyiid Galls Found on Tsushima Island

Gall-inducing cecidomyiids (Diptera: Cecidomyiidae) were not intensively surveyed on Tsushima until 2012 and only three species were recorded from the island, as follows (Yukawa 2014): *Asphondylia sphaera* Monzen on *Ligustrum japonicum* Thunberg (Oleaceae) (collecting record in Kim et al. 2015), *M. pustulae* on *Euonymus japonicus* Thunberg (Celastraceae) (Sunose 1981a) and *Thecodiplosis japonensis* Uchida & Inouye on *Pinus thunbergii* Parlatore and *Pinus densiflora* Siebold & Zuccarini (Pinaceae) (Inouye 1962). The most current surveys for cecidomyiid galls were conducted in various sites on Tsushima in October 2012, June 2013, May 2014 and September 2014. Through the field surveys, a total of 23 sorts of cecidomyiid gall were recognized on Tsushima in association with 23 plant species belonging to 17 families (Table 5.3) (Kim et al. 2015).

5.3.3 *Role of Tsushima as a Stepping Stone Island Between the Korean Peninsula and Kyushu, Japan*

So far, at least 210 sorts of cecidomyiid gall have been found in Kyushu (e.g. Nagai 2015, 2016, 2017, 2018, 2019; Elsayed et al. 2017; Elsayed et al. 2018a, b; see Kim et al. 2015 for the references before 2015). In addition, at least 70 sorts of newly found galls are waiting to be recorded from Kyushu (Yukawa et al. unpublished). Thus, a total of 280 (210 + 70) sorts of cecidomyiid gall are now known to occur in Kyushu. All 23 sorts (100%) of cecidomyiid gall that have been found on Tsushima also occur in Kyushu, which is equivalent to 8.2% of 280 sorts recorded from Kyushu (Table 5.3).

A total of 67 sorts of cecidomyiid gall induced by 36 identified and 30 unidentified species have been recorded from the Korean Peninsula including Jeju Island that is located in the East China Sea (ESK and KSAE 1994; Saitô 1932, 1933; Ganaha et al. 2004, 2007; Kim et al. 2019; Kodoi et al. 2003; Im and Ji 2015; Paik et al. 2004; Uechi et al. 2018; Yukawa et al. 2012) (Table 5.4). Among the 67 sorts, 16 (23.9%) have not yet been found in Japan although most host plants are distributed in Japan. From Kyushu side, 35 (12.5%) of the 280 sorts are commonly distributed in the Korean Peninsula. In future, much more cecidomyiid galls will be found in the peninsula when further intensive field surveys are conducted.

Among the 23 sorts found on Tsushima, 18 (78.3%) of them also occur in the Korean Peninsula, which is equivalent to 26.9% of the 67 sorts (Table 5.3). Five of the 23 sorts of cecidomyiid gall on Tsushima have not been found in the Korean Peninsula (Table 5.3) although *Celtis sinensis* Persoon (Cannabaceae), *Morus australis* Poiret (Moraceae) and *Isodon inflexus* (Thunberg) Kudo (Lamiaceae) are commonly distributed throughout, and *Eurya japonica* Thunberg (Pentaphylacacaeae) and *Trachelospermum asiaticum* (Siebold & Zuccarini) Nakai (Apocynaceae) are in southern parts of the Peninsula (Korea Forest Service 2010). This means that galling-cecidomyiids on these plant species have not yet expanded their distribution ranges to the Korean Peninsula. Neverthless, conical leaf galls were found on *C. sinensis* in Sunchon, South Korea (Yukawa et al. 2012). The galls were similar in shape to those of Japanese *Celticecis japonica* Yukawa & Tsuda but much

Table 5.3 Cecidomyiid galls found on Tsushima Island and their occurence in Kyushu and the Korean Peninsula (modified from Kim et al. 2015)

Family and species of host plant	Gall midge species	Gall shape	Collecting data on Tsushima	Distribution[a]	
				Kyushu	Korean Penn.
PINACEAE					
Pinus densiflora	*Thecodiplosis japonensis*	Needle gall	Inouye (1962)	○	○
FAGACEAE					
Quercus serrata, *Quercus* sp.	*Macrodiplosis* sp.	Downwardly rolled leaf-margin gall	Kim et al. (2015)	○	○
CANNABACEAE					
Humulus japonicus	*Schizomyia humuli*	Thick lenticular leaf gall	Kim et al. (2015)	○	○
Celtis sinensis	*Celticecis japonica*	Conical leaf gall	Kim et al. (2015)	○	×
MORACEAE					
Morus australis	An unidentified cecidomyiid	Subglobular leaf vein gall	Kim et al. (2015)	○	×
AMARANTHACEAE					
Achyranthes bidentata	*Lasioptera achyranthii*	Subglobular or fusiform stem gall	Kim et al. (2015)	○	○
LAURACEAE					
Lindera glauca	An unidentified cecidomyiid	Fusiform leaf gall	Kim et al. (2015)	○	○*
PENTAPHYLACACAEAE					
Eurya japonica	An unidentified cecidomyiid	Fusiform stem gall	Kim et al. (2015)	○	×
FABACEAE					
Pueraria lobata	An unidentified cecidomyiid	Downwardly globular leaf gall	Kim et al. (2015)	○	○
	Pitydiplosis puerariae	Thick lenticular leaf gall	Kim et al. (2015)	○	○
Lespedeza bicolor	An unidentified cecidomyiid	Inwardly folded leaf gall	Kim et al. (2015)	○	○
CELASTRACEAE					
Euonymus japonicus	*Masakimyia pustulae*	Suboval blister leaf gall	Kim et al. (2015), Yukawa et al. (2014)	○	○

(continued)

Table 5.3 (continued)

Family and species of host plant	Gall midge species	Gall shape	Collecting data on Tsushima	Distribution[a]	
				Kyushu	Korean Penn.
VITACEAE					
Ampelopsis glandulosa var. *heterophylla*	*Asphondylia baca*	Subglobular fruit gall	Kim et al. (2015)	○	○
CUCURBITACEAE					
Trichosanthes cucumeroides	*Lasioptera* sp.	Fusiform stem gall	Kim et al. (2015)	○	○*
CORNACEAE					
Swida macrophylla, *Swida controversa*	An unidentified cecidomyiid	Elongated pouch leaf vein gall	Kim et al. (2015)	○	○*
ARALIACEAE					
Hedera rhombea	*Asphondylia* sp.	Ellipsoidal flower bud gall	Kim et al. (2015)	○	○*
OLEACEAE					
Ligustrum japonicum	*Asphondylia sphaera*	Fruit gall	Kim et al. (2015), Yukawa et al. (2014)	○	○
APOCYNACEAE					
Trachelospermum asiaticum	*Schizomyia usubai*	Fusiform fruit gall	Kim et al. (2015)	○	×
LAMIACEAE					
Isodon inflexus	An unidentified cecidomyiid	Fusiform bud gall	Kim et al. (2015)	○	×
ASTERACEAE					
Chrysanthemum sp.	*Rhopalomyia* sp.	Globular leaf gall	Kim et al. (2015)	○	○*
Artemisia indica	*Rhopalomyia giraldii*	Large globular or subglobular stem gall	Kim et al. (2015)	○	○
	Rhopalomyia cinerarius	Globular leaf gall	Kim et al. (2015)	○	○
	Rhopalomyia yomogicola	Red conical leaf gall	Kim et al. (2015)	○	○

[a]○: Galls were found; ×: Galls were not found. * Collecting records in the Korean Penninsula were newly added from unpublished data

more slender and elongated than those of the Japanese species. Adult specimens have not been obtained but the gall shape was suggestive that a species of *Celticecis* is responsible for these galls. In Japan, at least three sorts of galls have been found on *Celtis* (Yukawa and Masuda 1996). In the light of diversification of *Celticecis* on

Table 5.4 List of galling-cecidomyiids, their host plants and distribution records in Korea together with distributional information in Japan for comparison

Gall midge	Host plant (family)	Galled organ	Distribution[a]		Reference to collecting record in Korea[b]
			Korea	Japan	
Aschistonyx eppoi	*Juniperus chinensis* (Cupressaceae)	Leaf bud	No data	Hn	ESK-KSAE (1994)
Asphondylia baca	*Ampelopsis glandulosa* var. *heterophylla* (Vitaceae)	Fruit	Gg, Gw, Cb, Gb, Gn	Hk, Hn, Sh, Ky	Saitô (1933)
Asphondylia baca	*Weigela hortensis* (Caprifoliaceae)	Leaf bud	Gg, Gb	Hk, Hn, Ky	ESK-KSAE (1994)
Asphondylia sphaera	*Ligustrum japonicum* (Oleaceae)	Fruit	Jn	Hn, Sh, Ky	Yukawa et al. (2012)
Asphondylia yushimai	*Glycine max* (Fabaceae)	Pod	Jn	Hk, Hn, Ky	Uechi et al. (2018)
Celticecis sp. (possibly)	*Celtis sinensis* (Cannabaceae)	Leaf	Jn	–	Yukawa et al. (2012)
Clinodiplosis rosaefoliae	*Rosa multiflora* (Rosaceae)	Leaf	No data	Hn, Sh, Ky	ESK-KSAE (1994)
Contarinia matusintome	*Pinus densiflora* (Pinaceae)	Leaf bud	Gg	Hn, Ky	Saitô (1932)
Daphnephila machilicola	*Machilus thunbergii* (Lauraceae)	Leaf	Jn, Jj	Hn, Ky	Paik et al. (2004); Yukawa et al. (2012); Im and Ji (2015)
Dasineura asteriae	*Aster scaber* (Asteraceae)	Leaf, petiole, stem	No data	Hn	ESK-KSAE (1994)
Dasineura viticola	*Vitis coignetiae* (Vitaceae)	Leaf	Gb	Hk, Hn, Ky	Saitô (1932)
Kiefferia pericarpiicola	Apiaceae spp. (Apiaceae)	Fruit	Gb	Hk	Kim et al. (2019)
Lasioptera achyranthii	*Achyranthes bidentata* (Amaranthaceae)	Stem	So, Gb, Jn, Jj	Hn, Sh, Ky	Paik et al. (2004)
Lasioptera astericola	*Aster scaber* (Asteraceae)	Stem	No data	Hn	ESK-KSAE (1994)
Lasioptera euphobiae	*Eupatorium makinoi* (Asteraceae)	Stem	Jn	Hk, Hn	Yukawa et al. (2012)

(continued)

Table 5.4 (continued)

Gall midge	Host plant (family)	Galled organ	Distribution[a] Korea	Japan	Reference to collecting record in Korea[b]
Lasioptera lespedezae	*Lespedeza bicolor* (Fabaceae)	Stem	Jn	Hn, Sh, Ky	Paik et al. (2004)
Lasioptera paederiae	*Paederia scandens* (Rubiaceae)	Vine	Jn	Hn, Ky	ESK-KSAE (1994); Paik et al. (2004)
Lasioptera rubi	*Rubus crataegifolius* (Rosaceae)	Stem	No data	Hn	ESK-KSAE (1994)
Lasioptera sp.	*Dioscorea opposite* (Dioscoreaceae)	Vine	So	Hn	Saitô (1933); Im and Ji (2015)
Masakimyia pustulae	*Euonymus japonicus* (Celastraceae)	Leaf	So, Jj	Hk, Hn, Ky	Yukawa et al. (2012); Im and Ji (2015)
Obolodiplosis robiniae	*Robinia pseudoacacia* (Fabaceae)	Leaf	So, Cb, Gb, Gn, Jn, Jj	Hk, Hn, Ky	Kodoi et al. (2003); Paik et al. (2004)
Orseolia miscanthi	*Miscanthus sinensis* (Poaceae)	Leaf bud	So	Hn, Sh, Ky	ESK-KSAE (1994); Im and Ji (2015)
Pseudasphondylia neolitseae	*Neolitsea sericea* (Lauraceae)	Leaf	Gn, Jn	Hn, Sh, Ky	Paik et al. (2004)
Pseudasphondylia rokuharensis	*Viburnum dilatatum* (Caprifoliaceae)	Fruit	Jj	Hn, Sh, Ky	Yukawa et al. (2012)
Rabdophaga rosaeformis	*Salix koreensis* (Salicaceae)	Leaf bud	No data	Hn	ESK-KSAE (1994)
Rabdophaga rosaria	*Salix koriyanagi* (Salicaceae)	Leaf bud	Gg, Jn	Hn, Sh	Saitô (1932); ESK-KSAE (1994)
Rabdophaga salicis	*Salix koreensis* (Salicaceae)	Stem	No data	Hk, Hn, Sh, Ky	ESK-KSAE (1994)
Rabdophaga salicivora	*Salix koreensis* (Salicaceae)	Stem	So, Gg, Cb, Gb, Gn, Jn	Hk, Hn, Sh, Ky	Saitô (1932); ESK-KSAE (1994)
Rhopalomyia caterva	*Artemisia japonica* (Asteraceae)	Leaf bud	No data	Hn	ESK-KSAE (1994)
Rhopalomyia cinerarius	*Artemisia indica* (Asteraceae)	Leaf, petiole, stem	Jj	Hk, Hn, Sh, Ky	Yukawa et al. (2012)

(continued)

Table 5.4 (continued)

Gall midge	Host plant (family)	Galled organ	Distribution[a]		Reference to collecting record in Korea[b]
			Korea	Japan	
Rhopalomyia giraldii	*Artemisia indica* (Asteraceae)	Stem	So, Gg, Gw, Cb, Gb, Gn, Jn	Hn, Sh, Ky	Saitô (1932); ESK-KSAE (1994); Im and Ji (2015)
Rhopalomyia longicauda	*Chrysanthemum* spp. (Asteraceae)	Leaf bud and leaf	Gw, Jj	–	Yukawa et al. (2012)
Rhopalomyia longitubifex	*Artemisia indica* (Asteraceae)	Leaf bud	Gw, Gb, Jn	Hn, Sh, Ky	Ganaha et al. (2004); Ganaha et al. (2007)
Rhopalomyia struma	*Artemisia indica* (Asteraceae)	Stem	So, Cb, Gb, Gn, Jn, Jj	Hn, Sh, Ky	Saitô (1932); ESK-KSAE (1994); Ganaha et al. (2004); Im and Ji (2015)
Rhopalomyia yomogicola	*Artemisia indica* (Asteraceae)	Leaf	Gw, So, Cb, Gb, Gn, Jn	Hn, Sh, Ky	Saitô (1932); ESK-KSAE (1994); Ganaha et al. (2004); Paik et al. (2004)
Rhopalomyia sp.	*Artemisia indica* (Asteraceae)	Leaf	So	Hk, Hn, Sh, Ky	Im and Ji (2015)
Schizomyia humuli	*Humulus japonicus* (Cannabaceae)	Leaf	Gw, So, Gb	Hn, Sh, Ky	Paik et al. (2004)
Schizomyia sasakii	*Ilex crenata* (Aquifoliaceae)	Axillary bud	No data	Hk, Hn, Sh, Ky	ESK-KSAE (1994)
Sitodiplosis mosellana	*Triticum aestivum* (Poaceae)	Grain	No data	Ky	ESK-KSAE (1994)
Thecodiplosis japonensis	*Pinus densiflora* (Pinaceae)	Needle	So, Gg, Gw, Cb, Cn, Gb, Gn, Bs, Jb, Jn, Jj	Hk, Hn, Ky	Saitô (1932)
Genus? *acutissima*[c]	*Quercus acutissima* (Fagaceae)	Twig	So	Hn, Ky	ESK-KSAE (1994); Im and Ji (2015)
Genus? *mori*[c]	*Morus alba* (Moraceae)	Leaf bud	No data	Hn, Sh	ESK-KSAE (1994)
Unidentified	*Achyranthes bidentata* (Amaranthaceae)	Fruit	So	Hk, Hn,	Im and Ji (2015)

(continued)

Table 5.4 (continued)

Gall midge	Host plant (family)	Galled organ	Distribution[a] Korea	Japan	Reference to collecting record in Korea[b]
				Sh, Ky	
Unidentified	*Aster tripolium* (Asteraceae)	Flower	So	Hk, Hn, Sh	Im and Ji (2015)
Unidentified	*Clematis apiifolia* (Ranunculaceae)	Stem	So	–	Im and Ji (2015)
Unidentified	*Deutzia glabrata* (Saxifragaceae)	Leaf	So	–	Im and Ji (2015)
Unidentified	*Dioscorea quinqueloba* (Dioscoreaceae)	Leaf	So	Hn	Im and Ji (2015)
Unidentified	*Elaeagnus umbellate* (Elaeagnaceae)	Leaf bud	So	Hn	Im and Ji (2015)
Unidentified	*Indigofera kirilowii* (Fabaceae)	Leaf	So	–	Im and Ji (2015)
Unidentified	*Lespedeza bicolor* (Fabaceae)	Leaf	Gw, Gb, Jn	Hk, Hn, Sh, Ky	Paik et al. (2004)
Unidentified	*Lespedeza cuneata*	Flower bud	So	–	Yukawa et al. (2012); Im and Ji (2015)
Unidentified	*Lespedeza maximowiczii* (Fabaceae)	Leaf	So	–	Im and Ji (2015)
Unidentified	*Lespedeza maximowiczii* (Fabaceae)	Petiole	So	–	Im and Ji (2015)
Unidentified	*Metaplexis japonica* (Asclepiadaceae)	Vine	So	Hk	Im and Ji (2015)
Unidentified	*Morus bombycis* (Moraceae)	Leaf	So	Hk, Hn, Ky	Im and Ji (2015)
Unidentified	*Mosla dianthera* (Lamiaceae)	Stem	So	–	Im and Ji (2015)
Unidentified	*Pelargonium inquinans* (Geraniaceae)	Stem	So	–	Im and Ji (2015)
Unidentified	*Phragmites communis* (Poaceae)	Stem	So	–	Im and Ji (2015)

(continued)

Table 5.4 (continued)

Gall midge	Host plant (family)	Galled organ	Distribution[a]		Reference to collecting record in Korea[b]
			Korea	Japan	
Unidentified	*Prunus padus* (Rosaceae)	Leaf	So	Hk	Im and Ji (2015)
Unidentified	*Pueraria lobata* (Fabaceae)	Leaf	Jn	Hk, Hn, Sh, Ky	Yukawa et al. (2012)
Unidentified	*Pueraria lobata* (Fabaceae)	Petiole	So	–	Im and Ji (2015)
Unidentified	*Sorbus alnifolia* (Rosaceae)	Leaf	So, Cn	Hk, Hn, Ky	Im and Ji (2015)
Unidentified	*Spiraea prunifolia* (Rosaceae)	Leaf	So	–	Im and Ji (2015)
Unidentified	*Styrax japonica* (Styracaceae)	Flower bud	So	Hn, Sh, Ky	Im and Ji (2015)
Unidentified	*Syringa reticulata* (Oleaceae)	Leaf	So	–	Im and Ji (2015)
Unidentified	*Vicia amoena* (Fabaceae)	Stem	So	–	Im and Ji (2015)
Unidentified	*Zelkova serrata* (Ulmaceae)	Leaf	So	–	Im and Ji (2015)

[a]Names of Korean and Japanese localities are abbreviated as follows; *[KOREA] So* Seoul, *Cb* Chungcheongbuk-do, *Cn* Chungcheongnam-do, *Gb* Gyeongsangbuk-do, *Gg* Gyeonggi-do, *Gn* Gyeongsangnam-do, *Bs* Busan, *Gw* Gangwon-do, *Jb* Jeollabuk-do, *Jn* Jeollanam-do, *Jj* Jeju-do, *[JAPAN] Hk* Hokkaido, *Hn* Honshu, *Ky* Kyushu, *Sh* Shikoku

[b]ESK-KSAE is an abbreviation for The Entomological Society of Korea & The Korean Society of Applied Entomology

[c]Generic position of these species should be reexamined based on recent taxonomic studies

Celtis species in North America (Gagné and Moser 2013), the origin and diversification processes of *Celticecis* species occurring in Asia including the Korean Peninsula and the Japanese Archipelago are interesting study subjects in the future.

It is remarkable that the gall-inducing cecidomyiid fauna of Tsushima is more similar to that of Kyushu (100%) than to that of the Korean Peninsula (78.3%) even though the island is more distant from Kyushu than from the peninsula (Fig. 5.3). Tsushima is more similar to Kyushu than the Korean Peninsula in the vegetation represented by plants of warm temperate forests due to the warm Tsushima Current (e.g. Nagasaki Biological Society 1976). Gall-inducing cecidomyiids are closely associated with their host plants, naturally resulting in a higher similarity in the cecidomyiid faunas between Tsushima and Kyushu.

Because Tsushima is located at latitudes different from those of the Korean Peninsula, insect populations invading from the peninsula seem to encounter more difficulty in establishing on Tsushima than those from Kyushu, probably because of different conditions of latitudinal factors, such as vegetation, temperature and daylength. Prevailing wind direction around Tsushima is northward in summer and southward in winter. Thus, the wind-dispersal of insects to Tsushima from the peninsula cannot be expected in summer. The warm Tsushima Current flows from southwest to northeast (e.g. Isoda 2011) and insect dispersal by rafting depends on the direction of currents (e.g. Abe 1984). These conditions indicate that Tsushima may play a more important role as a stepping stone island in the northward expansion of Japanese insects than the southward expansion of Korean insects.

5.4 Recolonization of the Krakatau Islands, Indonesia

5.4.1 The Location and History of the Krakatau Islands

The Krakatau Islands are situated in the Sunda Strait between Java and Sumatra, Indonesia, about 40 km away from the coast of Java and 35 km from Sumatra (Fig. 5.4). The islands previously composed of three islets, Krakatau (now called Rakata Besar), Lang (Rakata Kecil) and Verlaten (Sertung). As is well known, on 27 August 1883, volcanic activities of three vents on Krakatau gave rise to the world-famous eruption, which completely destroyed the northern two-third of the island. The remaining one-third of Krakatau, and its two neighboring islands, Rakata Kecil and Sertung, were almost certainly completely sterilized by a thick layer of hot ash and pumice (Docters van Leeuwen 1936; Richards 1952). In 1927–1930, submarine activities created a new volcanic island, Anak Krakatau, in the center of the above-mentioned three islands. Thus, the Krakataus now consist of four islets, Rakata, Panjang, Sertung and Anak Krakatau (Fig. 5.4). Successive activities of Anak Krakatau have been damaging biota on the Krakataus (Thornton et al. 1994). On 11 November 2007, Anak Krakatau lets out a massive roar as it blasts a gigantic cloud of smoke and flaming red rocks hundreds of meters into the night sky (Reuters World News 2007). On 22 December 2018, a sector collapse event occurred at Anak Krakatau volcano, triggering a deadly tsunami (Walter et al. 2019). The volcano remains restless and produces intermittent explosions every few days or so from its new crater (Volcano Discovery 2020).

After the 1883 eruption, recolonization of the Krakataus by arthropods has been studied on many occasions (e.g. Jacobson 1909; Docters van Leeuwen 1920, 1922; Dammerman 1922, 1948; Wheeler 1924, 1937; Roonwal 1969; Yamane 1983; Yukawa 1984a, b, c; Kanmiya and Yukawa 1985; Bush 1986; Yukawa and Yamane 1985; Evenhuis and Yukawa 1986; Thornton 1987, 1996; Compton et al. 1988; New and Sudarman 1988; New et al. 1988; Thornton and New 1988; Thornton and Rosengren 1988; Tsukaguchi and Yukawa 1988; Bush et al. 1990; Bush and Whittaker 1991; New and Thornton 1992a, b; Thakur and Thakur 1992; Yamane

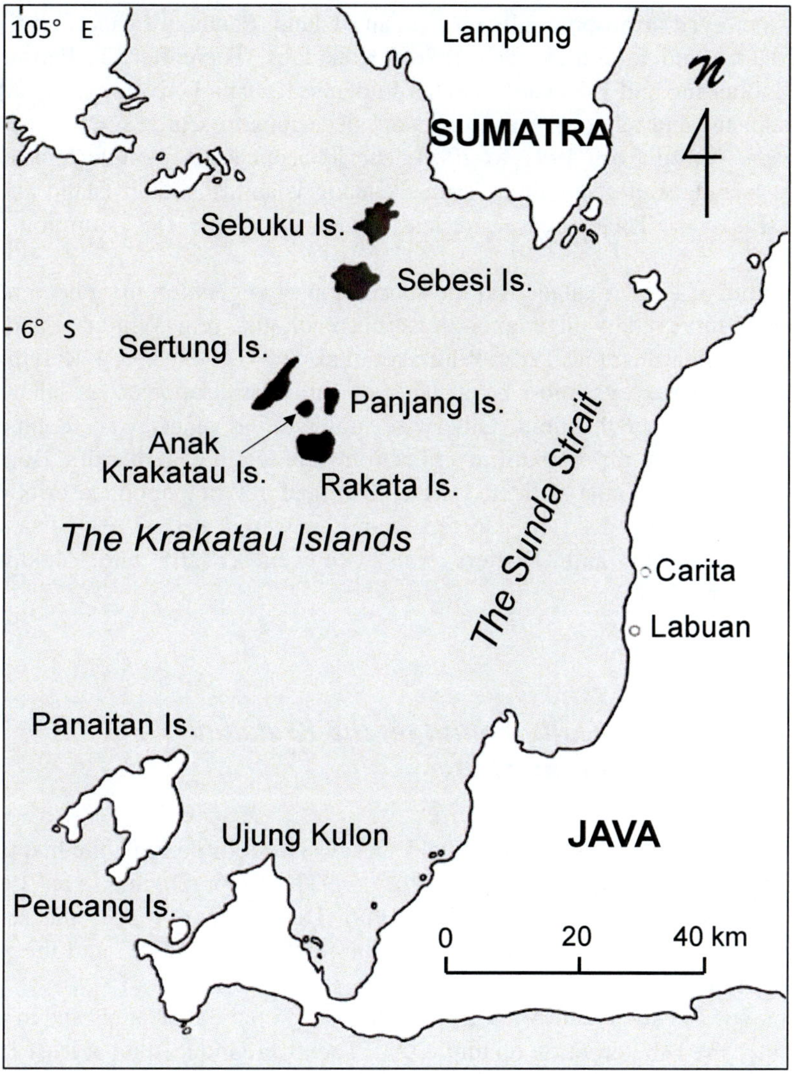

Fig. 5.4 Map of the Krakatau Islands and adjacent islands and areas (modified from Partomihardjo et al. 2011)

et al. 1992; Gathorne-Hardy et al. 2000; Yukawa et al. 2000, 2001; Thornton et al. 2002; Partomihardjo et al. 2011).

To search for arthropod galls, Docters van Leeuwen (1920, 1922) visited the Krakataus and adjacent islands several times during the period from 1919 to 1922 and compiled his collecting data in Docters van Leeuwen-Reijnvaan and Docters van Leeuwen (1926). Then, Dammerman (1948) listed 60 sorts of arthropod galls on 44 plant species based on the collecting records of Docters van Leeuwen (1920, 1936). Because three sorts of gall were omitted from the list in Dammerman (1948), the total number should be 63 in 1919–1922. In 1982, Yukawa and Partomihardjo

(1997) surveyed arthropod galls on Peucang Island, Panaitan Island and the Krakatau Islands, and found 15 sorts on the Krakataus. Thereafter, T. Partomihardjo (LIPI, Indonesia) and J. Yukawa and his students (Kyushu University, Japan) visited the Krakataus and adjacent areas on many occasions to search for arthropod galls during the period from 1991 to 2000. The adjacent areas include Sebesi Island, Sebuku Island, South Sumatra, Carita, Peucang Island, Panaitan Island and Ujung Kulon (Fig. 5.4). These areas were less influenced by the 1883 eruption than the Krakataus.

The flora of the Krakataus and the succession of vegetation have been relatively intensively surveyed by plant taxonomists and ecologists (e.g. Whittaker and Flenley 1982; Partomihardjo et al. 1985; Whittaker et al. 1984, 1989, 1992). Relying on the results of their surveys, most host plants of galling-arthropods were identified by T. Partomihardjo in the field. Otherwise, unidentified plants were compared for identification with the herbarium collection kept in the Herbarium Bogoriense, Bogor, Indonesia. Galls collected were identified relying upon descriptions and figures of galls provided by Docters van Leeuwen (1920, 1922), Docters van Leeuwen-Reijnvaan and Docters van Leeuwen (1926) and Yukawa and Partomihardjo (1997).

5.4.2 Arthropod Galls Found on the Krakataus and in Adjacent Areas

Partomihardjo et al. (2011) divided field surveys into four periods: the first period in the 1920s (Docters van Leeuwen 1920, 1922, 1936; Docters van Leeuwen-Reijnvaan and Docters van Leeuwen 1926; Dammerman 1948), the second in 1982 (Yukawa and Partomihardjo 1997), the third in 1991–1993 and the fourth in 2000 (Partomihardjo et al. 2011).

A total of 147 sorts of arthropod gall were found on the Krakataus and in adjacent areas since the 1883 eruption up until 2000. They were induced on at least 109 plant species belonging to 81 genera of 42 families including one species of Pteridophyta (Table 2 in Partomihardjo et al. 2011). Cecidomyiid galls were most abundant among them, 57 sorts having been found on 47 plant species belonging to 47 genera of 27 families (Table 5.5).

The 1991–1993 survey was the most intensive in recent years, and 45 different sorts of arthropod gall were collected from the islands (Table 3 in Partomihardjo et al. 2011). Among them, eriophyoid and cecidomyiid galls were dominant, and psylloid, thysanopteran, coccoid, hymenopteran, and lepidopteran galls were in a minority. In total, 52 different sorts of gall were collected from the islands during the period from 1982 to 2000 (Table 3 in Partomihardjo et al. 2011).

Table 5.5 Cecidomyiid galls collected from the Krakatau Islands and adjacent areas during four different survey periods (modified from Partomihardjo et al. 2011)

Family and species of host plant	Gall No.[a]	Rakata	Sertung	Panjang	Anak	Sebesi and Sebuku	South Sumatra[b]	Carita	Panaitan and Peucang	Ujung Kulon
OLEANDRACEAE										
Nephrolepis hirsutula	21,412	1	1	–	–	–	1	–	–	–
ACANTHACEAE										
Acanthus ilicifolius	20,006	–	–	–	–	–	–	–	–	3,4
ALANGIACEAE										
Alangium densiflorum	21,185	–	–	–	–	3	1	–	–	–
ANACARDIACEAE										
Buchanania arborescens	20,097	2,3,4	2	3,4	–	3	–	2	2	3,4
Mangifera indica	21,005	–	–	–	–	1,3	1	–	–	–
Mangifera indica	20,617	–	–	–	–	1,3	3	2	–	–
ANNONACEAE										
Pseuduvaria reticulata	PY001	–	–	–	–	–	–	–	2	3
ARISTOLOCHIACEAE										
Aristolochia tagala	No number	1	–	–	–	–	–	–	–	–
ASTERACEAE										
Wollastonia biflora	20,198	1	1	–	–	3	–	2	–	–
COMBRETACEAE										
Terminalia catappa	21,122	1,2,3,4	3,4	3,4	3,4	1,3	1,3	3	3	3
Terminalia catappa	PY004	3,4	3	3,4	3	3	3	3	–	3
CUCURBITACEAE										
Coccinia cordifolia	20,255	–	–	1	–	–	–	–	–	–
Trichosanthes bracteata	21,401	1,3	1	–	3	1,3	3	3	–	–
EUPHORBIACEAE										

(continued)

Table 5.5 (continued)

Family and species of host plant	Gall No.[a]	Rakata	Sertung	Panjang	Anak	Sebesi and Sebuku	South Sumatra[b]	Carita	Panaitan and Peucang	Ujung Kulon
Antidesma bunius	20,048	–	–	–	–	1	1	–	–	–
Antidesma sp.	PY006	–	–	–	–	–	–	3	–	–
Breynia cernua	PY007	3	–	–	–	–	–	–	–	–
Bridelia monoica	20,094	1,3	4	4	–	3	3	–	–	–
Drypetes macrophylla	PY008	–	–	–	–	–	–	–	2	–
Glochidion phillipicum	20,475	–	–	–	–	3	1,3	–	–	–
Mallotus richinoides	20,990	–	–	–	–	3	3	–	–	–
Melanolepis multiglandulosa	20,981	–	–	–	–	–	–	3	–	–
FABACEAE										
Derris heterophylla	20,321	1,3	–	–	–	–	–	–	–	–
Derris heterophylla	PY030	–	–	–	–	–	–	–	–	4
Derris heterophylla	PY031	–	–	–	–	–	–	–	–	4
Intsia bijuga	PY013	–	–	–	–	–	–	3	–	–
Millettia sericea	20,894	–	–	–	–	–	–	3	–	–
Spatholobus verugineus	PY014	–	–	–	–	–	–	–	2	3
Spatholobus verugineus	PY015	–	–	–	–	–	–	–	2	3
FICACEAE										
Ficus pubinervis	20,140	1,3	4	3	–	–	–	–	3	–
Ficus ribes	20,156	1	–	–	–	–	–	–	–	–
Ficus subulata	20,163	1,3	–	–	–	–	–	–	–	–
Ficus tinctoria ssp. *gibbosa*	20,120	1,3	1,3	3	–	–	–	–	3	–
FLACOURTIACEAE										
Flacourtia rukam	20,178	–	–	–	–	–	1	–	–	–

LEEACEAE										
Leea sambucina	21,032	1,2,3,4	1,3,4	1,3,4	–	1,3	1,3	3	2,3	3,4
MELIACEAE										
Dysoxylum densiflorum	20,342	–	–	–	–	–	–	–	–	3
Dysoxylum gaudichaudianum	PY016	–	2 N	–	–	–	–	–	–	–
MENISPERMACEAE										
Pericampylus glaucus	20,812	1	–	–	–	1,3	1	–	–	–
MYRTACEAE										
Syzygium polyanthum	20,418	3	–	3,4	–	–	–	–	3	3
Syzygium zollingerianum	21,134	–	–	–	–	1,3	1	–	–	–
ORCHIDACEAE										
Calanthe zollingeri	20,683	–	–	–	–	1,3	1	–	–	–
RUBIACEAE										
Anthocephalus chinensis	PY023	–	–	–	–	3	–	–	–	–
Neonauclea calycina	21,342	1,3	–	–	–	–	1	–	–	–
Neonauclea calycina	21,343	1,3	–	–	–	–	–	–	–	–
Neonauclea calycina	PY032	4	–	–	–	–	–	–	–	–
Paederia foetida	21,351	–	–	–	–	3	1	–	–	–
RUTACEAE										
Luvunga sp.	PY025	–	–	–	–	–	–	–	–	3
SAPOTACEAE										
Planchonella obovata	PY026	–	–	–	–	–	–	–	2	3
SONNERATIACEAE										
Sonneratia alba	21,111	–	–	–	–	3	–	–	–	–
STERCULIACEAE										
Heritiera littoralis	PY027	–	–	–	–	–	–	–	2	3
Melochia umbellata	21,080	1,3	–	–	–	–	–	–	–	–

(continued)

Table 5.5 (continued)

Family and species of host plant	Gall No.[a]	Rakata	Sertung	Panjang	Anak	Sebesi and Sebuku	South Sumatra[b]	Carita	Panaitan and Peucang	Ujung Kulon
Pterospermum diversifolium	PY033	–	–	–	–	–	–	–	–	4
URTICACEAE										
Poikilospermum suaveolens	PY028	–	–	–	–	3	–	–	–	–
VERBENACEAE										
Avicennia marina	20,074	–	–	–	–	1,3	1	–	–	–
Avicennia marina	20,075	–	–	–	–	1,3	1	–	–	–
Clerodendrum inerme	20,249	3	1,3	1	3	–	–	3	–	–
VITACEAE										
Cayratia trifolia	21,053	2,4	3	3	–	1,3	1	3	–	–
Cayratia trifolia	21,056	–	–	–	–	3	–	–	–	–
Tetrastigma lanceolarium	21,040	3	–	–	–	3	–	3	–	–

Numerals 1, 2, 3 and 4 in the table indicate, respectively, that each gall was found in the first (the 1920s) (Docters van Leeuwen 1920, 1922, 1936; Docters van Leeuwen-Reijnvaan and Docters van Leeuwen 1926; Dammerman 1948), second (1982) (Yukawa and Partomihardjo 1997), third (1991–1993) (Partomihardjo et al. 2011) or fourth expedition (2000) (Partomihardjo et al. 2011)

[a]Gall numbers are quoted from Docters van Leeuwen-Reijnvaan and Docters van Leeuwen (1926), except those with PY, which were found by T Partomihardjo and J Yukawa in the second, third, and fourth expeditions

[b]South Sumatra includes Lampung and its adjacent areas

5.4.3 Disharmony in the Relative Abundance of Sorts of Galls Among Galling-Arthropod Taxa

The accumulated number of sorts of arthropod gall found on the Krakataus was 74 in 1919–2000, being only 4.8% in the number of sorts for the whole of Indonesia (Table 5.6). The total number of butterfly species on the Krakataus was only 39, also much fewer than 583 in Java and 686 in Sumatra (Yukawa 1984a). Similar results were found for tephritid species of the genus *Dacus* (Yukawa 1984b) and for aculeate Hymenoptera (Yamane 1983).

Faunas of oceanic islands are frequently disharmonious (e.g. Carlquist 1974; MacArthur 1972), and remarkable examples of faunistic disharmony were found on the Krakataus for aculeate Hymenoptera (Yamane 1983) and termites (Abe 1984). As to the fauna of gall-inducing arthropods on the Krakataus, disharmony was evident in 1919–1922 (early stage of succession) with the relatively high Krakatau/Indonesia ratio (8.2% of 378 sorts) in eriophyoid galls compared to the other groups (Table 5 in Partomihardjo et al. 2011). The presence or absence of host plants are essential, then disharmony is referable chiefly to different dispersal means of arthropod groups. Eriophyoids would be able to disperse more easily with wind than other groups. In addition, synchronization of emergence or oviposition seasons with host plant phenology, such as time of shoot growth or flowering season, is also an important factor in the establishment of galling-arthropod populations (Yukawa 2000). Particularly for arthropods that induce flower and fruit galls, synchronization is more difficult, because the available season of host plant organ is limited. In contrast, it is easier for leaf or bud gall inducers because shoots are produced almost continuously in the tropics. Most eriophyoids are leaf or bud gall inducers and complete one generation within a relatively short period of time (Lindquist et al. 1996). Thus, eriophyoids might have more opportunities to colonize the islands than the other gall-inducing arthropods. In 1982–2000, however, differences in the percentage between Eriophyoidea and groups other than Thysanoptera reduced (Table 5 in Partomihardjo et al. 2011). As a result, disharmony became obscure in 1982–2000 following ecological succession. The statistical analysis of gall sorts accumulated from 1919 to 2000 indicates no significant difference between the Krakataus and the whole of Indonesia in the relative abundances of gall sorts induced by different arthropod taxa ($p < 0.01$; Fisher's exact test with simulated p-value based on 1e+05 replicates using the fisher.test function in R ver. 3.6.1) (Table 5.6).

In contrast, the statistical analysis indicates that the relative abundances of gall sorts are significantly different between Japan and the whole of Indonesia or the Krakataus (Table 5.6). The abundance of cecidomyiid and eriophyoid galls can be commonly seen in other faunistic studies of arthropod galls, for example in North America (Felt 1965), Japan (Yukawa and Masuda 1996) and Europe (Roskam 2019). However, the rates of galls induced by Aphidoidea and Hymenoptera were distinctly higher in the Holarctic Region than those in Indonesia (Tropics) (Table 5.6). This is caused by existence of many species of Aphidoidea (Hemiptera),

Table 5.6 Accumulated number of sorts of arthropod gall found from 1919 to 2000 on the Krakataus and in adjacent areas, and comparison between the Krakataus, the whole of Indonesia and Japan in the relative abundance of gall sorts induced by different arthropod taxa

Group of gall inducers	Krakatau Islands[a]	Sebesi and Sebuku[a]	South Sumatra[a]	Carita[a]	Panaitan and Peucang[a]	Ujung Kulon[a]	Whole of Indonesia[b]	Japan[c]
Thysanoptera	3 (4.1)	7	5	2	1	1	147 (9.6)	10 (0.7)
Psylloidea	6 (8.1)	7	5	0	4	1	145 (9.4)	73 (5.1)
Aphidoidea	5 (6.8)	2	1	0	0	0	66 (4.3)	189 (13.3)
Coccoidea	3 (4.1)	1	2	1	0	1	47 (3.1)	20 (1.4)
Other Hemiptera	0 (0.0)	0	0	0	0	0	11 (0.7)	11 (0.8)
Coleoptera	0 (0.0)	0	0	0	0	0	24 (1.6)	20 (1.4)
Hymenoptera	1 (1.4)	0	1	0	0	0	28 (1.8)	223 (15.8)
Cecidomyiidae	26 (35.1)	24	21	14	12	16	600 (39.1)	628 (44.2)
Tephritidae, etc.	1 (1.4)	1	1	1	0	0	12 (0.8)	20 (1.4)
Lepidoptera	1 (1.4)	3	2	2	1	1	66 (4.3)	58 (4.1)
Eriophyoidea	28 (37.8)	27	32	8	3	7	378 (24.6)	171 (12.0)
Unknown	0 (0.0)	0	1	0	0	0	12 (0.8)	0 (0.0)
Total	74 (100%)[a]	72	71	28	21	27	1536 (100%)[a]	1423 (100%)[b]

[a]Partomihardjo et al. (2011)

[b]Docters van Leeuwen-Reijnvaan and Docters van Leeuwen (1926)

[c]Yukawa and Masuda (1996). Relative abundances (numerals in parentheses) were compared between the Krakatau Islands, the whole of Indonesia and Japan. Different alphabetical letters after 100% of the total numbers indicate significant difference between different survey areas, $P < 0.01$ (Fisher's exact test with simulated p-value (based on 1e+05 replicates) using the fisher.test function in R ver. 3.6.1)

Tenthredinidae (Hymenoptera) on *Salix* (Salicaceae) and Cynipidae (Hymenoptera) on *Quercus* (Fagaceae) in the Holarctic Region.

5.4.4 Sebesi Island, Sebuku Island and South Sumatra

The islands of Sebesi and Sebuku locate in the Sunda Strait between South Sumatra and the Krakataus (Fig. 5.4). These islands have been paid special attention as stepping stones in the colonization of the Krakataus and were visited by various biologists for field surveys (e.g. Docters van Leeuwen 1923; Dammerman 1922, 1948; Bush et al. 1990; Thornton et al. 1994, 2002; Partomihardjo 1999; Yukawa et al. 2000).

From the 1920s to 2000, a total of 72 different sorts of arthropod gall were found on Sebesi-Sebuku, and 71 sorts in Lampung and its adjacent areas of South Sumatra (Table 5.6). Among 26 cecidomyiid galls found on the Krakataus, 13 (50%) were found also on Sebesi-Sebuku or in South Sumatra, and 15 (57.7%) were in Carita, Panaitan-Peucang and Ujung Kulon (Table 5.5). In addition, 10 (38.5%) of the 26 were commonly distributed on the Krakataus and both in Javan and Sumatran sides. These data seem to indicate that the galling-cecidomyiid fauna on the Krakataus have been derived both directly from Java and through the two stepping stone islands from Sumatra.

However, the butterfly and moth fauna of the Lampung district of South Sumatra, including Sebesi-Sebuku, contains Javanese species or subspecies in addition to true Sumatran elements (Toxopeus 1950; Yukawa et al. 2000). Among the ten Krakatau butterfly species that were identified to the subspecies level, nine were Javanese and only one was Sumatran subspecies (Yukawa 1984a). A similar tendency was also noted for aculeate Hymenoptera (Yamane 1983), in which four species derived from Java and one from Sumatra. In termites, two species of Javanese origin were recognized, and one of Sumatran origin (Yamane et al. 1992; Abe 1984). By examining Krakatau's resident land birds, reptiles, bats, land mollusca, termites, butterflies, pteridophytes and spermatophytes, Thornton et al. (2002) concluded that, overall, Sebesi had little impact as a stepping stone in the recolonization of the Krakataus, regardless of the fact that prevailing north-west monsoon and movement of sea current would tend to bring airborne and surface propagules, respectively, to the Krakataus from Sumatra and Sebesi-Sebuku (Thornton and Rosengren 1988; Gathorne-Hardy et al. 2000).

5.4.5 Colonization Curve

Preston (1962) and MacArthur & Wilson (1963) independently suggested that on islands there might be a balance of immigration by extinction so that the diversity of biotas could be understood as equilibrium. A dynamic equilibrium is reached when immigration and extinction rates are equal; namely colonization rate falls to zero (MacArthur and Wilson 1963, 1967).

Six sorts of cecidomyiid gall that had not been found in 1919–1922 on the Krakataus was newly detected in 1982–2000 possibly due to the arrival of their host plants along with vegetation succession (Table 5.7). In contrast, six other sorts of cecidomyiid gall that had been found in 1919–1922 were not detected in 1982–2000 although their host plants existed on the islands. Probably this was caused by overlook rather than extinction. The number of sorts of cecidomyiid gall seems to still increase, as has been noted for butterflies (Yukawa 1984a). The colonization curve for butterflies (Fig. 3 in Yukawa 1984a) indicates that the species number is still increasing on the islands with some turnover events along with vegetation succession. Recent arrival of vascular epiphytes at the Krakataus (Partomihardjo et al. 2004) provides lycaenid butterflies in particular with a variety of host plants, which will facilitate the increase of lycaenid species. Thus, the presence or absence of host plants is an essential factor influencing the establishment of newly arrived monophagous or oligophagous herbivores on the islands.

Whittaker et al. (1984) predict that, in the absence of major disturbance by humans, volcanicity, or other factors, the size of the flora of Rakata will continue to increase rather slowly by the addition of "primary" forest species, with the gradual decline (and sometimes loss) of some "secondary" forest species. A progression to the status of "primary rain forest" is clearly a very long way off. Therefore, colonization curves reflecting the species number of monophagous or oligophagous herbivores would not reach the stage of equilibrium even within the period of 100 years after the catastrophic eruption of Krakatau in 1883.

In contrast, colonization curves for polyphagous herbivores or predacious organisms seem to reach equilibrium earlier than monophagous or oligophagous herbivores because their establishment is not directly influenced by the vegetation succession. Actually, Yamane (1983) demonstrated that the colonization curve for aculeate Hymenoptera had already reached the equilibrium on the Krakataus.

5.4.6 Recent Progress in Island Biogeography

Whittaker et al. (2017) reviewed recent progress in integrating the largely ecological thinking of MacArthur & Wilson's theory of island biogeography (MacArthur and Wilson 1963, 1967) with insights on the longer-term dynamics of both the islands and their biotas, and showed new perspectives that predict how immigration, speciation, and extinction respond to the typical life cycle of hotspot islands, with phases of emergence, development, and submergence. For example, there was an "island progression rule" in which older landmasses donate colonists to younger islands (consistent with the generalization of islands as "sinks"), but recently evidence of "reverse colonization" from islands to continental regions has been increasing (e.g. Nicholson et al. 2005; Shaw and Gillespie 2016). A classic insular syndrome is the loss of dispersibility of formerly dispersive species following island

Table 5.7 Comparison of cecidomyiid galls found on the Krakatau Islands between two different survey periods

Family and species of host plant	Gall no.[a]	Survey period 1919–1922[b]	1982–2000[c]
OLEANDRACEAE			
Nephrolepis hirsutula	21,412	○	×
ANACARDIACEAE			
Buchanania arborescens	20,097	××	○
ARISTOLOCHIACEAE			
Aristolochia tagala	No number	○	×
ASTERACEAE			
Wollastonia biflora	20,198	○	×
COMBRETACEAE			
Terminalia catappa	21,122	○	○
Terminalia catappa	PY004	×	○
CUCURBITACEAE			
Coccinia cordifolia	20,255	○	×
Trichosanthes bracteata	21,401	○	○
EUPHORBIACEAE			
Breynia cernua	PY007	××	○
Bridelia monoica	20,094	○	○
FABACEAE			
Derris heterophylla	20,321	○	○
FICACEAE			
Ficus pubinervis	20,140	○	○
Ficus ribes	20,156	○	×
Ficus subulata	20,163	○	○
Ficus tinctoria ssp. *gibbosa*	20,120	○	○
LEEACEAE			
Leea sambucina	21,032	○	○
MELIACEAE			
Dysoxylum gaudichaudianum	PY016	××	○
MENISPERMACEAE			
Pericampylus glaucus	20,812	○	×
MYRTACEAE			
Syzygium polyanthum	20,418	××	○
RUBIACEAE			
Neonauclea calycina	21,342	○	○
Neonauclea calycina	21,343	○	○
Neonauclea calycina	PY032	×	○
STERCULIACEAE			
Melochia umbellata	21,080	○	○
VERBENACEAE			
Clerodendrum inerme	20,249	○	○
VITACEAE			

(continued)

Table 5.7 (continued)

Family and species of host plant	Gall no.[a]	Survey period	
		1919–1922[b]	1982–2000[c]
Cayratia trifolia	21,053	××	○
Tetrastigma lanceolarium	21,040	××	○
Total sort of galls found		18	20
Total sort of galls not found		8	6

[a]Gall numbers are quoted from Docters van Leeuwen-Reijnvaan and Docters van Leeuwen (1926), except those with PY, which were found by T. Partomihardjo and J. Yukawa in the second, third, and fourth expeditions
[b]Data of the first survey period (1919–1922) were based on Docters van Leeuwen (1920, 1922, 1936), Docters van Leeuwen-Reijnvaan and Docters van Leeuwen (1926) and Dammerman (1948)
[c]Data of the second survey period (1982–2002) were based on Yukawa and Partomihardjo (1997) and Partomihardjo et al. (2011). ○: Galls were found; ×: Galls were not found. ××: Plant species possibly did not exist on the islands (Docters van Leeuwen-Reijnvaan and Docters van Leeuwen 1926; Whittaker and Flenley 1982)

colonization, but now good evidence appears for this syndrome: at least for plants and invertebrates, many species lacking evident long-distance dispersal adaptations can reach quite remote archipelagos (e.g. Wada et al. 2012; Carvajal-Endara et al. 2017).

In the light of recent progress in island biogeography, classic insular syndromes, particularly those on hotspot islands like the Krakataus, are needed in future to be reexamined using expanded data sets, molecular biology, modern techniques of analysis, and Earth system science.

References

Abe T (1984) Colonization of the Krakatau Islands by termites (Insecta: Isoptera). Physiol Ecol Jpn 21:63–88

Bush MB (1986) The butterflies of Krakatoa. Entomol Mon Mag 122:51–58

Bush MB, Whittaker RJ (1991) Krakatau: colonization patterns and hierarchies. J Biogeogr 18:341–356

Bush MB, Bush DJB, Evans RD (1990) Butterflies of Krakatau and Sebesi. New records and habitat relations. In: Whittaker RJ, Asquith NM, Bush BM, Partomihardjo T (eds) Krakatau research project 1989 expedition report. School of Geography, University of Oxford, Oxford, pp 35–41

Carlquist S (1974) Island biology. Columbia University Press, New York

Carvajal-Endara S, Hendry AP, Emery NC, Davies TJ (2017) Habitat filtering not dispersal limitation shapes oceanic island floras: species assembly of the Galápagos archipelago. Ecol Lett 20:495–504

Compton SG, Thornton IWB, New TR, Underhill L (1988) T colonization of the Krakatau Islands by fig wasps and other chalcids (Hymenoptera, Chalcidoidea). Philos Trans R Soc London Ser B 322:459–470

Dammerman KW (1922) The fauna of Krakatau, Verlaten Island and Sebesy. Treubia 3:61–112

Dammerman KW (1948) The fauna of Krakatau 1883–1933. Verhandlingen der Koninklijke Nederlandsche Akademie van Wetenschappen Afdel Natuurkunde Part 2(44):1–594

Docters van Leeuwen WM (1920) The galls of Krakatau and Verlaten Eiland (Desert Island) in 1919. Annales du Jardin Botanique de Buitenzorg 31:83–92

Docters van Leeuwen WM (1922) The galls of the islands of the Krakatau-group and of the island of Sebesy. Bulletin du Jardin Botanique de Buitenzorg III 4:288–314

Docters van Leeuwen WM (1923) The vegetation of the island of Sebesy, situated in the Sunda-Strait, near the islands of the Krakatau group, in the year 1921. Annales du Jardin Botanique de Buitenzorg 32:135–192

Docters van Leeuwen WM (1936) Krakatau: 1883–1933. A Botany. Annales du Jardin Botanique de Buitenzorg 46(47):1–506

Docters van Leeuwen-Reijnvaan J, Docters van Leeuwen WM (1926) The Zoocecidia of the Netherlands east indies. Drukkerij de Unie, Batavia

Elsayed AK, Ogata K, Kaburagi K, Yukawa J, Tokuda M (2017) A new *Dasineura* species (Diptera: Cecidomyiidae) associated with *Symplocos cochinchinensis* (Loureiro) (Symplocaceae) in Japan. Jpn J Syst Entomol 23:81–86

Elsayed AK, Matsuo K, Kim W, Uechi N, Yukawa J, Gyoutoku N, Tokuda M (2018a) A new *Asphondylia* species (Diptera: Cecidomyiidae) and a eulophid wasp (Hymenoptera) inducing similar galls on leaf buds of *Schoepfia jasminodora* (Schoepfiaceae), with reference to their ecological traits and a description of the new gall midge. Entomol Sci 21:324–339

Elsayed AK, Yukawa J, Tokuda M (2018b) A taxonomic revision and molecular phylogeny of the eastern Palaearctic species of the genera *Schizomyia* Kieffer and *Asteralobia* Kovalev (Diptera: Cecidomyiidae: Asphondyliini), with descriptions of five new species of *Schizomyia* from Japan. ZooKeys 808:123–160

ESK & KSAE (The Entomological Society of Korea & The Korean Society of Applied Entomology) (1994) Check list of insects from Korea. Kon-Kuk University Press, Seoul

Evenhuis NL, Yukawa J (1986) Bombyliidae (Diptera) of Panaitan and the Krakatau Islands, Indonesia. Kontyû 54:450–459

Felt EP (1965) Plant galls and gall makers. Hafner Publishing Compa, New York and London

Forestry Administration Division and Forest Department Office, Agriculture and Forestry Department, Nagasaki Prefecture (2015) The 2012 report on the statistics of forest and forestry in Nagasaki prefecture. Agriculture and Forestry Department, Nagasaki Prefecture, Nagasaki. (In Japanese)

Gagné RJ, Moser JC (2013) The North American gall midges (Diptera: Cecidomyiidae) of hackberries (Cannabaceae: *Celtis* spp.). Mem Am Entomol Soc 49:1–103

Ganaha T, Yukawa J, Uechi N, Nohara M, Paik JC (2004) Identifications of some species of the genus *Rhopalomyia* (Diptera: Cecidomyiidae) inducing galls on *Artemisia* (Asteraceae) in South Korea. Esakia 44:45–55

Ganaha T, Nohara M, Sato S, Uechi YK, Yamauchi S, Yukawa J (2007) Polymorphism of axillary bud galls induced by *Rhopalomyia longitubifex* (Diptera: Cecidomyiidae) on *Artemisia princeps* and *A. montana* (Asteraceae) in Japan and Korea, with the designation of new synonyms. Entomol Sci 10:209–212

Gathorne-Hardy FJ, Jones DT, Mawdsley NA (2000) The recolonization of the Krakatau islands by termites (Isoptera), and their biogeographical origins. Biol J Linn Soc 71:251–267

Hachijojima Interpretation Association (2007) Field guide Hachijo-jima. Hachijojima Kankou Shinkou Jikkou Iinkai, Tokyo. (In Japanese)

Hasegawa M (2003) Ecological diversification of insular terrestrial reptiles: a review of the studies on the lizard and snakes of the Izu Islands. Glob Environ Res 7:59–67

Im H, Ji O (2015) A report on the plants galls. Jayeon-gwa-saengtae, Seoul. (In Korean)

Inoue K (1988) Pattern of breeding-system change in the Izu Islands in *Campanula punctata*: bumblebee-absence hypothesis. Plant Species Biol 3:125–128

Inoue K, Amano M (1986) Evolution of *Campanula punctata* lam. In the Izu Islands: changes of pollinators and evolution of breeding systems. Plant Species Biol 1:89–97

Inouye M (1962) Report on an inspection of the Japanese pine needle gall midge on Tsushima. Dantairin 17:16–19. (In Japanese)

Ishizawa J (1942) [Insects in the four seasons.] Ars, Tokyo. (In Japanese)
Isoda Y (2011) Climate change and physical process associated with the Tsushima warm current. Mem Fac Fish Sci 53:2–12. Hokkaido University (in Japanese, with English title and abstract)
Jacobson ER (1909) De nieuwe fauna van Krakatau. Jaarverslag Topographischen Dienst Nederlandsch-Indië 4:192–206
Kanmiya K, Yukawa J (1985) Chloropidae (Diptera) of Panaitan and the Krakatau Islands, Indonesia. Kontyû 53:461–474
Kim W, Tokuda M, Yukawa J (2015) Cecidomyiid galls found on Tsushima, a stepping stone island between the Korean peninsula and Kyushu, Japan. Acta Dipterol 26:21–37
Kim W, Minami T, Tokuda M, Matsuo K, Harris KM, Yukawa J (2019) Detection of two new cryptic species of *Kiefferia* (Diptera: Cecidomyiidae) by means of morphological, molecular and ecological studies. Entomol Sci 22:450–462
Kobayashi M (1985) On the Trichoptera from the island of Tsushima, with seven new species (Insecta). Bull Kanagawa Prefect Mus Nat Sci 16:7–22
Kodoi F, Lee HS, Uechi N, Yukawa J (2003) Occurrence of *Obolodiplosis robiniae* (Diptera: Cecidomyiidae) in Japan and South Korea. Esakia 43:35–41
Korea Forest Service (2010) Korea biodiversity information system. http://www.nature.go.kr/
Kurosawa Y (1978) Phylogeny and distribution of *Lucanus gamunus* Sawada et Y. Watanabe (Coleoptera, Lucanidae), endemic to the Izu Islands, Central Japan. Mem Nat Sci Mus, Tokyo 11:141–153
Kurosawa Y (1990) Insects fauna of the Izu Islands. Nihon no Seibutsu 4:23–28. (In Japanese)
Lindquist EE, Sabelis MW, Bruin J (1996) Eriophyoid mites, their biology, natural enemies and control. Elsevier, Amsterdam
MacArthur RH (1972) Geographical ecology. Patterns in the distribution of species. Harper and Row, New York
MacArthur RH, Wilson EO (1963) An equilibrium theory of insular zoogeography. Evolution 17:373–387
MacArthur RH, Wilson EO (1967) The theory of island biogeography. Princeton University Press, Princeton, NJ
Miyake T, Inoue K (2003) Character displacement in style length between pollinator-sharing *Clerodendrum trichotomum* and *C. izuinsulare* (Verbenaceae). Plant Syst Evol 243:31–38
Nagai A (2015) Arthropod galls in Miyazaki prefecture VI. Nishimoro no Seibutsu 8:45–65. (In Japanese)
Nagai A (2016) Arthropod galls in Miyazaki prefecture VII. Nishimoro no Seibutsu 9:26–49. (In Japanese)
Nagai A (2017) Arthropod galls in Miyazaki prefecture VIII. Nishimoro no Seibutsu 10:22–37. (In Japanese)
Nagai A (2018) Arthropod galls in Miyazaki prefecture IX. Nishimoro no Seibutsu 11:10–17. (In Japanese)
Nagai A (2019) Arthropod galls in Miyazaki prefecture X. Nishimoro no Seibutsu 12:18–26. (In Japanese)
Nagasaki Biological Society (1976) Biota on Tsushima. The Nagasaki Biological Society, Isahaya. (In Japanese)
Naito A, Tokuda M (2018) Predation by birds on larvae of *Schizomyia* sp. (Diptera: Cecidomyiidae) leaving infructescence galls on *Castanopsis sieboldii* (Fagaceae). Miyakensis 21:47–53. (In Japanese)
New TR, Sudarman HK (1988) Neuroptera (Insecta) of the Krakatau Islands, Indonesia. Phil Trans R Soc London Ser B 322:481–485
New TR, Thornton IWB (1992a) Colonization of the Krakatau Islands by invertebrates. In: Thornton IWB (ed) Krakatau – a century of change. GeoJournal, vol 28, pp 219–224
New TR, Thornton IWB (1992b) The butterflies of Anak Krakatau, Indonesia: faunal development in early succession. J Lepid Soc 46:83–96

New TR, Bush MB, Thornton IWB, Sudarman HK (1988) The butterfly fauna of the Krakatau Islands after a century of recolonization. Phil Trans R Soc London Ser B 322:445–457

Nicholson KE, Glor RE, Kolbe JJ (2005) Mainland colonization by island lizards. J Biogeogr 32:929–938

Oiki S, Kawahara T, Inoue K, Ohara M, Maki M (2001) Random amplified polymorphic DNA (RAPD) variation among populations of the insular endemic plant *Campanula microdonata* (Campanulaceae). Ann Bot 87:661–667

Okada Y (1927) A study on the distribution of tailless batrachians of Japan. Annot Zool Jpn 11:137–143

Osozawa S, Shinjo R, Armid A, Watanabe Y, Horiguchi T, Wakabayashi J (2012) Palaeogeographic reconstruction of the 1.55 ma synchronous isolation of the Ryukyu Islands, Japan, and Taiwan and inflow of the Kuroshio warm current. Int Geol Rev 54:1369–1388

Paik JC, Yukawa J, Uechi N, Sato S, Ganaha T (2004) Gall-inducing species of the family Cecidomyiidae (Diptera) recorded from the Korean peninsula and surrounding islands, in comparison with the gall-midge fauna of Japan. Esakia 44:57–66

Partomihardjo T (1999) Flora Sebesi, peran dan potensinya dalam proses hunian kembali Kepulautan Krakatau [Flora of Sebesi Island: its role and potential in the recolonization proses of the Krakatau Islands]. In: Proceedings, National Sminor on Indonesian Plant Conservation. LIPI, Jakarta, pp 123–129

Partomihardjo T, Riswan S, Abdulhadi R (1985) Plant communities on the Anak Krakatau. In: Sastrapradja D, Soemodihardjo S, Soemartadipura A, Soegiarto A, Adisoemarto S, Kusumadinata K, Birowo S, Riswan S, Lapian AB (eds) Proceedings of the symposium on 100 years development of krakatau and its surroundings, I. Indonesian Institute of Sciences (LIPI), Jakarta, pp 411–416

Partomihardjo T, Suzuki E, Yukawa J (2004) Development and distribution of vascular epiphytes communities of the Krakatau Islands, Indonesia. South Pac Stud 25:7–26

Partomihardjo T, Yukawa J, Uechi N, Abe J (2011) Arthropod galls found on the Krakatau Islands and in adjacent areas of Indonesia, with reference to faunistic disharmony between the islands and the whole of Indonesia. Esakia 50:9–21

Preston FW (1962) The canonical distribution of commonness and rarity: part I. Ecology 43:185–432. Part II *Ibid* 43:410–432

R Core Team (2018) R: a language and environment for statistical computing. R Foundation for Statistical Computing, Vienna. http://www.R-project.org/

Reuters World News (2007) Indonesia's Krakatau Roars, dazzles with fireworks. November 11, 2007/5:29 pm/12 Years AGO

Richards PW (1952) The tropical rain Forest: an ecological study. Cambridge University Press, Cambridge

Roonwal ML (1969) The pioneer termite fauna of the recent volcanic island, Anak Krakatau, Sunda Strait, Indonesia. Bull Syst Zool 1:39–44

Roskam JC (2019) Plant galls of Europe. KNNV Publishers, Zeist

Saitô K (1932) Studies of insect galls in Korea. Bull Sci Res Alumni Assoc Morioka Imperial Coll Agric For 7:93–107. (In Japanese)

Saitô K (1933) Studies of insect galls in Korea II. J Chosen Nat Hist Soc 15:8–10. (In Japanese)

Sasa M, Suzuki H (1999) Studies on the chironomid midges of Tsushima and Iki Islands, western Japan. Part 2. Species of Orthocladiinae and Tanypodinae collected on Tsushima. Trop Med 41:75–132

Shaw KL, Gillespie RG (2016) Comparative phylogeography of oceanic archipelagos: hotspots for inferences of evolutionary process. Proc Natl Acad Sci USA 113:7986–7993

Shiuchi T, Hotta M (2015) Flora of Tokara Islands. Kagoshima Univ Mus Bull 7:1–367. (In Japanese)

Sunose T (1981a) Biogeography of two color types of *Masakimyia pustulae*, – the origin of the white color type and its range expansion. Panmixia 4:1–7. (In Japanese)

Sunose T (1981b) A study of biogeography on the Izu Islands with notes on the distribution of gall midges. Panmixia 4:8–12. (In Japanese)

Sunose T, Yukawa J (1979) Interrelationship between the leaf longevity of the evergreen spindle tree, *Euonymus japonicus* Thunb. and the euonymus gall midge, *Masakimyia pustulae* Yukawa and Sunose (Diptera, Cecidomyiidae) in different environments. Jpn J Ecol 29:29–34

Taira A (1990) The dawn of the Japanese archipelago. Iwanami-shoten, Tokyo. (In Japanese)

Taira A, Tashiro M (1987) Historical biogeography and plate tectonic evolution of Japan and eastern Asia. Terra Scientific Pub. Co., Tokyo

Takaoka H, Saito K (2005) A new species of *Simulium* (*Nevermannia*) from Izu Islands, Japan (Diptera: Simuliidae). Med Entomol Zool 56:309–317

Thakur ML, Thakur RK (1992) Termite fauna of Krakatau and associated islands, Sunda Strait, Indonesia. Treubia 30:213–317

Thornton IWB (1987) Zoological expedition to the Krakataus. Preliminary Report La Trobe University Department of Zoology Miscellaneous Series 3, pp 1–59

Thornton IWB (1996) Krakatau. The destruction and reassembly of an island ecosystem. Harvard University Press, Cambridge, MA

Thornton IWB, New TR (1988) Krakatau invertebrates: the 1980s fauna in the context of a century of recolonization. Phil Trans R Soc London Ser B 322:493–522

Thornton IWB, Rosengren NJ (1988) Zoological expeditions to the Krakatau Islands, 1984 and 1985: general introduction. Phil Trans R Soc London Ser B 332:273–526

Thornton IWB, Partomihardjo T, Yukawa J (1994) Observations on the effects, up to July 1993, of the current eruptive episode of Anak Krakatau. Glob Ecol Biogeogr Lett 4:88–94

Thornton IWB, Runciman D, Cook S, Lumsden LF, Partomihardjo T, Schedvin NK, Yukawa J, Ward SA (2002) How important were stepping stones in the colonization of Krakatau? Biol J Linn Soc 77:275–317

Tokuda M (2012) Biology of Asphondyliini (Diptera: Cecidomyiidae). Entomol Sci 15:361–383

Tokuda M (2014) Gall midges (Diptera: Cecidomyiidae) on the Izu Islands, Tokyo, Japan. Nat Insects 49(3):26–29. pl. 1. (In Japanese)

Tokuda M (2017) Arthropod galls found on Kodakarajima and Takarajima, the Tokara Islands, Japan. Jpn J Entomol (new Series) 20:101–108. (In Japanese with English summary)

Tokuda M (2018) Arthropod galls found in northern parts of the Tokara Islands, Japan. Jpn J Syst Entomol 24:99–107

Tokuda M, Kawauchi K (2013) Arthropod galls found on Toshima and Shikinejima Islands, the Izu Islands, Japan. Jpn J Syst Entomol 19:261–274

Tokuda M, Uechi N, Yukawa J (2002) Distribution of *Asteralobia* gall midges (Diptera: Cecidomyiidae) causing axillary bud galls on *Ilex* species (Aquifoliaceae) in Japan. Esakia 42:19–31

Tokuda M, Tabuchi K, Yukawa J, Amano H (2004) Inter- and intraspecific comparisons between *Asteralobia* gall midges (Diptera: Cecidomyiidae) causing axillary bud galls on *Ilex* species (Aquifoliaceae): species identification, host range, and mode of speciation. Ann Entomol Soc Am 97:954–970

Tokuda M, Matsuo K, Yukawa J (2012a) Insect galls found on Mikurajima and Aogashima, the Izu Islands, Tokyo Japan. Jpn J Entomol (New Series) 15:75–84. (in Japanese with English summary)

Tokuda M, Matsuo K, Yukawa J (2012b) Insect galls found on Miyakejima and Hachijojima, the Izu Islands, Tokyo, Japan. Esakia 52:59–66

Tokuda M, Matsuo K, Kiritani K, Yukawa J (2013) Insect galls found on Ohshima, Kozushima and Niijima Islands, the Izu Islands, Tokyo, Japan. Acta Dipterol 25:1–16

Tokuda M, Kawauchi K, Kikuchi T, Iwasaki Y (2015) Arthropod galls newly found on the Izu Islands, Tokyo, Japan. Jpn J Syst Entomol 21:363–365

Toxopeus LJ (1950) Over de pioneer-fauna van Anak Krakatau, met enige beschouwingen over het onstaat van de Krakatau-fauna. Chronica Nat 106:27–34

Tsukaguchi S, Yukawa J (1988) Neuroptera collected from the Krakatau Islands, Indonesia. Kontyû 56:481–490

Uechi N, Kim W, Tokuda M, Fujii T, Kikuchi H, Kakizaki M, Iwasaki A, Paik JC, Yukawa J (2018) Genetic and ecological differences between *Asphondylia yushimai* and the ivy gall midge, *Asphondylia* sp. (Diptera: Cecidomyiidae), with a new distribution record of the former from Hokkaido and South Korea. Appl Entomol Zool 53:363–371

Volcano Discovery (2020) Krakatoa (Krakatau). Volcano News & Eruption Updates. 25 May 2019

Wada S, Kawakami K, Chiba S (2012) Snails can survive passage through a bird's digestive system. J Biogeogr 39:69–73

Walter TR, Haghighi MH, Schneider FM, Coppola D, Motagh M, Saul J et al (2019) Complex hazard cascade culminating in the Anak Krakatau sector collapse. Nat Comm 10. https://doi.org/10.1038/s41467-019-12284-5

Wheeler WM (1924) Ants of Krakatau and other islands in the Sunda Strait. Treubia 5:239–258

Wheeler WM (1937) Additions to the ant-fauna of Krakatau and Verlaten Island. Treubia 16:21–24

Whittaker RJ, Flenley JR (1982) The flora of Krakatau. In: Flenley JR, Richards K (eds) The Krakatoa Centenary Expedition (1983: final report. Miscellaneous series number 33, 9–53.). University of Hull, Hull

Whittaker RJ, Richards K, Wiriadinata H, Flenley JR (1984) Krakatau I883 to 1983: a biogeographical assessment. Prog Phys Geogr 8:61–81

Whittaker RJ, Bush MB, Richards K (1989) Plant recolonization and vegetation succession on the Krakatau Islands, Indonesia. Ecol Monogr 59:59–123

Whittaker RJ, Bush MB, Partomihardjo T, Asquith NM, Richards K (1992) Ecological aspects of plant colonisation of the Krakatau Islands. GeoJournal 28:201–211

Whittaker RJ, Fernández-Palacios JM, Matthews TJ, Borregaard MK, Triantis KA (2017) Island biogeography: taking the long view of nature's laboratories. Science 357:eaam8326. https://doi.org/10.1126/science.aam8326

Yamane SK (1983) The aculeate fauna of the Krakatau Islands (Insecta, Hymenoptera). Rep Fac Sci Kagoshima Univ (Earth Sci Biol) (16), 75–107

Yamane SK, Abe T, Yukawa J (1992) Recolonization of the Krakataus by Hymenoptera and Isoptera (Insecta). GeoJournal 28:213–218

Yamauchi S, Ikenaga H, Yukawa J (1982) Midge galls collected from the south-west islands of Japan. Satsuma 31:1–23. (in Japanese with English summary)

Yukawa J (1971) A revision of the Japanese gall midges (Diptera: Cecidomyiidae). Mem Fac Agric Kagoshima Univ 8:1–203

Yukawa J (1984a) Geographical ecology of the butterfly fauna of the Krakatau Islands, Indonesia. Tyô to Ga 35:47–74

Yukawa J (1984b) Fruit flies of the genus *Dacus* (Diptera: Tephritidae) on the Krakatau Islands in Indonesia, with special reference to an outbreak of *Dacus albistrigatus* De Meijere. Jpn J Ecol 34:281–288

Yukawa J (1984c) An outbreak of *Crypticerya jacobsoni* (green) (Homoptera: Margarodidae) on Rakata Besar of the Krakatau Islands in Indonesia. Appl Entomol Zool 19:175–180

Yukawa J (1988) Midge galls of Amami-ôshima. Mem Fac Agric Kagoshima Univ 24:141–145

Yukawa J (2000) Synchronization of gallers with host plant phenology. Popul Ecol 42:105–113

Yukawa J (2014) Family Cecidomyiidae. In: Nakamura T, Saigusa T, Suwa M (eds) Catalogue of the insects of Japan. Entomological Society of Japan, Fukuoka, pp 126–160. (In Japanese)

Yukawa J, Masuda H (1996) Insect and mite galls of Japan in colors. Zenkoku Nôson Kyôiku Kyôkai, Tokyo. (In Japanese with English explanation for color plates)

Yukawa J, Partomihardjo T (1997) Insect and mite galls collected from Peucang, Panaitan, and the Krakatau Islands, Indonesia. Tropics 7:141–152

Yukawa J, Rohfritsch O (2005) Biology and ecology of gall-inducing Cecidomyiidae (Diptera). In: Raman A, Schaefer CW, Withers TM (eds) Biology, ecology, and evolution of gall-inducing arthropods, vol 1. Science Publications, Enfield, pp 273–304

Yukawa J, Sunose T (1976) Description of a new gall midge (Diptera, Cecidomyiidae) on *Euonymus*, with notes on its bionomics. Kontyû 44:159–168

Yukawa J, Yamane SK (1985) Odonata and Hemiptera collected from the Krakatau Islands, Indonesia. Kontyû 53:690–698

Yukawa J, Partomihardjo T, Yata O, Hirowatari T (2000) An assessment of the role of Sebesi Island as a stepping-stone for the colonisation of the Krakatau Islands by butterflies. Esakia 40:1–10

Yukawa J, Ogata K, Yata O, Tadauchi O, Kamitani S, Simbolon H, Partomihardjo T, Ngakan PO, Yamaguchi D (2001) An interim report of the 2000 survey of entomofauna on Lombok Island, Bali Island, the Krakatau Islands, and in Ujung Kulon, Indonesia. Esakia 41:1–10

Yukawa J, Uechi N, Ganaha-Kikumura T, Paik JC (2012) Cecidomyiid galls found on Jeju Island and in Sunchon and its vicinity, South Korea. Esakia 52:45–49

Yukawa J, Tokuda M, Yamagishi K (2014) Host plant ranges and distribution records of identified and unidentified species of the genus *Lasioptera* (Diptera: Cecidomyiidae) in Japan. Esakia 53:1–15

Yukawa J, Moriya T, Kanmiya K (2019) Comparison in the flight ability between the soybean pod gall midge, *Asphondylia yushimai* and the aucuba fruit gall midge, *A. aucubae* (Diptera: Cecidomyiidae). Appl Entomol Zool 54:167–174

Part IV
Ecology and Behavior

Chapter 6
Life History Traits

Junichi Yukawa and Nami Uechi

Abstract Life history strategy of univoltine gall-inducing cecidomyiids has been divided into four types based on their overwintering sites and larval developmental stadia. The four types can be applied also to multivoltine gall midges. We show that lower developmental threshold temperature and thermal constant are important factors to determine life history patterns and 50% emergence date, of which the latter is essential to evaluate the degree of synchronization of emergence and host plant phenology. We emphasize the importance of using non-linear relationship between rearing temperature and developmental speed in determining the lower developmental threshold temperature to avoid higher estimation. Based on the life history strategies, we further demonstrate examples of gall midges exhibiting long-term diapause, host organ alternation and host plant alternation, and discuss the adaptive significance of such life history traits.

Keywords Developmental threshold temperature · Host organ alternation · Host plant alternation · Life history strategy · Long-term diapause · Thermal constant

6.1 Introduction

As usual in other insects, gall-inducing cecidomyiids are also divided into uni-, bi- or multivoltine species according to the number of generations repeated in a year, which is fundamentally determined by the lower developmental threshold temperature, thermal constant required for the development from the egg to adult stage (Sect. 6.3), photoperiod and the presence or absence of diapausing period. Some species

J. Yukawa (✉)
Entomological Laboratory, Faculty of Agriculture, Kyushu University, Fukuoka, Japan
e-mail: galling-cecidomyiids@kyudai.jp

N. Uechi
Institute of Plant Protection, National Agriculture and Food Research Organization, Tsukuba, Japan
e-mail: uechinmi@affrc.go.jp

J. Yukawa, M. Tokuda (eds.), *Biology of Gall Midges*, Entomology Monographs,
https://doi.org/10.1007/978-981-33-6534-6_6

require two or more years to complete one generation by entering extended diapause (Sect. 6.4). Because galls are induced through the growth reaction of host plants, their life history patterns are closely linked to their host-plant phenology, particularly to timing when the plant parts are available for oviposition. Extending leaf buds and shoots, fresh leaves, flower buds, flowers and young fruit are targets of oviposition. Most galling-cecidomyiids are mono- or oligophagous species and can complete their annual life cycle on one or some congeneric plant species. A few of them seasonally alternate galling-sites to be bivoltine on a single host plant species. The remaining species are polyphagous, utilizing for galling more than one plant species across different plant families. Some of the polyphagous species alternate host plants seasonally (Sect. 6.5). Overwintering sites and stages are also important elements to determine life history patterns depending on deciduous or evergreen plant species (Sect. 6.2). Thus, the life history traits of galling-cecidomyiids are characterized by the combinations of the number of generations per year, host plant range, oviposition sites and overwintering sites.

6.2 Life History Strategy

Life-history patterns of gall midges are closely linked to their host-plant phenology. In temperate areas, most trees produce new shoots and flowers once a year, and therefore tree-inhabiting gall midges are inevitably univoltine because they need fresh host organs to induce galls. In contrast, gall midges on herbaceous plants that produce new shoots repeatedly can be multivoltine, such as *Dasineura affinis* (Kieffer) on *Viola odorata* Linnaeus (Violaceae) (Birch et al. 1992) and many species of *Rhopalomyia* on *Artemisia*. Even on trees, when new shoots are available in seasons other than spring, the gall midges could become bivoltine, like *Taxodiomyia cupressiananassa* (Osten Sacken) on *Taxodium distichum* (Linnaeus) Richard (Taxodiaceae) (Chen and Appleby 1984) and *Asphondylia tojoi* Elsayed & Tokuda on *Schoepfia jasminodora* Siebold & Zuccarini (Schoepfiaceae) (Elsayed et al. 2018a). A few species have two generations per year by alternating different organs of the same host. For example, *Asphondylia sarothamni* (Loew) utilizes pods of *Cytisus* (= *Sarothamnus*) *scoparius* (Linnaeus) Link (Fabaceae) from May to July and overwintering buds of the same plant from July to May in the following year (Parnell 1964) (see Sect. 6.5). Some species remain univoltine even though new shoots and leaves are produced for a long time within the season. For example, *Pitydiplosis puerariae* Yukawa et al. on *Pueraria lobata* (Willdenow) Ohwi (Fabaceae) is univoltine in Japan and South Korea, while that on *Pueraria montana* (Loureiro) Merrill in Yuanshan, Taiwan is naturally multivoltine (Yukawa et al. 2012). *Tokiwadiplosis matecola* Simbolon & Yukawa exhibits a unique life history strategy, using the lammas shoots (late season shoots usually produced as compensatory reactions) of *Lithocarpus edulis* (Makino) (Fagaceae) for galling instead of normal spring shoots that are more abundant and predictable food resources than lammas shoots (Simbolon and Yukawa 1992; Okuda and Yukawa 2000).

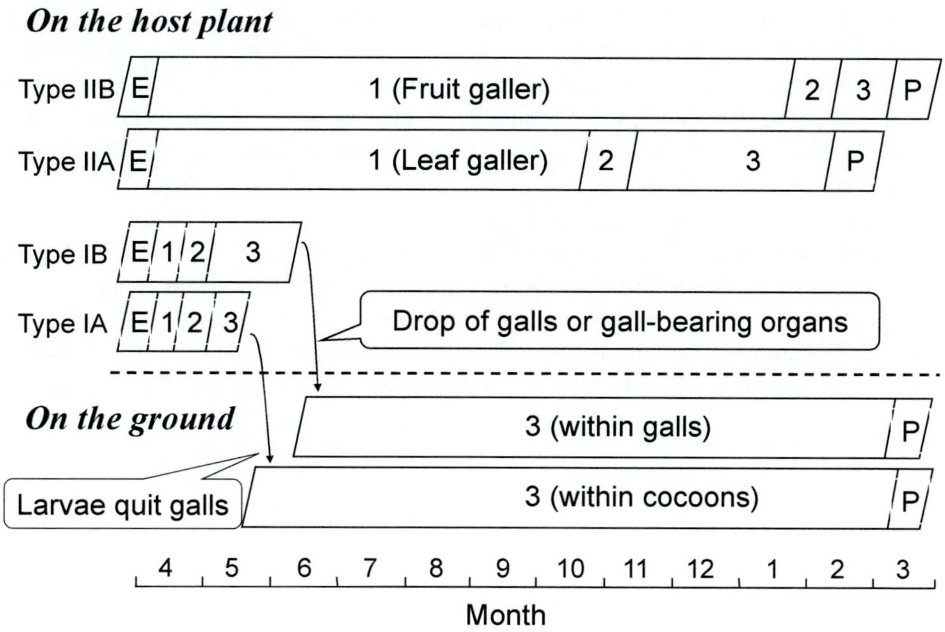

Fig. 6.1 Schematic representation of four types of life history strategies for univoltine gall-inducing cecidomyiids

At least 75 univoltine gall-midge species exist in Japan. Based on their overwintering sites and larval developmental stadia, their life-history patterns can be divided into two main types, each subdivided into two further categories (Yukawa 1987) (Fig. 6.1).

6.2.1 Type IA

Larvae mature quickly, exit the galls, and drop to the ground, mostly before summer. They remain in cocoons in the summer, autumn, and winter and pupate the following spring. Larvae of conifer-inhabiting species tend to remain within the gall until autumn. This life-history pattern is common in *Dasineura*, *Contarinia*, and many other genera (Skuhravý et al. 1993). Most gall midges exhibiting this pattern are associated with deciduous trees or herbaceous plants, but some occur on coniferous trees, and rarely on evergreen broad-leaved trees. Their life history strategy has an ecologically adaptive significance as it shortens the duration of the third larval stadium on the host plant to avoid the attack of ectoparasitoids (Yukawa 1987) (see Chap. 11 for Parasitic strategies). Instead, they are exposed to various sorts of mortality factors on the ground for relatively long period of time although the intensity of the factors has hardly been evaluated.

The life history strategy of *Macrodiplosis selenis* Kim & Yukawa belongs to type IA in a broad sense. However, *M. selenis* has a relatively long period of the second larval stadium from July to October on the host trees, *Quercus* spp. (Fagaceae) (Kim et al. 2015) unlike Palaearctic congeners, *Macrodiplosis roboris* (Hardy) and *Macrodiplosis pustularis* (Bremi) of which full-grown larvae drop to the ground before summer (e.g. Skuhravá et al. 1998; Skuhravá et al. 2006; Alford 2012). The strategy of *M. selenis* seemed to be dangerous for the second instars that were exposed to parasitoids on the host tree for a relatively long period of time. However, the field survey revealed that the second instars were not so severely attacked by ectoparasitoids before they became third instars, which dropped to the ground soon or later in autumn. This life history strategy is significant to avoid mortality factors such as predation and aridity on the ground (Kim et al. 2015).

6.2.2 Type IB

Mature larvae do not exit from galls, but galls or gall-bearing organs drop to the ground usually in early summer, or in late autumn. These larvae overwinter within the galls on the ground and pupate inside the galls in the following spring. This life history strategy has also an adaptive significance to avoid the attack of ectoparasitoids by shortening the duration of the third larval stadium on the host plants (Yukawa 1987).

Gall midges of this type induce mostly leaf galls or fruit galls on deciduous trees or herbaceous plants. For example, conical leaf galls induced by *Celticecis japonica* Yukawa & Tsuda on *Celtis sinensis* Persoon (Cannabaceae) drop to the ground in May and June and the full-grown larvae overwinter in the galls (Yukawa and Tsuda 1987). The genus *Mikiola* contains four Palaearctic species, *Mikiola fagi* (Hartig) in Europe, *Mikiola orientalis* (Kieffer) in Turkey, *Mikiola bicornis* Sato & Yukawa and *Mikiola glandaria* Sato & Yukawa in Japan. They all induce acorn-shaped galls on the leaves of *Fagus* species (Fagaceae). Galls become conspicuous in late May. In autumn, larvae molt into the final instar in the galls, which drop to the ground before the fall of leaves (e.g. Kampichler and Teschner 2002; Sato and Yukawa 2008).

Three known Palaearctic species of *Kiefferia* induce fruit galls on various plant species of Apiaceae. Because both galled and ungalled fruit of Apiaceae cannot remain on plants during the winter, *Kiefferia* species inevitably adopt type IA or type IB strategy. It is remarkable that two Japanese species, *Kiefferia ezoensis* Kim & Yukawa and *Kiefferia olla* Kim & Yukawa have type IB while a European *Kiefferia pericarpiicola* (Bremi) has type IA life history strategy (Kim et al. 2019).

6.2.3 Type IIA

Type IIA larvae grow less rapidly than those of types IA and IB but mature before winter. However, full-grown larvae do not exit from galls, nor do galls and gall-bearing organs drop to the ground. Full-grown larvae remain within the galls on the plant through the winter and pupate inside the galls in the following spring. This behavior is exhibited by various species of *Bruggmanniella* (Tokuda and Yukawa 2006), *Daphnephila* (Yukawa 1974; Tokuda et al. 2008), *Illiciomyia* (Tokuda 2004; Yukawa et al. 2016c), *Pseudasphondylia* (Yukawa 1974; Tokuda and Yukawa 2005; Tokuda et al. 2007), *Masakimyia* (Yukawa and Sunose 1976), etc. Such species are associated with evergreen broad-leaved trees. Most members of *Lasioptera* induce stem galls on various plants and overwinter in the galls as full-grown larvae (e.g. Yukawa et al. 2014). Species of *Rabdophaga* that are responsible for stem and bud galls on *Salix* exhibit type IIA pattern (e.g. Nijveldt and Yukawa 1982).

Schizomyia sasakii (Monzen) and *Schizomyia soyogo* Kikuti overwinter as full-grown larvae in axillary bud galls on *Ilex* species (Aquifoliaceae) (Tabuchi and Amano 2003; Tokuda et al. 2004) although many other congeners have type IA pattern (e.g. Elsayed et al. 2018b) and *Schizomyia broussonetiae* Elsayed & Tokuda that induces hairy leaf galls on *Broussonetia papyrifera* (Linnaeus) Ventenat (Moraceae) exhibits type IB pattern (Elsayed et al. 2019a).

Gall midges with type IIB strategy avoid attacks of endoparasitoids by enlarging the gall size relatively quickly. Attacks of ectoparasitoids are restricted to short periods before and after overwintering.

6.2.4 Type IIB

This pattern is basically similar to type IIA, but larvae do not mature until the following spring, and the duration of the first larval stadium is the longest among the four types. Many univoltine *Asphondylia* species that induce flower buds or fruit galls are of this type (e.g. Sunose 1983; Yukawa and Ohsaki 1988; Uechi and Yukawa 2004, 2006; Tokuda 2012; Yukawa et al. 2019b).

Gall midges with type IIB strategy lay their eggs into flower buds or young fruit, by which they can scape to some extent from endoparasitoids that attack eggs or first instars of host gall midges. In addition, they overwinter as first instars and emerge later than leaf gall midges because flower buds can be available for oviposition distinctly later than leaf bud extension or opening of fresh leaves. Thus, activities of ectoparasitoids are limited to a short period after overwintering.

6.2.5 *Other Types of Life History Strategy*

Exceptions to the aforementioned four types of strategies may exist but still they can be referred essentially to one of the four types, regardless of uni-, bi- or multivoltine, by considering overwintering stages and pupation sites.

Extended diapause and host plant alternation (including host organ alternation and short-term host plant-alternation) will be treated in later sections of this chapter.

6.2.6 *Paedogenesis by Non-Gall Inducing Cecidomyiids*

Although Mamaev (1968) provided general information on biology of non-gall inducing cecidomyiids, their life history patterns have seldom been intensively studied because of difficulties in the filed survey of mycophagous or saprophagous free-living larvae inhabiting mushrooms, rotten woods, or other decayed matters. This book does not intend to cite detailed life history traits of non-gall inducing cecidomyiids but here we briefly refer to larval and pupal paedogenesis (Type I paedomorphosis in McMahon and Hayward 2016) exhibited by some of non-gall inducing species.

Paedogenesis is a unique way of facultative larval or pupal reproduction and has evolved at least six times independently in insects, twice in Cecidomyiidae (Hodin and Riddiford 2000). Holometabolous insects that have lost complete metamorphosis by Type I paedomorphosis have been known in Micromalthidae (Coleoptera), Cecidomyiidae, Chironomidae, and Syrphidae (Diptera) (McMahon and Hayward 2016). In Cecidomyiidae, at least 11 mycophagous species belonging to eight genera have been known to reproduce multiple first instars by paedogenesis at the second or third larval stadium or at the hemi-pupal stage (Wyatt 1967). Some of paedogenetic species are pests of cultivated mushrooms (Wyatt 1959). In 1984, *Mycophila speyeri* (Barnes) was found infesting the oyster mushroom *Pleurotus ostreatus* (Jacquin ex Franchet) P. Kumun (Pleurotaceae) in Miyazaki Prefecture, Japan (Sanui and Yukawa 1986; Yukawa 1996). Mother larvae (possibly third instars) exhibited variations in body length and color, and in the presence or absence of sternal spatula. Red mature larvae with the spatula developed to pupae, while yellow or yellowish-orange mother larvae without the spatula produced two to 33 young larvae, depending on different breeding conditions (Sanui and Yukawa 1986). The time required to complete one paedogenetic cycle was 4–5 days at 25 °C, and one larva increased to 2000 individuals within 3 weeks at 27 °C. Sex ratio was female biased, and each female has only four to six large ovarian eggs.

McMahon and Hayward (2016) discussed significance in the loss of complete metamorphosis during the evolution of life history strategy and are hoping that their review will stimulate further studies of this field. Actually, paedogenesis includes exciting study subjects in various aspects of biology, such as cytology, ecology, embryology, evolutionary genetics, etc. Applied entomologists need to establish effective control measures against mushroom pests that suddenly increase to an outbreak situation.

6.3 Lower Development Threshold Temperature and Thermal Constant

6.3.1 Lower Developmental Threshold Temperature

Among various life history traits, lower developmental threshold temperature (hereafter LDT) and thermal constant are important factors to determine voltinism, diapausing season, and synchronization with host plant phenology. Therefore, LDT and thermal constant are essential to understand the life history strategy of respective species. The LDT has been determined for many arthropods by assuming a linear relationship between developmental speed and temperature. In Japan, for example, the LDT has been determined for at least 400 arthropod species and about 25% of them have the LDT ranging from 10 to 12 °C (Kiritani 2012). Bergant and Trdan (2006) argued that the LDT and thermal constants, based on laboratory experiments, commonly suffer from a great amount of uncertainty, and should be used with caution in practice. In particular, they emphasized that a sufficient temperature range should be covered in the experiment to approach the borders of the linear response as closely as possible. Otherwise, the LDT that was estimated with linear relationship between temperature and developmental speed can be frequently overestimated than actual temperature (Fig. 6.2).

As to gall-inducing cecidomyiids, the LDT has been determined for at least 17 species (Table 6.1). It should be remarked here that the LDT of most species was determined without assuming the linear relationship between developmental speed and temperature. In order to decrease the uncertainty in determining LDT around the borders, gall midge larvae were reared under various temperature conditions as has been practiced by Yukawa et al. (2013), Kim et al. (2015), Yukawa et al. (2016a), Uechi et al. (2018), and Yukawa et al. (2019a, b). The results of their experiments indicated the border and non-linear relationships between developmental speed and temperature around the LDT.

The following is one of the examples of methods in determining the LDT of the *Alpinia* fruit gall midge *Asphondylia* sp. that has the type IIB life history strategy (see Yukawa et al. 2019a, b for details). *Alpinia* fruit galls were collected from the field in February and some of them dissected before incubation to confirm that the galls contained only overwintered first instars. The dissection of galls is essential because, unlike non-gall inducing insects, larval developmental stadium of gall midges cannot be confirmed from outside. To determine the LDT, the remaining galls were incubated for 21 days at 11, 13, 15, 17, 19, 21, or 23 °C. Lighting conditions were set at LD 10:14, which was day-length at the time of gall collection.

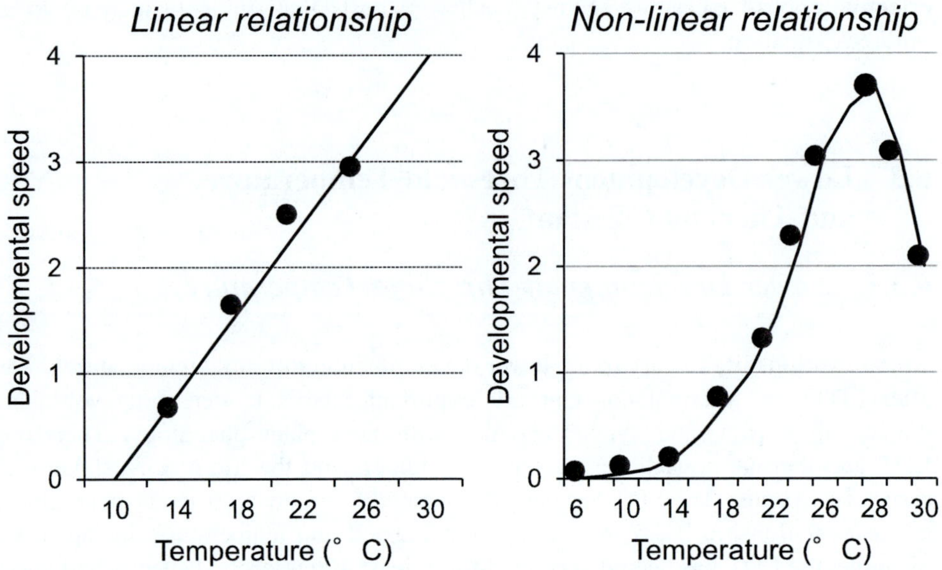

Fig. 6.2 Schematic representation of linear and non-linear relationship between temperature and developmental speed

Developmental stages were divided into seven categories; first, second, third instar, early, mid, late pupal stadium, and adult emergence that was indicated by a pupal case left on an emergence hole of the gall. The seven categories from first instar to adult were scored respectively as: 1.0, 2.0, 3.0, 4.0, 4.3, 4.7, and 5.0. Average developmental scores under respective temperature conditions were calculated after gall dissection. Then, scores were compared before and after incubation to determine LDT. These rearing and scoring methods have been successfully used to determine the LDT and the rate of development for some gall-inducing cecidomyiids (e.g. Ohtani et al. 1983; Yukawa et al. 2016a; Uechi et al. 2018).

As has been demonstrated for *Asphondylia* species with type IIB life history strategy, both field survey and laboratory experiment are easier for gall midges with type IIB life history strategy than those with other types because they overwinter as first instars in the galls on the host plants. In the other types, surveys and experiments are usually limited to the development from third instars to adults because they overwinter as third instars (Table 6.1). The LDT of the pine needle gall midge *Thecodiplosis japonensis* Uchida & Inouye (type IIB species) was finely determined to be 6.1 °C with model construction and validation on the basis of long-term field data and laboratory experiments (Nam and Choi 2014). The LDT of the wheat blossom midge *Sitodiplosis mosellana* (Géhin) was determined to be 9 °C with laboratory experiments (Wise and Lamb 2004). The blueberry gall midge *Dasineura oxycoccana* (Johnson) was estimated as 9.8 °C with linear and nonlinear regression models, and both models fit the data well (Roubos and Liburd 2010). Foster and Taylor (1975) showed, using linear model, differences in the LDT of the Hessian fly

Table 6.1 Known lower development threshold temperatures and thermal constants for gall-inducing cecidomyiids

Gall midge	Locality	Host plant (family)	LDT (°C)	Thermal constant[a]	Developmental stage	References
Asphondylia aucubae	Japan	*Aucuba japonica* (Garryaceae)	14	150	First instar to pupa	Ohtani et al. (1983)
Asphondylia baca	Japan	*Weigela hortensis* (Caprifoliaceae)	17	164.9	First instar to pupa	Yukawa et al. (2016a, b, c)
Asphondylia sphaera	Japan	*Ligustrum japonicum* (Oleaceae)	13	138.9	First instar to pupa	Ohtani et al. (1983)
Asphondylia yushimai	Japan	*Glycine max* (Fabaceae)	15	47.4	First instar to pupa	Yukawa et al. (2016a, b, c)
Asphondylia sp. 1	Japan	*Alpinia intermedia* (Zingiberaceae)	14	220	First instar to pupa	Yukawa et al. (2019a, b)
Asphondylia sp. 2	Japan	*Hedera rhombea* (Araliaceae)	11	84	First instar to pupa	Uechi et al. (2018)
Contarinia nasturtii	USA UK	*Brassica oleracea* (Brassicaceae)	7.2 12	Unknown	Third instar to pupa	Noll (1959) Readshaw (1966)
Dasineura oxycoccana	N America	*Vaccinium corymbosum* (Ericaceae)	9.8	134	Third instar to pupa	Roubos and Liburd (2010)
Dasineura tetensi	Sweden	*Ribes nigrum* (Grossulariaceae)	7	44–69	Third instar to pupa	Hellqvist (2001)
Illiciomyia yukawai	Japan	*Illicium anisatum* (Schisandraceae)	14	33.1–68.7	Third instar to pupa	Yukawa et al. (2013)
Macrodiplosis selenis	Japan	*Quercus serrata* (Fagaceae)	10	Unknown	Egg to first instar	Kim et al. (2015)
Mayetiola destructor	USA	*Triticum aestivum* (Poaceae)	12.2 1.6	27 343	Egg to first instar First to adult	Foster and Taylor (1975)
Rhopalomyia sp.	Switzerland	*Matricaria recutita* (Asteraceae)	7.02	478.5	Egg to adult	Hinz (1998)
Stenodiplosis sorghicola	USA	*Sorghum bicolor* (Poaceae)	15[b]	Unknown	Third instar to pupa	Baxendale et al. (1984)

(continued)

Table 6.1 (continued)

Gall midge	Locality	Host plant (family)	LDT (°C)	Thermal constant[a]	Developmental stage	References
Sitodiplosis mosellana	USA	*Triticum aestivum* (Poaceae)	9	306	Third instar to pupa	Wise and Lamb (2004)
Thecodiplosis japonensis	S Korea	*Pinus thunbergii* and *Pinus densiflora* (Pinaceae)	5.1 6.1	626.7	Third instar to pupa	Son et al. (2007) Nam and Choi (2014)
Tokiwadiplosis matecola	Japan	*Lithocarpus edulis* (Fagaceae)	>16	Variable[c]	Third instar to pupa	Okuda and Yukawa (2000)

Modified from Yukawa et al. (2016a)

[a]Thermal constant to 50% emergence date is expressed as "degree-days"

[b]Possibly between 10 and 15 °C

[c]80–387 when LDT was estimated to be 16 °C and 40–301 when LDT to be 18 °C

Mayetiola destructor Say between the egg stage and the subsequent developmental stages.

To determine the LDT is desirable to predict emergence and other life history events of gall midges. Information on temperature-dependent development is useful in searching for effective control measures against pest gall midges (e.g. Foster and Taylor 1975; Wise and Lamb 2004: Roubos and Liburd 2010; Uechi et al. 2018) and also in considering diversification process of life history traits in relation to host range expansion or host plant shift (e.g. Yukawa et al. 2019a, b).

6.3.2 *Emergence Curve*

To obtain emergence curves, the daily number of adult emergences should be recorded by counting pupal cases left on the galls on selected census trees. Otherwise, galls are incubated at temperature higher than the LDT and daily number of emergences should be monitored similarly. This survey is needed to perform every day, if possible, or at three- to five-day intervals during the emergence season of the target species. Then, the emergence data obtained are transformed into emergence curves in cumulative percentage. Before determining 50% emergence date (hereafter ET_{50}), we need to test if the emergence curves can be applied to the gamma distribution model because an insect emergence curve frequently has a long tail that is caused by delayed emergence of some individuals (e.g. Yukawa and Akimoto 2006). Gamma distribution has a desired shape characteristic of a long tail and is convenient to use for statistical inference (Matis et al. 1992). After we confirm that the emergence curves are applied to the gamma distribution model by comparing actual and gamma distribution data with the method of regression analysis, we can regard the nearest date when the half of adults emerged to be ET_{50}.

6.3.3 *Thermal Constant*

After determining LDT, we calculate thermal constant required for the overwintered first instars to become adults using the following equation:

$$\mathrm{TT} = (\mathrm{RT} - \mathrm{LDT}) \times \mathrm{D}$$

where TT is thermal constant, RT is rearing temperature, LDT is lower developmental threshold temperature, and D is incubation period in days from the first day to ET_{50}.

In the case of field data, we calculate the thermal constant from March to ET_{50} using the following eq.

$\mathrm{TT} = \sum (\mathrm{DMT} - \mathrm{LDT})$ where DMT is daily mean temperature higher than LDT. The daily mean temperatures from March to ET_{50} can be obtained from

Meteorological Observatory near by the census field. Thus, the thermal constant is calculated by accumulating temperatures above LDT for the period when overwintered first instars started to develop until the ET_{50} of respective gall midge populations.

There are many examples of a tradeoff between LDT and thermal constant, in which insects with relatively low LDT require larger amounts of thermal constant than those with high LDT (Honěk 1996; Kiritani and Yamashita 2008). In other cases, LDT is not correlated with thermal constant because the thermal constants are variable with food quality, sexes, and other factors (Kiritani 2012). In the case of *Asphondylia* species (Table 6.1), for example, *A. yushimai* that has lower LDT, 15 °C, requires smaller amount of thermal constant, 47.4 day-degrees, while *A. baca* that has higher LDT, 17 °C, requires greater amount of thermal constant, 164.9 day-degrees. These different combinations of LDT and thermal constant resulted in a great difference in the number of annual generations between the two species (Yukawa et al. 2016a).

6.3.4 Heat Stress

Effects of temperature change are frequently contradictory (Bale et al. 2002). Higher temperatures can help to accelerate the development of individuals and increase the survival rate but can be accompanied by lower adult body mass and lower fertility. Heat stress affected insects in various ways. In the southern green stinkbug, *Nezara viridula* (Linnaeus) (Hemiptera: Pentatomidae), for example, the body size became smaller, the development was delayed, and the rate in the failure of molt increased under 2.5 °C higher temperature conditions than usual (Musolin et al. 2010) and its natural populations suffered heat stress during the summer in spite of its subtropical origin (Kiritani 2011). Later, Kikuchi et al. (2016) found that high temperature negatively affects *N. viridula* indirectly through the heat-vulnerable obligate bacterial symbiont.

In recent years, the heat stress has been receiving more attention in relation to global warming (e.g. Kiritani and Yukawa 2010; Kiritani 2012; Kingsolver et al. 2013). Kiritani and Yukawa (2010) emphasized that heat stress affects insect at lower temperatures than that we have previously predicted. As to gall midges, Yukawa et al. (2016a) demonstrated that the larval development of the soybean pod gall midge *Asphondylia yushimai* Yukawa & Uechi, the ampelopsis fruit gall midge *A. baca* Monzen, the aucuba fruit gall midge *Asphondylia aucubae* Yukawa & Ohsaki and the ligustrum fruit gall midge *Asphondylia sphaera* Monzen was delayed at 28–29 °C. Unfortunately, further effects of heat stress on the failure of molt, the body size, life span, fecundity, flight ability, etc. have never been studied for any species of gall midges. Although higher developmental threshold temperature has been hardly determined because it varies with ecological and physiological characters tested (Kiritani and Yukawa 2010), we need to accumulate further data as to the higher developmental threshold temperature to evaluate influence of global warming on insect populations.

6.4 Long-Term Diapause

In some insect species, some portions of the population are known to enter extended diapause, resulting in polymodal emergence patterns. Waldbauer (1978) divided the emergence patterns into three types. In type A polymodality, diapause does not extend over more than one winter. All individuals that enter diapause in the same year terminate diapause the following growing season, exhibiting a bimodal or a trimodal emergence curve. For example, *Chaoborus americanus* (Johannsen) (Diptera: Chaoboridae), *Eurytides marceluus* (Cramer) (Lepidoptera: Papilionidae) and *Helicoverpa armigera* (Hübner) (= *Chloridea obsoleta*) (Lepidoptera: Noctuidae) have a bimodal emergence curve (e.g. Waldbauer 1978). In type B polymodality, insects exhibit bimodal emergence patterns within one season, which are represented by individuals belonging to different year classes. The emperor dragonfly *Anax imperator* Leach (Odonata: Aeshnidae) is one of the examples (Corbet 1963). Many insects exhibit type C polymodality, in which the emergence of a year class extends over two or more years (e.g. Barnes 1958; Danks 1987; Powell 1987; Turgeon et al. 1994; Corley et al. 2004; Yukawa and Rohfritsch 2005). Hanski (1988) reviewed theoretical predictions about 'extra-long diapause' in a perspective of density dependence and independence.

At least 14 species of galling-cecidomyiid have been known to exhibit type C polymodality (Table 6.2 and Fig. 6.3). Their larval developmental stadia entering extended diapause are different between two groups of gall midges in relation to their life history strategies (Sect. 6.2). In one group that exhibits IA or IB life history strategy (Yukawa 1987), third instars or occasionally prepupae enter extended diapause in cocoons or desiccated galls on the ground (hereafter type C-I polymodality). The other group of gall midge that exhibits IIA or IIB life history strategy (Yukawa 1987) enters extended diapause at the first or second larval stadium in the galls on host plants (hereafter type C-II polymodality). In the latter group, the duration of diapause cannot extend beyond three winters because of the limited longevity of host leaves (Takasu and Yukawa 1984; Yukawa et al. 2018). Therefore, their life history patterns are divided into 1-year (univoltine), 2-year and at most 3-year types.

In this section, examples of type C polymodality exhibited by galling cecidomyiids are demonstrated in relation to life history strategy and the significance of the extended diapause is discussed.

6.4.1 Type C-I Polymodality

Examples of type C-I polymodality have been known for eight galling-cecidomyiid species (Table 6.2). They mainly belong to Cecidomyiini or Dasineurini and are associated with grasses or conifers except *Contarinia pisi* (Loew) that induces galls on fabaceous plants. In this group, the most remarkable example of extended

Table 6.2 List of galling-cecidomyiids that exhibit polymodal emergence patterns

Polymodality	Tribe[a]	Species	Host plants (family)	References
Type C-I	Cecidomyiini	*Contarinia pisi* (Loew)	*Pisum* spp. (Fabaceae)	Barnes and Arnold (1960)
	Cecidomyiini	*Contarinia tritici* (Kirby)	*Triticum vulgare* (Poaceae)	Barnes (1952)
	Cecidomyiini	*Plemeliella abietina* Seitner	*Picea* spp. (Pinaceae)	Annila (1981)
	Clinodiplosini	*Sitodiplosis mosellana* Géhin	*Triticum* spp. (Poaceae)	Barnes (1958)
	Dasineurini	*Kaltenbachiola canadensis* (Felt)	*Picea glauca* (Pinaceae)	Tripp (1955)
	Dasineurini	*Kaltenbachiola rachiphaga* (Tripp)	*Picea glauca* (Pinaceae)	Prévost (1990)
	Probably Dasineurini	*Hasegawaia sasacola* Monzen	*Sasa* spp. (Poaceae)	Sunose (1978)
	(Unplaced to tribe)	*Strobilotia carpophaga* (Tripp)	*Picea glauca* (Pinaceae)	Tripp (1955)
Type C-II	Asphondyliini	*Daphnephila machilicola* Yukawa	*Machilus thunbergii* (Lauraceae)	Maeda et al. (1982)
	Asphondyliini	*Illiciomyia yukawai* Tokuda	*Illicium anisatum* (Illiciaceae)	Yukawa et al. (2013)
	Asphondyliini	*Pseudasphondylia neolitseae* Yukawa	*Neolitsea sericea* (Lauraceae)	Takasu and Yukawa (1984)
	Asphondyliini	*Schizomyia soyogo* Kikuti[b]	*Ilex integra* (Aquifoliaceae)	Tabuchi and Amano (2003)
	Lasiopterini	*Lasioptera camelliae* Ohno & Yukawa	*Camellia japonica* (Theaceae)	Ohno and Yukawa (1984)
	(Unplaced to tribe)	*Taxomyia taxi* (Inchbald)	*Taxus baccata* (Taxodiaceae)	Redfern (1975)

[a]Tribal position followed Gagné and Jaschhof (2017)

[b]In Tabuchi and Amano (2003), this species was identified as *Asteralobia sasakii* (Monzen) but later identified to be *A. soyogo* Kikuti (Tokuda et al. 2004) and recently combined with the genus *Schizomyia* (Elsayed et al. 2018b)

diapause is that of the wheat blossom midge *S. mosellana*, of which adults, as well as its parasitoid, emerge for as long as 13 years (Barnes 1958). The percentage emergence is not always proportionately lower as the years go by. The overwhelming proportion of lemon wheat blossom midge *Contarinia tritici* (Kirby) emerges the following year but some do in the second, third and even fourth year (Barnes 1952, 1958). In the pea midge *Contarinia pisi* (Loew), more adults of the 1956 year-class emerged in 1957 than in 1958 but further emergences were not confirmed thereafter (Barnes and Arnold 1960). The spruce-cone axis midge *Kaltenbachiola rachiphaga* (Tripp) (= *Dasineura rachiphaga*) enters extended diapause as third instars in galls for two or more winters (Prévost 1990). Prepupae of the bamboo-bud midge

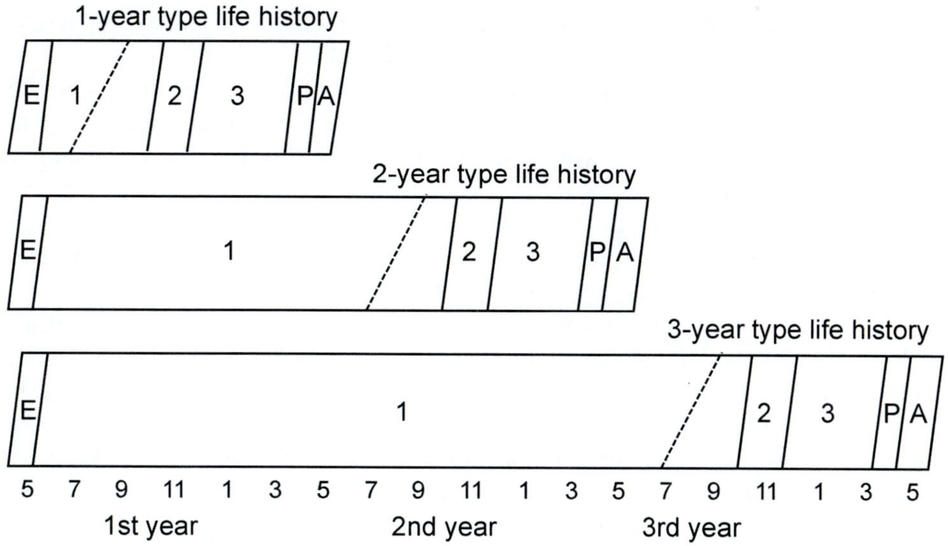

Fig. 6.3 Schematic representation of 1-year, 2-year and 3-year life history patters for gall-inducing cecidomyiids. Dashed lines indicate the maturity of galls

Hasegawaia sasacola Monzen show extended diapause from one to four years in desiccated galls on the host plant or on the ground (Sunose 1978).

In the case in which full-grown larvae overwinter in cocoons on the ground, the existence of larvae that entering extended diapause is hardly detected in the field. Rearing of full-grown larvae belonging to the same year-class should be observed in an insectary as has been demonstrated for *S. mosellana* in Barnes (1958). Further galling-cecidomyiid species that exhibit the polymodal emergence will be detected in the future.

6.4.2 *Type C-II Polymodality*

So far, this type of polymodality has been known for six species of galling-cecidomyiids (Table 6.2). Five of them belong to the tribe Asphondyliini and are associated with evergreen broad-leaved trees. *Daphnephila machilicola* Yukawa induces galls on the under leaf surface of *Machilus thunbergii* (Blume) Koidzumi (Lauraceae) (Fig. 3.1b) (Yukawa 1974; Sect. 3.2). Some first instars enter extended diapause before the galls become prominent and the emergence of a year class extends over two rarely three years (Maeda et al. 1982). The ratio of the 2-year type population to the 1-year type varied largely (5.6–100%) from year to year and at different localities. *Pseudasphondylia neolitseae* Yukawa induces leaf galls on *Neolitsea sericea* (Blume) Koidzumi (Lauraceae) (Fig. 2.1a) (Yukawa 1974). Some first instars enter extended diapause after the galls become mature and the emergence of a year class extends over two years (Takasu and Yukawa 1984). The ratio of the 2-year type population to the 1-year type was less than 25% although it varies largely from year to year and at different localities.

Illiciomyia yukawai Tokuda induces leaf galls on *Illicium anisatum* Linnaeus (Schisandraceae) (Fig. 10.1a) (Tokuda 2004). Some first instars enter extended diapause. Galls of the 1-year type become conspicuous in early June and mature in August–September, while those of the 2-year type remain as feeble swellings until early June of the following year, then they become conspicuous and mature in August–September (Yukawa et al. 2013). The ratio of the 2-year type population to the 1-year type varies largely from year to year and at different localities.

Schizomyia soyogo Kikuti was previously identified as *Asteralobia sasakii* (Monzen) in Tabuchi and Amano (2003) but later identified to be *A. soyogo* (Tokuda et al. 2004) and recently combined with the genus *Schizomyia* (Elsayed et al. 2018b). This species is responsible for axillary bud galls on *Ilex integra* Thunberg, *Ilex chinensis* Sims, *Ilex leucoclada* (Maximowicz) Makino and *Ilex pedunculosa* Miquel (Aquifoliaceae) (Fig. 2.10c) (Tokuda et al. 2004). Tabuchi and Amano (2003) found 21.8% (in 2000) and 23.9% (in 2001) of first instars on *I. integra* entered extended diapause and spent two or more winters.

Lasioptera camelliae Ohno & Yukawa that is responsible for leaf vein galls on *Camellia japonica* Linnaeus (Theaceae) enter extended diapause at the second larval stadium and the emergence of a year class extends over two rarely three years (Ohno and Yukawa 1984). The ratio of the 2-year type population to the 1-year type varies largely from year to year.

Taxomyia taxi (Inchbald) induces artichoke galls on leaf bud of *Taxus baccata* Linnaeus (Taxaceae) in the second year (Redfern 1975) but its tribal position has not yet been determined (Gagné and Jaschhof 2017). In *T. taxi*, most individuals have a 2-year life cycle, but some develop in one year. The life cycle rarely (<2%) takes longer than two years. The percentage of 1-year type varied considerably with trees and from year to year, ranging from 1.1 to 4.0% in 1969, 0 to 70% in 1970, and 0 to 18.4% in 1971 (Redfern 1975).

6.4.3 Significance of Polymodality

Extended diapausing populations are considered to face higher risks of mortality than univoltine individuals because of their longer generation time. Nevertheless, extended diapausing individuals exist as shown in the aforementioned examples. Sunose (1978) considered that the rate of extended diapausing individuals determined by the balance of three factors: (1) mortality in the diapausing prepupae, (2) mortality of the univoltine fraction of a population at a catastrophe (food shortage caused by withering of bamboo after mast flowering) and (3) frequency of catastrophic happenings.

Sunose (1978) demonstrated that the extended diapause by *H. sasacola* has an adaptive significance to avoid mortality of the univoltine fraction of a population during a catastrophe (food shortage caused by mast-flowering of bamboo). Other possible adaptive significance of the extended diapause in cecidomyiids has been referred to population density and parasitism. The extended diapause by *K. rachiphaga* is ecologically significant in avoiding intraspecific competition for oviposition sites (Prévost 1990). Extended diapausing individuals of *D. machilicola* and *P. neolitseae* have advantages to diversify risks against the attack of ectoparasitoids because 2-year and 3-year type mature larvae are attacked by the parasitoids of different generations in the subsequent years (Maeda et al. 1982; Takasu and Yukawa 1984). In *T. taxi*, the parasitism by *Mesopolobus diffinis* (Walker) (Hymenoptera: Pteromalidae) is a vital mortality factor in 1-year life cycles, but not in 2-year cycles (Redfern and Cameron 1978, 1993). However, Redfern and Hunter (2005) demonstrated, based on further population studies over 35 years, that parasitism had no effect on the dynamics of univoltine *T. taxa.*

Despite the accumulation of practical examples and theoretical predictions, an adaptive significance of the extended diapause has not been sufficiently explained. Besides, causes of the initiation and termination of extended diapause have been little studied from mechanistic (physiological) viewpoints. Takasu and Yukawa (1984) referred to geographical and annual variations in the proportion of 2-year to 1-year type individuals of *P. neolitseae* but did not provide further analytical data. The analysis of such variations may implicate an adaptive significance of extended diapause.

To find an adaptive significance for *I. yukawai* to produce 2-year type individuals, Yukawa et al. (2013) counted catastrophic events during the 18-year field surveys and compared the relative frequencies between different catastrophes. They regarded high mortalities caused by factors, such as a serious shortage of host buds available for oviposition (the mean number of host buds per female was less than 0.5), a high percentage parasitism (more than 80%) and an extremely low survival rate of host leaves (less than 20%), as catastrophic events. Then, they concluded that the existence of 2-year type individuals of *I. yukawai* has effective ecological advantages to overcome the high mortality caused by the fall of galled leaves, because the other catastrophic events exceeded the frequency of catastrophe caused by the fall of galled leaves.

At the moment there is no clear explanation for an adaptive ecological significance of very long-term diapause exhibited by the wheat blossom midge *S. mosellana* under agricultural field conditions. Barnes (1958) considered that free mating of individuals from different year classes might strengthen the gene pool. This gall midge possibly expanded its host range from wild plants of Poaceae to cultivated wheat quite recently in terms of evolutionally history. The questions are what catastrophic events for *S. mosellana* on the wild host plants were and how frequently the events happened. Otherwise, the gall midge might obtain a long-term diapausing habit within an extremely short period of time after it came to wheat fields, which are certainly disturbed frequently in various artificial ways. Such a short-term adaptation, if any, would be an interesting study subject in future.

6.5 Host Plant Alternation

Host plant alternation is one of the life history traits of herbivorous insects and has been known for several insect taxa, such as gall wasps (e.g. Stone et al. 2002), aphids (e.g. Dixon and Kundu 1994) and some whiteflies (e.g. Miyatake 1980). Host plant alternation and utilization of occasional alternative hosts have adaptive significance to increase of voltinism that enhances reproductive potential, to escape from parasitoid attacks by changing habitats and to seek fresher and more nutritious host plant (Uechi and Yukawa 2006).

6.5.1 Host Plant-Alternating Species of Asphondylia

In Diptera, host-alternating habit and overwintering-host plants have been successively confirmed by means of genetic analysis for gall midges of the genus *Asphondylia* (Cecidomyiidae: Asphondyliini: Asphondyliina). Five species are now known to exhibit host plant-alternation (Table 6.3; Fig. 6.4). The soybean pod gall midge *A. yushimai* is one of the major pests of soybean *Glycine max* (Linnaeus) Merrill (Fabaceae) in Japan and South Korea (Yukawa et al. 2003; Uechi et al. 2018). To seek its overwintering hosts was a long-term issue to establish control measures against *A. yushimai*. Finally the DNA sequencing data obtained from larvae inhabiting fruit galls on *Laurocerasus zippeliana* (Miquel) Browicz (Rosaceae) was confirmed to coincide with the data of *A. yushimai* (Yukawa et al. 2003) and successively *Osmanthus heterophyllus* (G. Don) P.S. Green (Oleaceae) was found to be an overwintering host (Uechi et al. 2005) (Fig. 6.4a–c).

Based on morphological studies and biological information, the carob gall midge *Asphondylia gennadii* (Marchal) in Cyprus has been known to utilize carob *Ceratonia siliqua* Linnaeus (Fabaceae) as an overwintering host, and many other plants including pepper *Capsicum annuum* Linnaeus (Solanaceae), caper *Capparis spinosa* Linnaeus (Capparidaceae), and sea squill *Urginea maritima* (Linnaeus) Baker (Liliaceae), as summer hosts (Harris 1975; Orphanides 1975; Gagné and Orphanides 1992). However, host plant alternation by *A. gennadii* had not been confirmed at the DNA level, but Uechi et al. (2004) revealed that individuals reared from galls on carob, caper, and sea squill were included in a clade of *A. gennadii* in a Neighbor-joining tree based on mtDNA COI gene.

Sunose (1992) first suspected that the ampelopsis fruit gall midge *A. baca* moves from *Ampelopsis glandulosa* var. *heterophylla* (Thunberg) Momiyama (Vitaceae) to *Weigela* spp. (Caprifoliaceae) for overwintering. Later its host plant alternation was confirmed by genetic analysis (Uechi et al. 2004) (Fig. 6.4d, e). The ivy fruit gall midge *Asphondylia* sp. move from the overwintering host *Hedera rhombea* (Miquel) Bean (Araliaceae) to summer-autumn hosts *Phytolacca americana* Linnaeus (Phytolaccaceae) and *Paederia foetida* Linnaeus (Rubiaceae) (Uechi et al. 2018)

Table 6.3 *Asphondylia* and *Pseudasphondylia* species that alternate host plants or host organs seasonally

Species of *Asphondylia* or *Pseudasphondylia*	Host plant and organ[a]		Literatures
	Summer-Autumn season	Winter-Spring season	
Host plant-alternating species			
A. yushimai	*Glycine max, G. soya, Lespedeza* spp., *Desmodium* spp. and others (Fabaceae) (PD), *Chamaecrista nomame* (Caesalpiniaceae) (PD)	*Laurocerasus zippeliana* (Rosaceae) (FR) *Osmanthus heterophyllus* (Oleaceae) (FR)	Yukawa et al. (2003), Uechi et al. (2005)
A. gennadii	*Capsicum annuum* (PD), *Solanum tuberosum* (Solanaceae) (FB), *Capparis spinosa* (Capparidaceae) (LB), *Urginea maritima* (Liliaceae) (FB), *Eruca sativa, Sinapis* spp. (Brassicaceae) (PD), *Asphodelus fistulosus* (Asphodelaceae) (FB), *Hypericum crispum* (Hypericaceae) (FB)	*Ceratonia siliqua* (Fabaceae) (PD)	Harris (1975), Orphanides (1975), Gagné and Orphanides (1992), Uechi and Yukawa (2006)
A. baca	*Ampelopsis glandulosa* var. *heterophylla* (Vitaceae) (FR)	*Weigela* spp. (Caprifoliaceae) (LB)	Sunose (1992), Uechi et al. (2004)
A. sp. (Ivy fruit gall midge)	*Phytolacca americana* (Phytolaccaceae) (FR), *Paederia foetida* (Rubiaceae) (FB)	*Hedera rhombea* (Araliaceae) (FR)	Uechi et al. (2018)
A. monacha	*Solidago juncea, S. erecta, S. uliginosa* (Asteraceae) (LB)	*Solidago altissima* (Asteraceae) (RB)	Dorchin et al. (2015)
Possible host plant-alternating species			
Asphondylia capsicicola	*Capsicum* spp. (Solanaceae) (PD)	Unknown	Uechi et al. (2017)
A. uvarum	*Vitis* spp. (Vitaceae) (FR)	Unknown	Gagné et al. (2018)
A. sp. (Grape berry gall midge)	*Vitis* spp. (FR)	Unknown	Sasaki and Uechi (2018)
A. sp. (Helwingia fruit gall midge)	*Helwingia japonica* (Helwingiaceae) (FR)	Unknown	Yukawa and Masuda (1996)
A. sp. (Ardisia fruit gall midge)	*Ardisia japonica* (Primulaceae) (FL, FR)	Unknown	Uechi et al. (2004)

(continued)

Table 6.3 (continued)

Species of *Asphondylia* or *Pseudasphondylia*	Host plant and organ[a]		Literatures
	Summer-Autumn season	Winter-Spring season	
A. sp. (Pieris flower bud gall midge)	*Pieris japonica* (Ericaceae) (FB)	Unknown	Usuba (1982)
P. matatabi	*Actinidia polygama* (Actinidiaceae) (FB)	Unknown	Yukawa and Masuda (1996)
P. kiritanii	*Cornus controversa* (Cornaceae) (FB)	Unknown	Tokuda and Yukawa (2005)
P. tominagai	*Eleutherococcus spinosus* (Araliaceae) (FB)	Unknown	Elsayed et al. (2019b)
Short-term host plant-alternating species			
A. sphaera	*Ligustrum* spp. (Oleaceae) (FB, FR)	*Rhus* spp. (Anacaridiaceae) (FB)[b]	Uechi and Yukawa (2006)
Host organ-alternating species			
A. sarothamni	*Cytisus scoparius* (Fabaceae) (PD)	*C. scoparius* (Fabaceae) (LB)	Parnell (1964)
A. rudbeckiaeconspicua	*Rudbeckia laciniata* (Asteraceae) (FL)	*R. laciniata* (Asteraceae) (LB)	Plakidas (1988), Gagné (1989)
A. acaciae	*Acacia* spp. (Mimosaceae) (FR)	*Acacia* spp. (Mimosaceae) (LB)	Kolesik et al. (2010)

[a]Abbreviations indicate host organs as follows: *BT* branch tip, *FB* flower bud, *FL* flower, *FR* fruit, *LB* leaf bud, *PD* pod, *RB* rhizome bud
[b]*Rhus* spp. are not overwintering hosts but used as short-term alternate hosts in June–July

(Fig. 6.4f–h). A North American species *Asphondylia monacha* (Osten Sacken) alternates congeneric host plants of *Solidago* (Asteraceae) (Dorchin et al. 2015).

6.5.2 *Possible Host Plant-Alternating Species of* Asphondylia

The following two identified and four unidentified *Asphondylia* gall midges are suspected to be host-alternating species (Table 6.3). The Asian chili pod gall midge *Asphondylia capsicicola* Uechi et al. infests pods of *Capsicum annuum* Linnaeus and *Capsicum frutescens* Linnaeus (Solanaceae) in Indonesia and Vietnam. Previously this gall midge had been identified as *A. gennadii* (= *A. capsici*) but was distinguishable from it by the morphological features of pupa and DNA sequencing data (Uechi et al. 2017). Like *A. gennadii*, it is highly possible that *A. capsicicola* utilizes overwintering hosts other than *Capsicum* species. The North American

Fig. 6.4 Selected examples of galls induced by host-alternating species of *Asphondylia*. (**a**), pod galls of *A. yushimai* on *Glycine max* (Summer-Autumn host); (**b**), fruit galls of *A. yushimai* on *Laurocerasus zippeliana* (Winter-Spring host); (**c**), ibid. on *Osmanthus heterophyllus* (Winter-Spring host) (reproduced with permission from Uechi et al. 2018); (**d**), fruit galls of *A. baca* on *Ampelopsis glandulosa* var. *heterophylla* (Summer-Autumn host); (**e**), bud galls of *A. baca* on *Weigela* spp. (Winter-Spring host); (**f**), a male and a flower bud gall of the ivy fruit gall midge *Asphondylia* sp. on *Paederia foetida* (Summer-Autumn host); (**g**), fruit galls of the ivy fruit gall midge on *Phytolacca americana* (Summer-Autumn host); (**h**), fruit gall of the ivy fruit gall midge on *Hedera rhombea* (Winter-Spring host)

Asphondylia uvarum Gagné (Gagné et al. 2018) and the Japanese grape berry gall midge *Asphondylia* sp. (Sasaki and Uechi 2018) are pests of grape but their overwintering hosts have never been detected. Overwintering host plants of the three unidentified Japanese species of *Asphondylia* on *Helwingia japonica* (Thunberg) FG Dietrich (Helwingiaceae), *Ardisia japonica* (Thunberg) Blume (Primulaceae) and *Pieris japonica* (Thunberg) D. Don ex G. Don (Ericaceae) are also unknown.

Asphondylia contains about 300 nominal species in the world that mostly form galls on flower buds or flowers and prevent fruiting in various plant species and families (Gagné and Jaschhof 2017). Like other gall-inducing cecidomyiids, many *Asphondylia* species are known from one plant species (tentatively defined as "monophagous"), and some are from one plant genus or one plant family ("oligophagous"). Except host-alternating species, polyphagy across different plant families has been known only for a few species, such as *Asphondylia websteri* Felt that induces galls on plants belonging to Fabaceae, Lauraceae and Simmondsiaceae (Gagné and Wuensche 1986). Because annual life cycle may have not been intensively studied for some of the "monophagous" or "oligophagous" species, the number of host-alternating species will increase in future when their overwintering host plants are detected.

6.5.3 Possible Host Plant-Alternating Species of Pseudasphondylia

The genus *Pseudasphondylia* (Cecidomyiidae: Asphondyliini: Asphondyliina) consists of 10 species (Gagné and Jaschhof 2017). At least three of them are possibly host-alternating species (Table 6.3) because adults emerge before winter, but no oviposition sites are available on their original host plants at the time of emergence. They inevitably search overwintering host plants. However, their overwintering host plants have not yet been detected. They are *Pseudasphondylia matatabi* Yuasa & Kumazawa on *Actinidia polygama* (Siebold & Zuccarini) Maximowicz (Actinidiaceae) (Yukawa and Masuda 1996), *Pseudasphondylia kiritanii* Tokuda & Yukawa on *Cornus controversa* Hemsley (Cornaceae) (Tokuda and Yukawa 2005), and *Pseudasphondylia tominagai* Elsayed & Tokuda on *Eleutherococcus spinosus* (Linnaeus) S. Y. Hu (Araliaceae) (Elsayed et al. 2019b).

6.5.4 Short-Term Host Plant-Alternating Species

The ligustrum fruit gall midge *A. sphaera* induces fruit galls on *Ligustrum obtusifolium* Siebold & Zuccarini, *Ligustrum japonicum* Thunberg, *Ligustrum lucidum* Aiton, and *Ligustrum ovalifolium* Hasskarl (Oleaceae) in Japan (Monzen

1937; Yukawa 1971; Yukawa and Miyamoto 1979). *Asphondylia sphaera* is mostly univoltine entirely on fruit galls of *Ligustrum* but a very few individuals emerge in June from flower bud galls on *Ligustrum* and *Rhus* (Yukawa and Masuda 1996). Ecological surveys and molecular analysis confirmed that a part of the population of *A. sphaera* diverted into a short-season generation on flower buds of *Rhus succedanea* Linnaeus and *Rhus sylvestris* Siebold & Zuccarini (Anacardiaceae) and flower buds of *L. japonicum* before coming to young fruit of *Ligustrum* (Uechi and Yukawa 2006). These are short-term alternative host plants or host organs other than fruit of *Ligustrum* species that is the main host organ. This situation is another type of polyphagy. By using short-term alternative hosts, *A. sphaera* effectively adds a generation and increases the population size but only a small portion of the whole population of spring emergent adults can contribute to the additional generation, because of the following reasons: (1) high mortality of eggs, first and second instars in the flower-bud galls, of which at least 70% drop to the ground in May (Yukawa et al. 2016b), (2) difficulties in synchronization with the developmental stage of flower bud, which is much shorter than that of the young fruit stage. In addition, flower bud production by *L. japonicum* fluctuates greatly from year to year, almost on a biennial cycle (Yukawa et al. 2016b) and (3) phylogenetic constraint: species exhibiting a type IIB life-history strategy (Yukawa 1987) overwinter as first instars, which require a longer period of time to mature after overwintering, in comparison with type IIA species that overwinter as mature larvae and induce galls on young leaves in early spring. To synchronize with flower bud phenology, the short-season generation larvae must mature more rapidly than individuals that use young fruit for galling (Uechi and Yukawa 2006).

6.5.5 Host Organ Alternation

Host organ alternation is another type of host plant alternation. Instead of alternating plant species, some species alternate host organ on the same plant species. In Europe, *Asphondylia sarothamni* (Loew) exhibits such pattern of host alternation. This species alternatively utilizes pods for galling in summer and leaf bud for overwintering (Parnell 1964). A North American species *Asphondylia rudbeckaeconspicua* Osten Sacken utilizes flowers and leaf buds alternatively (Plakidas 1988; Gagné 1989). *Asphondylia acaciae* Kolesik, utilizes fruit and buds of *Acacia* spp. (Mimosaceae) in Australia (Kolesik et al. 2010). *Asphondylia punica* Marchal (= *conglomerata*) uses stems and flower buds of *Atriplex halimus* Linnaeus (Chenopodiaceae) in Egypt (Elsayed et al. 2015).

6.5.6 Some Behavioral Traits Supporting Host Plant Alternation

Most *Asphondylia* species form galls on flower buds or flowers (Gagné and Jaschhof 2017). Such plant organs are available for a short and limited period within a year. Therefore, they inevitably search alternative plant species for oviposition and gall induction if they attempt to repeat further generations. Some behavioral traits that have been inherent in *Asphondylia* might enable them to alternate seasonally host plants across different plant families.

All species of Asphondyliini are associated with fungal symbiont and many species induce ambrosia galls, in which the fungal mycelium develops on the inner wall of the gall (Tokuda 2012, reviews in Bissett and Borkent 1988; Yukawa and Rohfritsch 2005). Females convey fungi with setae of the seventh abdominal sternite, which constitute a mycangial structure (Borkent and Bissett 1985; Kobune et al. 2012) and introduce them together with eggs into host meristematic tissues (Yukawa and Rohfritsch 2005). Such an association with fungal symbiont possibly enables *Asphondylia* to utilize several plant species belonging to different families. We need to confirm a possibility that the fungal association may help to overcome chemical barrier between different plant taxa.

Strong flight ability is essential in search for alternative host plants because they grow in different habitats that are relatively distant from each other. For example, the soybean pod gall midge *A. yushimai* flies to 4 km distant soybean plants from an overwintering-host *L. zippeliana* that grows in an evergreen broad-leaved forest (Yukawa et al. 2003). Yukawa et al. (2019a, b) compared flight ability between a host-alternating and multivoltine *A. yushimai* and a non-host alternating and univoltine *A. aucubae* that induces fruit galls on *A. japonica*. In terms of flying speed (22.9 m/min) and the maximum continuous flying time (13.7 km within 10 h), *A. yushimai* females carried stronger potential flight ability than *A. aucubae*. Generally, *Asphondylia* species have relatively large body size, large wing, tough thorax and long adult life span, which may also help their long distant flight. Nocturnal habit exhibited by *Asphondylia* species is an additional important trait to avoid desiccation from long-distant flight during the daytime.

6.5.7 How to Determine Host Plant Combinations

First, to collect various cecidomyiid galls from the field in winter season is essential. Larval specimens for genetic analysis can be easily obtained by dissecting the galls and the tribal or generic position of the gall midge can be identified by examining larval sternal spatula (e.g. Gagné 1989). Adults of *Asphondylia* species can be reared from the galls because they emerge directly from galls. However, to determine the combination of their summer-autumn hosts and winter-spring hosts is difficult since congeners are morphologically quite similar to each other. Therefore, intensive behavioral observation, phenological field survey on possible host plants, oviposition experiments in cages and precise morphological comparison have been attempted (e.g. Orphanides 1975 for *A. gennadii*, Sunose 1992 for *A. baca*, Yukawa et al. 1983; Yukawa et al. 1986, for *A. yushimai*). From the late 1990s, molecular technique started to be used for the comparison of inter- and intraspecific differences between cecidomyiid species (e.g. Shirota et al. 1999; Yukawa et al. 2003; Dorchin et al. 2004; Tokuda et al. 2004; Tokuda et al. 2008). This technique has been effectively applied to combine unknown overwintering-hosts with already known host plants in the other seasons by using larval specimens obtained easily from various *Asphondylia* galls.

Another way of molecular technique is to use the plant DNA barcoding region amplified from insect DNA extracts (Jurado-Rivera et al. 2009). We can analyze plant DNA remained in the female intestine and identify the plant species or genus that the gall midge fed during its larval stage. For this purpose, we need to catch females that flew to a known host plant for oviposition from an unknown host plant although there are some difficulties in catching females in the night (nocturnal habit) and at the good phenological timing. In addition, DNA sequencing data have not registered in GenBank for some plant species.

References

Alford DV (2012) Family Cecidomyiidae (gall midges). In: Northcott J (ed) Pests of ornamental trees, shrubs and flowers a colour handbook, 2nd edn. Manson Publishing Ltd, London, pp 398–400

Annila E (1981) Fluctuations in cone and seed insect populations in Norway spruce. Commun Inst Forest Fenn 101:1–32. (In Finnish with English summary)

Bale JS, Masters GJ, Hodkinson ID, Wmack CA, Bezemer TM, Brown VK, Butterfield J, Buse A, Coulson JC, Farrar J, Good JEG, Harrington R, Hartley S, Jones TH, Lindroth RL, Press MC, Symrnioudis I, Watt AD, Whittaker JB (2002) Herbivory in global climate change research: direct effects of rising temperature on insect herbivores. Glob Chang Biol 8:1–16

Barnes HF (1952) Studies of fluctuations in insect populations. XII further evidence of prolonged larval life in the wheat-blossom midges. Ann Appl Biol 39:370–373

Barnes HF (1958) Wheat blossom midges on Broadbalk, Rothamstead experiment station, 1927–56. Proceedings of 10th International Congress of Entomology, Montreal, vol. 3, pp 367–374

Barnes HF, Arnold MK (1960) Delayed emergence of the pea midges. Plant Pathol 9:52–54

Baxendale FP, Teetes GL, Sharpe PJH, Wu H (1984) Temperature-dependent model for development of nondiapausing sorghum midges (Diptera: Cecidomyiidae). Environ Entomol 13:1572–1576

Bergant K, Trdan S (2006) How reliable are thermal constants for insect development when estimated from laboratory experiments? Entomol Exp Appl 120:251–256

Birch ML, Brewer JW, Rohfritsch O (1992) Biology of *Dasineura affinis* (Cecidomyiidae) and influence of its gall on *Viola odorata*. In: Shorthouse JD, Rohfritsch O (eds) Biology of insect-induced galls. Oxford University Press, New York, NY, pp 171–184

Bissett J, Borkent A (1988) Ambrosia galls: the significance of the fungal nutrition in the evolution of the Cecidomyiidae (Diptera). In: Pirozynski KA, Hawksworth DL (eds) Coevolution of fungi with plants and animals. Academic Press, New York, NY, pp 203–225

Borkent A, Bissett J (1985) Gall midges (Diptera: Cecidomyiidae) are vectors for their fungal symbionts. Symbiosis 1:185–194

Chen C, Appleby JE (1984) Biology of the cypress twig gall midge, *Taxodiomyia cupressiananassa* (Diptera: Cecidomyiidae), in Central Illinois. Ann Entomol Soc Am 77:203–207

Corbet PS (1963) A biology of dragonflies. Quadrangle Books, Chicago

Corley JC, Capurro AF, Bernstein C (2004) Prolonged diapause and the stability of host-parasitoid interactions. Theor Popul Biol 65:193–203

Danks HV (1987) Insect dormancy: an ecological perspective. Biological Survey of Canada, Ottawa

Dixon AFG, Kundu R (1994) Ecology of host alternation in aphids. Eur J Entomol 91:63–70

Dorchin N, Freidberg A, Mokady O (2004) Phylogeny of the Baldratiina (Diptera: Cecidomyiidae) inferred from morphological, ecological and molecular data sources, and evolutionary patterns in plant-galler relationships. Mol Phylogenet Evol 30:503–515

Dorchin N, Joy JB, Hilke LH, Wise MJ, Abrahamson WG (2015) Taxonomy and phylogeny of the *Asphondylia* species (Diptera: Cecidomyiidae) of north American goldenrods: challenging morphology, complex host associations, and cryptic speciation. Zool J Linnean Soc 174:265–304

Elsayed AK, Skuhravá M, Karam HH, Elminshawy A, Al-Eryan MA (2015) New records and new species of gall midges (Diptera: Cecidomyiidae) developing on Chenopodiaceae in Egypt. Zootaxa 3904:105–115

Elsayed AK, Matsuo K, Kim W, Uechi N, Yukawa J, Gyoutoku N, Tokuda M (2018a) A new *Asphondylia* species (Diptera: Cecidomyiidae) and a eulophid wasp (Hymenoptera) inducing similar galls on leaf buds of *Schoepfia jasminodora* (Schoepfiaceae), with reference to their ecological traits and a description of the new gall midge. Entomol Sci 21:324–339

Elsayed AK, Yukawa J, Tokuda M (2018b) A taxonomic revision and molecular phylogeny of the eastern Palaearctic species of the genera *Schizomyia* Kieffer and *Asteralobia* Kovalev (Diptera: Cecidomyiidae: Asphondyliini), with descriptions of five new species of *Schizomyia* from Japan. ZooKeys 808:123–160

Elsayed AK, Yukawa J, Tokuda M (2019a) Two new species of *Schizomyia* (Diptera: Cecidomyiidae) from Japan, with an updated key to larval, pupal and adult *Schizomyia* in Japan. Zootaxa 4688:348–360

Elsayed AK, Yukawa J, Tokuda T (2019b) *Pseudasphondylia tominagai,* a new gall midge species (Diptera: Cecidomyiidae) inducing flower bud galls on *Eleutherococcus spinosus* (Araliaceae) in Japan. Biodivers Data J 7:e35673. https://doi.org/10.3897/BDJ.7.e35673

Foster JE, Taylor PL (1975) Thermal-unit requirements for development of the hessian fly under controlled conditions. Environ Entomol 4:195–202

Gagné RJ (1989) The plant-feeding gall midges of North America. Cornell University Press, Ithaca

Gagné RJ, Jaschhof M (2017) A catalog of the Cecidomyiidae (Diptera) of the world, 4th edn. Digital. [Cited 10 May 2020] Available from URL: https://www.ars.usda.gov/ARSUserFiles/80420580/Gagne_2017_World_Cat_4th_ed.pdf

Gagné RJ, Orphanides GM (1992) The pupa and larva of *Asphondylia gennadii* (Diptera: Cecidomyiidae) and taxonomic implications. Bull Entomol Res 82:313–316

Gagné RJ, Wuensche AL (1986) Identity of the *Asphondylia* (Diptera: Cecidomyiidae) on guar, *Cyamopsis tetragonoloba* (Fabaceae), in the southwestern United States. Ann Entomol Soc Am 79:246–250

Gagné RJ, Kim JW, Uechi N, Yukawa J (2018) A new pest *Asphondylia* (Diptera: Cecidomyiidae) on grape berries (Vitaceae) in southwestern North America with descriptive notes on the genus. Proc Entomol Soc Wash 120:779–790

Hanski I (1988) Four kinds of extra long diapause in insects: a review of theory and observations. Ann Zool Fenn 25:37–53

Harris KM (1975) The taxonomic status of the carob gall midge, *Asphondylia gennadii* (Marchal), comb. n. (Diptera, Cecidomyiidae), and of other *Asphondylia* species recorded from Cyprus. Bull Entomol Res 65:377–380

Hellqvist S (2001) Phenology of the blackcurrant leaf midge (*Dasineura tetensi*) in Northern Sweden. Acta Agr Scand B 51:84–90

Hinz H (1998) Life history and host specificity of *Rhopalomyia* n. sp. (Diptera: Cecidomyiidae), a potential biological control agent of scentless chamomile. Environ Entomol 27:1537–1547

Hodin J, Riddiford LM (2000) Parallel alterations in the timing of ovarian ecdysone receptor and ultraspiracle expression characterize the independent evolution of larval reproduction in two species of gall midges (Diptera: Cecidomyiidae). Dev Genes Evol 210:358–372

Honěk A (1996) The relationship between thermal constants for insect development: a verification. Acta Soc Zool Bohem 60:115–152

Jurado-Rivera J, Vogler A, Reid C, Petitpierre E, Gómez-Zurita J (2009) DNA barcoding insect-host plant associations. Proc R Soc B Biol Sci 276:639–648

Kampichler C, Teschner M (2002) The spatial distribution of leaf galls of *Mikiola fagi* (Diptera: Cecidomyiidae) and *Neuroterus quercusbaccarum* (Hymenoptera: Cynipidae) in the canopy of a central European mixed forest. Eur J Entomol 99:79–84

Kikuchi Y, Tada A, Musolin DL, Hari N, Hosokawa T, Fujisaki K, Fukatsu T (2016) Collapse of insect gut Symbiosis under simulated climate change. mBio 7:e01578–e01516. https://doi.org/10.1128/mBio.01578-16

Kim W, Matsunaga K, Gyoutoku N, Matsuo L, Minami T, Yukawa J (2015) Life history strategy and adult and larval behavior of *Macrodiplosis selenis* (Diptera: Cecidomyiidae), a species that induces leaf-margin fold galls on deciduous *Quercus* (Fagaceae). Entomol Sci 18:470–478

Kim W, Minami T, Tokuda M, Matsuo K, Harris KM, Yukawa J (2019) Detection of two new cryptic species of *Kiefferia* (Diptera: Cecidomyiidae) by means of morphological, molecular and ecological studies. Entomol Sci 22:450–462

Kingsolver JG, Diamond SE, Buckley LB (2013) Heat stress and the fitness consequences of climate change for terrestrial ectotherms. Funct Ecol 27:1–9

Kiritani K (2011) Impacts of global warming on *Nezara viridula* and its native congeneric species. J Asia Pac Entomol 14:221–226

Kiritani K (2012) The low development threshold temperature and the thermal constant in insects and mites in Japan (2nd edition). Bull Nat Inst Agro-Environ Sci 31:1–74. (In Japanese with English summary)

Kiritani K, Yamashita H (2008) Phenology of reproduction and the lower development threshold in *Carabus lewisianus* Breuning. Jpn J Appl Entomol Zool 52:77–86. (In Japanese with English abstract)

Kiritani K, Yukawa J (eds) (2010) Effects of global warming on insects. Zenkoku Nôson Kyôiku Kyôkai, Tokyo. (In Japanese)

Kobune S, Kajimura H, Masuya H, Kubono T (2012) Symbiotic fungal flora in leaf galls induced by *Illiciomyia yukawai* (Diptera: Cecidomyiidae) and in its mycangia. Microb Ecol 63:619–627

Kolesik P, Adair RJ, Eick G (2010) Six new species of *Asphondylia* (Diptera: Cecidomyiidae) damaging flower buds and fruit of Australian *Acacia* (Mimosaceae). Syst Entomol 35:250–267

Maeda N, Sato S, Yukawa J (1982) Polymodal emergence pattern of the machilus leaf gall midge, *Daphnephila machilicola* Yukawa (Diptera, Cecidomyiidae). Kontyû 50:44–50

Mamaev BM (1968) Evolution of gall forming insects: gall midges. In Akademia Naul, Leningrad. (in Russian, translated into English in 1975, Harris KM (ed) British library, Boston Spa, UK)

Matis JH, Rubink WL, Makela M (1992) Use of the gamma distribution for predicting arrival times of invading insect populations. Environ Entomol 21:436–440

McMahon DP, Hayward A (2016) Why grow up? A perspective on insect strategies to avoid metamorphosis. Ecol Entomol 41:505. https://doi.org/10.1111/een.12313

Miyatake Y (1980) A list of the whiteflies of Japan with their host plant and distribution data (Homoptera: Aleyrodidae). Rostria 32:291–330. (In Japanese)

Monzen K (1937) On some new gall midges. Kontyû 11:180–194. (In Japanese)

Musolin DL, Tougou D, Fujisaki K (2010) Too hot to handle? Phenological and life-history responses to simulated climate change of the southern green stink bug *Nezara viridula* (Heteroptera: Pentatomidae). Glob Chang Biol 16:73–87

Nam Y, Choi W (2014) An empirical predictive model for the spring emergence of *Thecodiplosis japonensis* (Diptera: Cecidomyiidae): model construction and validation on the basis of 25 years of field observation data. J Econ Entomol 107:1136–1141

Nijveldt W, Yukawa J (1982) A taxonomic study on *Salix*-inhabiting gall midges in Japan (Diptera, Cecidomyiidae). Bull Kitakyushu Mus Nat Hist 4:23–56

Noll J (1959) Über den Einfluss von Temperatur und Bodenfeuchtigkeit auf die Larven und Puppen der Kohldrehherzmücke *Contarinia nasturtii* (Kieffer) als Grundlage für die Vorausberechnung der ersten Schjlupftermins im Frühjahr. Arch Garten 7:362–415

Ohno K, Yukawa J (1984) Description of a new gall midge (Diptera: Cecidomyiidae) causing leaf galls on *Camellia japonica* L., with notes on its bionomics. Kontyû 52:427–434

Ohtani T, Yazawa Y, Yukawa J (1983) Developmental zero point of the ligustrum fruit midge, *Asphondylia sphaera* Monzen and the aucuba fruit midge, *Asphondylia* sp. (Diptera: Cecidomyiidae). Proc Assoc Plant Prot Kyushu 29:118–121. (In Japanese with English summary)

Okuda S, Yukawa J (2000) Life history strategy of *Tokiwadiplosis matecola* (Diptera: Cecidomyiidae) relying upon the lammas shoots of *Lithocarpus edulis* (Fagaceae). Entomol Sci 3:47–56

Orphanides GM (1975) Biology of the carob midge complex, *Asphondylia* spp. (Diptera, Cecidomyiidae), in Cyprus. Bull Entomol Res 65:381–390

Parnell JR (1964) Investigation on the biology and larval morphology of the insects associated with the galls of *Asphondylia sarothamni* H. Loew (Diptera: Cecidomyiidae) on broom (*Sarothamnus scoparius* (L.) Wimmer). Trans R Entomol Soc Lond 116:255–273

Plakidas JD (1988) The newly discovered spring crown gall of *Asphondylia rudbeckiaeconspicua* (Diptera: Cecidomyiidae) on *Rudbeckia laciniata* (Asteraceae) in Pennsylvania. Proc Entomol Soc Wash 90:393–393

Powell JA (1987) Records of prolonged diapause in Lepidoptera. J Res Lepidoptera 25:83–109

Prévost YH (1990) Spruce cone axis midge, *Dasineura rachiphaga* Tripp (Diptera: Cecidomyiidae), in cones of black spruce, *Picea mariana* (mill.) B.S.P. Canadian Entomol 122:441–447

Readshaw JL (1966) The ecology of the swede midge, *Contarinia nasturtii* (Kieffer) (Diptera: Cecidomyiidae). 1. Life history and influences of temperature on development. B Entomol Res 56:685–700

Redfern M (1975) The life history and morphology of the early stages of the yew gall midge *Taxomyia taxi* (Inchbald) (Diptera: Cecidomyiidae). J Nat Hist 9:513–533

Redfern M, Cameron RAD (1978) Population dynamics of the yew gall midge *Taxomyia taxi* (Inchbald) (Diptera: Cecidomyiidae). Ecol Entomol 3:251–263

Redfern M, Cameron RAD (1993) Population dynamics of the yew gall midge *Taxomyia taxi* and its chalcid parasitoids: a 24-year study. Ecol Entomol 18:365–378

Redfern M, Hunter M (2005) Time tells: long-term patterns in the population dynamics of the yew gall midge, *Taxomyia taxi* (Cecidomyiidae), over 35 years. Ecol Entomol 30:86–95

Roubos CR, Liburd OE (2010) Pupation and emergence of blueberry gall midge, *Dasineura oxycoccana* (Diptera: Cecidomyiidae), under varying temperature conditions. Fla Entomol 93:283–290

Sanui T, Yukawa J (1986) Discovery of a paedogenetic gall midge (Diptera: Cecidomyiidae) in Japan and notes on its reproduction in vitro and infestation in a cultivating factory of the oyster mushroom. Jpn J Appl Entomol Zool 30:50–54. (In Japanese with English summary)

Sasaki M, Uechi N (2018) Occurrence of the grape gall midge, *Asphondylia* sp. (Diptera: Cecidomyiidae), and biological characteristics of the genus *Asphondylia*. Plant Prot 72:103–107. (In Japanese)

Sato S, Yukawa J (2008) Descriptions of two new *Mikiola* species (Diptera: Cecidomyiidae) that induce leaf galls on *Fagus crenata* (Fagaceae) in Japan. Stud Dipterol 15:151–164

Shirota Y, Iituka K, Asano J, Abe J, Yukawa J (1999) Intraspecific variations of mitochondrial cytochrome oxidase I sequence in an aphidophagous species, *Aphidoletes aphidimyza* (Diptera: Cecidomyiidae). Entomol Sci 2:209–215

Simbolon H, Yukawa J (1992) Description of a new gall midge (Diptera, Cecidomyiidae) causing galls on *Lithocarpus edulis* (Fagaceae) in Japan. Jpn J Entomol 60:609–617

Skuhravá M, Skuhravý V, Dengler K (1998) Gall inducing and other gall midge species (Diptera: Cecidomyiidae) associated with oaks (*Quercus* spp.) (Fagaceae) in the Palaearctic region. In: Csóka G, Mattson W, Stone GN, Price PW (eds) The biology of gall-inducing arthropods, pp 1–11. USDA Forest Service, North Central Research Station, General Technical Report NC-199

Skuhravá M, Skuhravý V, Jørgensen J (2006) Gall midges (Diptera: Cecidomyiidae) of Denmark. Entomologiske Meddeleser 74:94

Skuhravý V, Skuhravá M, Brewer JW (1993) The saddle gall midge *Haplodiplosis marginata* (Diptera: Cecidomyiidae) in Czech Republic and Slovak Republic from 1971–1989. Acta Societatis Zoologica Bohemoslovaca 57:117–137

Son Y, Lee JH, Chung YJ (2007) Temperature-dependent post-diapause development and prediction of spring emergence of the pine needle gall midge (Diptera: Cecidomyiidae). J Appl Entomol 131:674–683

Stone G, Schonrogge K, Atkinson R, Bellido D, Pujade-Villar J (2002) The population biology of oak gall wasps (Hymenoptera: Cynipidae). Annu Rev Entomol 47:633–668

Sunose T (1978) Studies on extended diapause in *Hasegawaia sasacola* Monzen (Diptera, Cecidomyiidae) and its parasites. Kontyû 46:400–415

Sunose T (1983) Redescription of *Asphondylia morivorella* (Naito), comb. n. (Diptera: Cecidomyiidae), with notes on its bionomics. Appl Entomol Zool 18:22–29

Sunose T (1992) Life cycle of the ampelopsis fruit gall midge *Asphondylia baca*. Nat Insec 27:29–34. (In Japanese)

Tabuchi K, Amano H (2003) Polymodal emergence pattern and parasitoid composition of *Asteralobia sasakii* (Monzen) (Diptera: Cecidomyiidae) on *Ilex crenata* and *I. integra* (Aquifoliaceae). Appl Entomol Zool 38:493–500

Takasu K, Yukawa J (1984) Two-year life history of the neolitsea leaf gall midge, *Pseudasphondylia neolitseae* Yukawa (Diptera, Cecidomyiidae). Kontyû 52:596–604

Tokuda M (2004) *Illiciomyia* Tokuda, a new genus for *Illiciomyia yukawai* sp. n. (Diptera: Cecidomyiidae: Asphondyliini) inducing leaf galls on *Illicium anisatum* (Illiciaceae) in Japan. Esakia 44:1–11

Tokuda M (2012) Biology of Asphondyliini (Diptera: Cecidomyiidae). Entomol Sci 15:361–383

Tokuda M, Yukawa J (2005) Two new and three known Japanese species of genus *Pseudasphondylia* Monzen (Diptera: Cecidomyiidae: Asphondyliini) and their life history strategies. Ann Entomol Soc Am 98:259–272

Tokuda M, Yukawa J (2006) First records of genus *Bruggmanniella* (Diptera: Cecidomyiidae: Asphondyliini) from Palaearctic and oriental regions, with descriptions of two new species that induce stem galls on Lauraceae in Japan. Ann Entomol Soc Am 99:629–637

Tokuda M, Tabuchi K, Yukawa J, Amano H (2004) Inter- and intraspecific comparisons between *Asteralobia* gall midges (Diptera: Cecidomyiidae) causing axillary bud galls on *Ilex* species (Aquifoliaceae): species identification, host range, and mode of speciation. Ann Entomol Soc Am 97:954–970

Tokuda M, Yukawa J, Goukon K (2007) Life history traits of *Pseudasphondylia rokuharensis* (Diptera: Cecidomyiidae) affecting emergence of adults and synchronization with host plant phenology. Environ Entomol 36:518–523

Tokuda M, Yang MM, Yukawa J (2008) Taxonomy and molecular phylogeny of *Daphnephila* gall midges (Diptera: Cecidomyiidae) inducing complex leaf galls on Lauraceae, with descriptions of five new species associated with *Machilus thunbergii* in Taiwan. Zool Sci 25:533–545

Tripp HA (1955) Description and habits of Cecidomyiidae (Diptera) from white spruce cones. Canadian Entomol 87:253–263

Turgeon JJ, Roques A, Groot P (1994) Insect fauna of coniferous seed cones: diversity, host plant interactions and management. Annu Rev Entomol 39:179–212

Uechi N, Yukawa J (2004) Description of *Asphondylia itoi* sp. n. (Diptera: Cecidomyiidae) inducing fruit galls on *Distylium racemosum* (Hamamelidaceae) in Japan. Esakia 44:27–43

Uechi N, Yukawa J (2006) Life history patterns and host ranges of the genus *Asphondylia* (Diptera: Cecidomyiidae). In: Ozaki K, Yukawa J, Ohgushi T, Price PW (eds) Galling arthropods and their associates: ecology and evolution. Springer-Verlag, Tokyo, pp 275–285

Uechi N, Yukawa J, Yamaguchi D (2004) Host alternation by gall midges of the genus *Asphondylia* (Diptera: Cecidomyiidae). In: Evenhuis NL, Kaneshiro KY (eds) Contributions to the systematics and evolution of diptera, pp 53–66. D. Elmo Hardy Memorial Volume Of Bishop Museum Bulletin In Entomology 12

Uechi N, Yukawa J, Usuba S (2005) Discovery of an additional winter host of the soybean pod gall midge, *Asphondylia yushimai* (Diptera: Cecidomyiidae) in Japan. Appl Entomol Zool 40:597–607

Uechi N, Yukawa J, Tokuda M, Maryana N, Ganaha-Kikumura T, Kim W (2017) Description of the Asian chili pod gall midge, *Asphondylia capsicicola* sp. n., with comparative notes on *A. gennadii* (Marchal) (Diptera: Cecidomyiidae) that induces the same sort of pod gall on the same host plant species in the Mediterranean region. Appl Entomol Zool 52:113–123

Uechi N, Kim W, Tokuda M, Fujii T, Kikuchi H, Kakizaki M, Iwasaki A, Paik JC, Yukawa J (2018) Genetic and ecological differences between *Asphondylia yushimai* and the ivy gall midge, *Asphondylia* sp. (Diptera: Cecidomyiidae) with a new distribution record of the former from Hokkaido and South Korea. Appl Entomol Zool 53:363–371

Usuba S (1982) Miscellaneous notes on insect galls VI. Insekuto 32:69–74. (In Japanese)

Waldbauer GP (1978) Phenological adaptation and the polymodal emergence patterns of insects. In: Dingle H (ed) Evolution of insect migration and diapause. Springer-Verlag, Heidelberg, pp 127–144

Wise IL, Lamb RJ (2004) Diapause and emergence of *Sitodiplosis mosellana* (Diptera: Cecidomyiidae) and its parasitoid *Macroglenes penetrans* (Hymenoptera: Pteromalidae). Canadian Entomol 136:77–90

Wyatt IJ (1959) Some aspects of Cecid biology of importance in mushroom culture. Mush Sci 4:271–279

Wyatt IJ (1967) Pupal paedogenesis in the Cecidomyiidae (Diptera) 3 – a reclassification of the Heteropezini. Trans R Entomol Soc London 119:71–98

Yukawa J (1971) A revision of the Japanese gall midges (Diptera: Cecidomyiidae). Mem Fac Agric Kagoshima Univ 8:1–203

Yukawa J (1974) Descriptions of new Japanese gall midges (Diptera, Cecidomyiidae, Asphondyliidi) causing leaf galls on Lauraceae. Kontyû 42:293–304

Yukawa J (1987) Life history strategies of univoltine gall-making Cecidomyiidae (Diptera) in Japan. Phytophaga 1:121–139

Yukawa J (1996) Identification of paedogenetic gall midge, *Mycophila speyeri* (Diptera: Cecidomyiidae) and possibility of accidental introduction to Japan. Jpn J Appl Entomol Zool 40:135–143. (In Japanese with English summary)

Yukawa J, Akimoto K (2006) Influence of synchronization between adult emergence and host plant phenology on the population density of *Pseudasphondylia neolitseae* (Diptera: Cecidomyiidae) inducing leaf galls on *Neolitseae sericea* (Lauraceae). Popul Ecol 48:13–21

Yukawa J, Masuda H (1996) Insect and mite galls of Japan in colors. Zenkoku Nôson Kyôiku Kyôkai, Tokyo. (In Japanese with English explanation for color plates)

Yukawa J, Miyamoto K (1979) Redescription of *Asphondylia sphaera* Monzen (Diptera, Cecidomyiidae), with notes on its bionomics. Mem Fac Agric Kagoshima Univ 15:99–106

Yukawa J, Ohsaki N (1988) Separation of the aucuba fruit midge, *Asphondylia aucubae* sp. nov. from the *ampelopsis* fruit midge, *Asphondylia baca* Monzen (Diptera: Cecidomyiidae). Kontyû 56:365–376

Yukawa J, Rohfritsch O (2005) Biology and ecology of gall-inducing Cecidomyiidae (Diptera). In: Raman A, Schaefer CW, Withers TM (eds) Biology, ecology, and evolution of gall-inducing arthropods. Science Publishers, Inc, Enfield, NH and Plymouth, UK, pp 273–304

Yukawa J, Sunose T (1976) Description of a new gall midge (Diptera, Cecidomyiidae) on *Euonymus*, with notes on its bionomics. Kontyû 44:159–168

Yukawa J, Tsuda K (1987) A new gall midge (Diptera: Cecidomyiidae) causing conical leaf galls on *Celtis* (Ulmaceae) in Japan. Kontyû 55:123–131

Yukawa J, Ohtani T, Yazawa Y (1983) Host-change experiments from wild plants to soybean in *Asphondylia* species (Diptera: Cecidomyiidae). Proc Assoc Plant Prot Kyushu 29:115–117. (In Japanese with English summary)

Yukawa J, Nakawatase A, Horikiri M (1986) Age structure and survival rate of the soybean pod gall midge, *Asphondylia* sp. (Diptera: Cecidomyiidae) occurring on soybean cultivated in early winter. Proc Assoc Plant Prot Kyushu 32:127–129. (In Japanese with English summary)

Yukawa J, Uechi N, Horikiri M, Tuda M (2003) Description of the soybean pod gall midge, *Asphondylia yushimai* sp. n. (Diptera: Cecidomyiidae), a major pest of soybean and findings of host alternation. Bull Entomol Res 93:73–86

Yukawa J, Haitsuka S, Miyaji K, Kamikado T (2006) Influence of the population dynamics of a gall-inducing cecidomyiid and its parasitoids on the abundance of a successor, *Lasioptera yadokariae* (Diptera: Cecidomyiidae). In: Ozaki K, Yukawa J, Ohgushi T, Price PW (eds) Galling arthropods and their associates: ecology and evolution. Springer-Verlag, Tokyo, pp 241–249

Yukawa J, Ikenaga H, Sato S, Tokuda M, Ganaha-Kikumura T, Uechi N, Matsuo K, Mishima M, Tung GH, Paik JC, Ren BQ, Don XO (2012) Description and ecological traits of a new species of *Pitydiplosis* (Diptera: Cecidomyiidae) that induces leaf galls on *Pueraria* (Fabaceae) in East Asia, with a possible diversification scenario of intraspecific groups. Entomol Sci 15:81–98

Yukawa J, Nakagawa K, Saigou T, Awa T, Fukuda T, Higashi M (2013) Adult behavior of an ambrosia gall midge *Illiciomyia yukawai* (Diptera: Cecidomyiidae) and synchronization between its emergence and host plant phenology. Entomol Sci 16:400–412

Yukawa J, Tokuda M, Yamagishi K (2014) Host plant ranges and distribution records of identified and unidentified species of the genus *Lasioptera* (Diptera: Cecidomyiidae) in Japan. Esakia 54:1–15

Yukawa J, Ichinose M, Wanggyu K, Uechi N, Gyotoku N, Fujii T (2016a) Lower development threshold temperatures and thermal constants for four species of *Asphondylia* (Diptera: Cecidomyiidae) in Japan and their larval developmental delay caused by heat stress. Appl Entomol Zool 57:71–80

Yukawa J, Miyamoto K, Yamaguchi T, Takesaki K, Uechi N, Matsuo K (2016b) Key factor/key stage analysis of long-term life table data for a fruit gall midge, *Asphondylia sphaera* (Diptera: Cecidomyiidae). Ecol Entomol 41:516–526

Yukawa J, Nakagawa K, Kohno A, Tokuda M, Kiritani K, Matsuo K, Mitsui H, Fujii T (2016c) Geographical and annual variations in the proportion of extended diapausing individuals of *Illiciomyia yukawai* (Diptera: Cecidomyiidae) with reference to an adaptive significance of its bimodal emergence pattern. Entomol Sci 19:275–289

Yukawa J, Yamamura K, Fujimoto K, Tokuda M, Muroya S, Maki T, Sadoshima T, Fukuda T (2018) Key-factors and key-stages that determine the leaf longevity of an evergreen broad-leaved tree, *Neolitsea sericea* (Lauraceae) at different growing sites in southern Japan. Ecol Res 33:175–190

Yukawa J, Moriya T, Kanmiya K (2019a) Comparison in the flight ability between the soybean pod gall midge, *Asphondylia yushimai* and the aucuba fruit gall midge, *A. aucubae* (Diptera: Cecidomyiidae). Appl Entomol Zool 54:167–174

Yukawa J, Tokuda M, Uechi N, Yasuda K, Ganaha-Kikumura T, Matsuo K, Shimizu Y, Yamaguchi D (2019b) Ecological divergence among morphologically and genetically related *Asphondylia* species (Diptera: Cecidomyiidae), with new life history data for three congeners including the *Alpinia* fruit gall midge. Entomol Sci 22:437–449

Chapter 7
Behavior

Junichi Yukawa and Wanggyu Kim

Abstract Adult life span of gall-inducing cecidomyiids is very short, usually one, sometimes two to three days although they live longer under humid laboratory conditions if they do not swarm or lay eggs. In this chapter, various patterns of adult activity including emergence, swarming, mating, and oviposition are demonstrated and compared between diurnal and nocturnal species and between species with different life history strategies. We refer to the results of comparison in the flight ability between a host-alternating and multivoltine *Asphondylia yushimai* and a monophagous and univoltine *A. aucubae*. It is remarkable that *A. yushimai* females carry the potential strong flight ability, i.e. continuously 13.7 km within 10 h. Some aspects of larval behavior are also referred to in this chapter.

Keywords Daily activity · Emergence · Fecundity · Flight ability · Mating · Oviposition · Sex ratio

7.1 Emergence Time of Day

Barnes (1930) first intensively studied the emergence of gall midges from various behavioral and ecological aspects such as danger at the time of emergence, emergence time of day and year, daily fluctuations, and factors governing the emergence. In this chapter, we refer mainly to the emergence time of day, and in Chap. 8 the emergence time of year will be treated together with temperature conditions.

Patterns of daily activity are divided into diurnal and nocturnal, and daily emergence patterns are into unimodal and bimodal (Fig. 7.1). The unimodal

J. Yukawa (✉)
Entomological Laboratory, Faculty of Agriculture, Kyushu University, Fukuoka, Japan
e-mail: galling-cecidomyiids@kyudai.jp

W. Kim
Animal Resources Division, National Institute of Biological Resources, Incheon, South Korea
e-mail: entowk@korea.kr

J. Yukawa, M. Tokuda (eds.), *Biology of Gall Midges*, Entomology Monographs,
https://doi.org/10.1007/978-981-33-6534-6_7

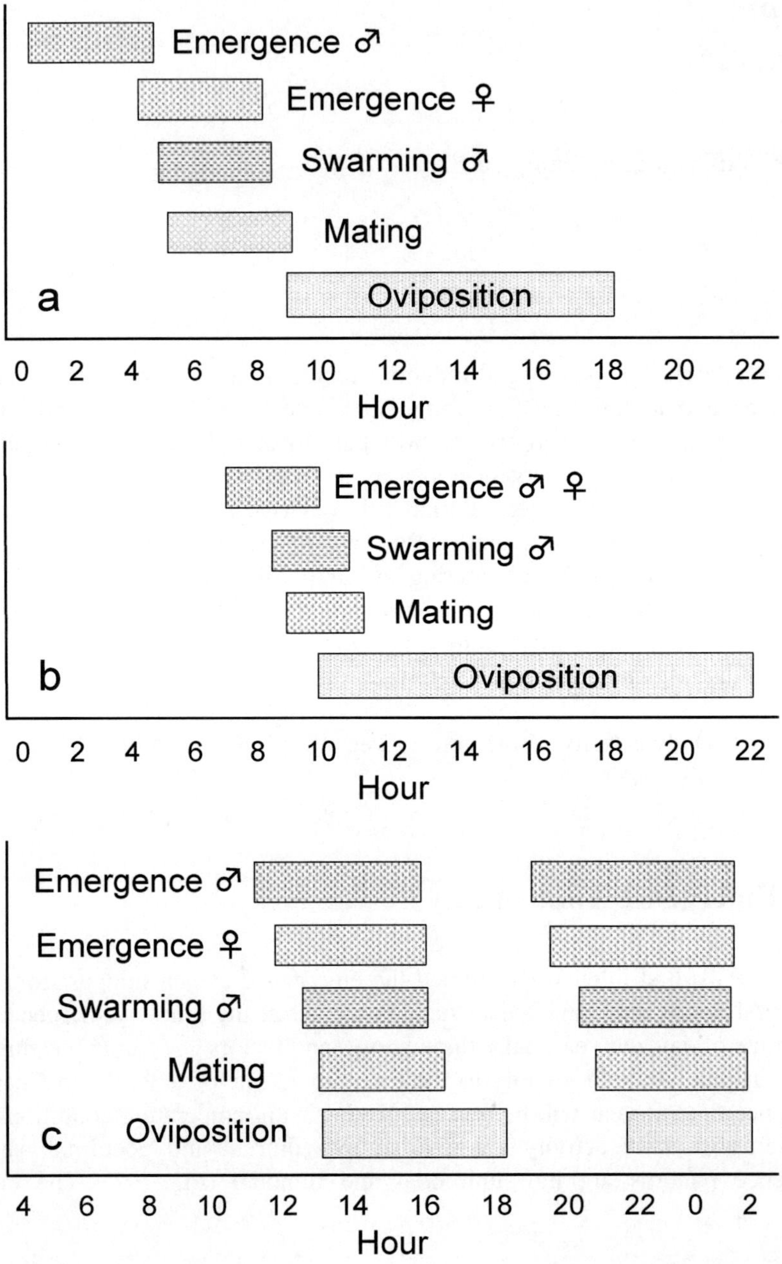

Fig. 7.1 Schematic representation of three types of daily activity exhibited by gall-inducing cecidomyiids. (**a**), unimodal emergence type in which males emerge earlier than females; (**b**), unimodal emergence type in which males and females emerge almost simultaneously; (**c**), bimodal emergence type that is mainly exhibited by nocturnal species

emergence pattern can be seen in various diurnal gall midges. Usually males emerge earlier than females (Fig. 7.1a) as has been noted for *Contarinia merceri* Barnes, *Contarinia tritici* Kirby, *Dasineura alopecuri* Reuter, *Dasineura pyri* Bouché, *Rabdophaga terminalis* Loew, *Stenodiplosis geniculate* Reuter (Barnes 1930), *Stenodiplosis sorghicola* (Coquillett) (Summers 1975), *Haplodiplosis marginata* (Roser) (Skuhravý et al. 1993), and *Dasineura mali* Kieffer (Harris et al. 1999). In *Masakimyia pustulae* Yukawa & Sunose, males emerge in midnight between 11:00 pm and 2:00 am, and female emergence reaches a crest 7 h or so later (Sunose 1983). In the laboratory, the males of *Illiciomyia yukawai* Tokuda start emergence from 9:00 pm. The emergence reaches a peak between 0:00 am and 1:00 am and continue intermittently until 11:00 am, while the females emerge from 5:00 am until 5:00 pm, with a peak between 5:00 am and 6:00 am (Yukawa et al. 2013).

Both males and females of *Pseudasphondylia neolitseae* Yukawa emerge at the same time between 7:30 am and 10:00 am (Yukawa et al. 1976) (Fig. 7.1b). The emergence of males and females occurs between 9:00 am and 11:00 am in *Macrodiplosis selenis* Kim & Yukawa (Kim et al. 2014, 2015). It is remarkable that females emerge before males in the Douglas-fir cone gall midge *Contarinia oregonensis* Foote (Miller and Borden 1984).

In some species, an hourly emergence pattern is more or less bimodal (the first peak usually in the morning and the second one after sunset) and extends over several hours (Fig. 7.1c). In *Asphondylia aucubae* Yukawa & Ohsaki, the hourly emergence of individuals extended over 16 hours, but there were two peaks for both sexes, respectively. At 10:00 am to 11:00 am, the males reached the first peak of emergence about 4 h before the females. The emergence of both the sexes attained the second peak almost simultaneously around 9:00 pm to 11:00 pm. This pattern is frequently seen in nocturnal species of *Asphondylia* (Yukawa and Miyamoto 1979; Yukawa and Ohsaki 1988b), but the bimodality sometimes becomes unclear because of the influence of light intensity and daytime temperature.

7.2 Time Required for Emergence

As pointed out by Barnes (1930), the emergence of insects from the pupal stage is a delicate process attended by many dangers, and dependent upon various external influences. Actual emergence takes 5–15 min in *Lasioptera rubi* (Schrank) (Barnes 1930), 20–45 min in *P. neolitseae* (Yukawa et al. 1976), about 105 min in *M. pustulae* (Yukawa and Sunose 1976), about 80 min in *M. selenis* (Kim et al. 2015). In the case of *Asphondylia sarothamni* (Loew), females take 24 min to emerge and males 30 min (Barnes 1930). Laboulbéne (1873) observed that males of *Monarthropalpus buxi* (Laboulbéne) required longer time for emergence than females probably because of the difference in the length of the antennae.

7.3 Mortality at the Time of Emergence

During the time of emergence, insects are in a helpless condition. Their enemies take advantage of this. For example, the common flowerbug *Anthocoris nemorum* Linnaeus (Hemiptera: Anthocoridae) greedily seizes emerging *R. terminalis* and *D. pyri* (Barnes 1930). Spiders have been observed sucking *D. alopecuri* in the act of emergence (Barnes 1930). Eleven species of spiders have been known as predators of *P. neolitseae* at the time of emergence or oviposition (Yukawa 1983). Heavily galled leaves of its host plant *Neolitsea sericea* (Blume) Koidzumi (Lauraceae) are frequently folded outward in various degrees, providing favorable webbing conditions to the spiders. Because *P. neolitseae* emerges from galls on under surface of the host leaves, adult gall midges are killed by the direct predation and are also caught in the webs. More than 50% of galled leaves are folded under high gall midge population density (Yukawa 1983).

7.4 Emergence Projections Induced by Fig Gall Midges

Gall induction and leaf mining are major examples in the manipulation of plant development, including morphological and physiological changes of host plant tissues. Giron et al. (2016) demonstrated various examples of host plant manipulation by gall-inducing organisms and leaf-mining insects, including other herbivorous insects that share the plant manipulation with inducers.

There are morphologically distinct examples of host plant manipulation by fig gall midges. Yafuso et al. (2013) found that species-specific projections developed before emergence from female flower galls induced by three unidentified cecidomyiid species in the syconia of *Ficus microcarpa* Linnaeus (Moraceae) on Okinawa and Amami Islands, Japan (Fig. 7.2). The projection was derived from

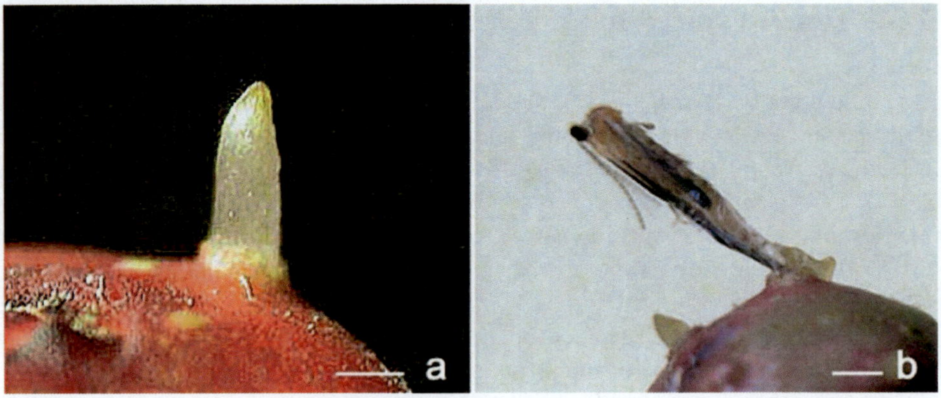

Fig. 7.2 Emergence projections of fig gall midges. (**a**), horn-like projection of sp. 1 of Yafuso et al. (2013); (**b**), female emergence of sp. 1. Scale bars, 2 mm

plant tissues consisting of a mass of small square cells in the basal and distal portions and regularly arranged long cells in the middle portion. Yafuso et al. (2013) considered that the gall midges manipulate the fig plant to develop the projection before emergence, so that the pupa can easily pass through the sticky epidermis of the syconium.

A similar way of cecidomyiid departure from fig fruit galls has been observed for *Ficiomyia perarticulata* Felt on *Ficus citriflora* Miller in Florida, USA (Roskam and Nadel 1990) and unidentified gall midges on *Ficus benjamina* Linnaeus in Xishuangbanna, southwestern China (Bai et al. 2008; Miao et al. 2011). In 2013, N Wachi (University of the Ryukyus, Japan) informed us that he found similar projections on fruit galls of *Ficus sundaica* Blume in Cambodia. Such a way of departure from fig fruit galls seems to be common in tropical and subtropical areas and to be an adaptive habit for fig gall midges to reduce risks under helpless conditions at the time of emergence.

7.5 Sex Ratio

The sex ratio is usually 1:1 (Barnes 1931) but is frequently biased in favor of females. For example, the sex ratio is approximately 1:2 in *P. neolitseae* (Yukawa et al. 1976), *Asphondylia sphaera* Monzen (Yukawa and Miyamoto 1979), *A. aucubae* (Yukawa and Ohsaki 1988b), *I. yukawai* (Yukawa et al. 2013), and 70% in *M. selenis* (Kim et al. 2015) although the ratio varies with locality and from year to year (Yukawa and Sunose 1976; Yukawa et al. 1976) and with season and host plant quality (Dorchin and Fredberg 2004). In *C. oregonensis*, females constitute 53–58% of the total number emerged (Miller and Borden 1984).

In addition, 'unisexual' families exist, in which female offspring of all-female families can produce all males, all females, or both males and females (Barnes 1931). Elimination of paternally derived chromosomes during spermatogenesis explains the incidence of unisexual families for the Hessian fly *Mayetiola destructor* (Say) (Gallum and Hatchett 1969). Dorchin and Fredberg (2004) demonstrated that *Izeniola obesula* Dorchin exhibited strict monogeny, resulting in galls that contain either all female or all male progeny and that the skewed sex ratio among galls, ranging from 1:1 to 1:4, possibly caused by higher mortality occurred among male galls. Both *Dasineura folliculi* Felt and *Dasineura carbonaria* Felt exhibit monogeny, and gall sex ratios differed between generations of *D. folliculi* on *Solidago rugosa* Miller (Asteraceae) but not on *Solidago gigantean* Aiton (Dorchin et al. 2007).

7.6 Swarming and Mating

Swarming mass has been observed for various non-gall inducing cecidomyiids of Micromyinae such as *Micromya lucorum* Rondani, *Ansifera gombakensis* Jaschhof, and *Campylomyza flavipes* Meigen and Lestremiinae such as *Conarete calcuttaensis* Nayar. Swarming of *Anarete pritchardi* Kim and *Anarete* sp. (Lestremiinae) was most intensively studied by Chiang and his group (see Kanmiya and Yukawa 2020 for the references as to Micromyinae and Lestremiinae).

Some other non-gall inducing cecidomyiids mate by hanging on abandoned spider-webs or other substrates in a face-to-face position. This way of mating has been seen in predacious or flower-visiting cecidomyiids, such as *Aphidoletes aphidimyza* (Rondani) (Yukawa et al. 2008), an aphidophagous species, and *Resseliella kadsurae* Yukawa, Sato & Xu, a pollinator of *Kadsura longipedunculata* Finet & Gagnepain (Schisandraceae) (Yukawa et al. 2011).

Gall-inducing cecidomyiids with type IA or IB life history strategy overwinter as mature larvae on the ground (Yukawa 1987; see also Chap. 6). Therefore, males usually fly over the ground under the host plants in search for females that locate on fallen leaves or short grasses after emergence (e.g. Yukawa and Tsuda 1987 for *Celticecis japonica* Yukawa & Tsuda; Kim et al. 2015 for *M. selenis*). However, some females of *Hartigiola faggalli* (Monzen) fly from the fallen leaves and alight on fresh leaves or old twigs of the host tree, *Fagus crenata* Blume (Fagaceae), and mating takes place there as well as on the ground (Mishima et al. 2014).

Kim et al. (2015) observed emergence (Fig. 7.3a, b), swarming and mating of *M. selenis* under natural conditions. The males started swarming from 9:00 am over the ground under the host trees *Quercus serrata* Thunberg ex Murray (Fagaceae). The swarming consisted of two to ten males and continued until noon. Some males were walking on the ground in search for females. Females rested on the fallen leaves or grasses after emergence, waiting for males to locate them. To call males, females protruded the ovipositor until males arrived. Females seemed to emit sex pheromones during this time because some males, sensing the presence of females, left the swarm and flew to the females. Mating took place without ceremony and lasted for about 40 seconds. After mating, the males returned to the swarm. Mating was observed mainly between 9:00 am and 10:00 am soon after the emergence of females. Males disappeared after swarming and mating. This pattern of hourly activity of *M. selenis* varied to some extent depending on weather conditions of the day, as has been noted for other gall midges (e.g. Barnes 1930; Yukawa et al. 1976; Skuhravá et al. 1998; Yukawa et al. 2013).

Gall-inducing cecidomyiids with type IIA or IIB life history strategy overwinter in the galls on evergreen host plants (Yukawa 1987; see also Chap. 6). Males exhibit swarming around the host plant from which they emerged because females are expected to emerge from galls on the same plant. After emergence, the females locate on the galls from which they emerged and wait for the males to come. Some of swarming males fly towards females in response to pheromonal calling. Such mating behavior has been observed for *P. neolitseae* (Yukawa et al. 1976), *Asphondylia*

Fig. 7.3 *Macrodiplosis selenis* and its galls on the host plant *Quercus serrata*. (**a**), male; (**b**), female (reproduced with permission from Kim et al. 2015); (**c**), newly extending shoot of *Q. serrata*; (**d**), eggs deposited on the upper surface of a host leaf; (**e**), eggs deposited on the under surface of a host leaf; (**f**), first instar crawling toward leaf margin on the upper surface of a host leaf; (**g**), leaf margin folded galls. Scale bars, mm

yushimai Yukawa & Uechi (Yukawa et al. 1983) and *I. yukawai* (Yukawa et al. 2013).

The followings are detailed observation records for swarming and mating behavior of *P. neolitseae* (Yukawa et al. 1976). Under favorable conditions, the males started swarming around the host plant *N. sericea*, from which they had emerged, at 8:30 am and continued until 11:00 am. The females were usually resting on the undersurface of the host leaves and waiting for males to locate them. One or more males, sensing the presence of the females, left the swarm and flew to the leaves to mate. Mating took place without ceremony and lasted for 99 ± 8 s (average ± s.d.). Under favorable weather conditions, at least 75 % of females mated on each census data. After mating the males returned to the warm.

According to Yukawa et al. (1976), mating failures were caused by: (1) too many males attempting to mate with one female, which made the female flying far away from both the host plant and swarming site (in this case the female could not be recognized again by any of swarming males); (2) the contrary case where females were not located by any of the males during the swarming period. The first case occurred when the male swarm size was large and the second when it was too small or there was no available swarm near the females. The second situation was caused mainly by the small number of daily emergences, which was frequently observed during several days before and after main emergence period of the year. The large sized swarm was naturally observed during the main period of emergence.

When the temperature was over 6.5 °C at the time of emergence, the daily emergence normally occurred, otherwise it did not occur even during the main emergence period of the year. The adults that had postponed their emergence were then added to those that were to emerge on the next favorable morning. If the temperature was not warm enough (below 16 °C) at swarming time, the emerged males could not fly off and were compelled to stay on the under surface of the leaves and were unable to mate. Thus, mating was seriously impaired by unfavorable temperature when it might be warm enough for emerging, but not for swarming. Once the females mated, they could fly and oviposit even if the temperature was considerably lower than 16 °C in the afternoon.

7.7 Flight Ability

Insects with wings fly to find food resources, to locate places to live, to hold territories, to search mating partners, to escape from danger and for other purposes (e.g., Pringle 1957). Therefore, an evaluation of their potential flight ability is essential in predicting their movement and dispersal. Therefore, various sorts of apparatus have been devised to evaluate the flight ability (Cloudsley-Thompson 1955, 1961; Miller 1979; Kanmiya 1994). Among these apparatuses, the thread tethered flight method is relatively simple, convenient and has been used to evaluate the flight ability of various insects (see references in Yukawa et al. 2019). In particular, a flight-mill together with various subordinate devices, (a high-speed video camera, for example), is an effective tool to evaluate the potential flight ability of insects (e.g. Fyodorova and Lapshin 1990; Ribak et al. 2017).

As to gall-inducing cecidomyiids, however, their flight ability had seldom been studied because the flight-mill has difficulties in measuring the flight ability of insects that do not have a strong power to pull the rotor of the flight-mill. Kanmiya (1994) developed a thread tethered flight method using strain gauge transducers to overcome the difficulties and successfully analyzed the flight power, flight direction and temporal progress of *Oraesia* spp. (Lepidoptera: Noctuidae).

Hao et al. (2013) studied the active flight potential of the orange wheat blossom midge *S. mosellana* under various environmental factors using a 26-channel computer-monitored flight-mill system. They demonstrated that the most suitable temperature for flight and mean flight distance was 16–24 °C (438 m in male, 735 in female); mean flight duration peaked at 16 °C (70 min in male, 113 in female) while mean speed peaked at 28 °C (9.0 m/min in male, 10.2 in female). Flight performance gradually declined between 10 and 400 lux light intensity.

Yukawa et al. (2019) compared the flight ability between *A. yushimai* and *A. aucubae* using the thread tethered flight method by modifying devices with a CCD camera and an apparatus for analyzing tethered flight sounds to measure the flight power of tiny gall midges that have a weight of about 2 mg (see Figs. 2 and 3 in Yukawa et al. 2019 for the devices and apparatus). *Asphondylia yushimai* is a host-alternating, multivoltine gall midge and one of the major pests of soybean in Japan and South Korea (e.g., Uechi et al. 2018; Yukawa et al. 2003; see also Chaps. 6 and 12). *Asphondylia aucubae* is a monophagous, univoltine species, and induces fruit galls on *Aucuba japonica* Thunberg (Garryaceae) in Japan (Yukawa and Ohsaki 1988a; Yukawa et al. 2016).

According to Yukawa et al. (2019), the total flying time of females of the two *Asphondylia* species was more than 2 h on average. The maximum continuous flying time was nearly 10 h in *A. yushimai* and 5 h in *A. aucubae*. The speed of wing-beat stroke was higher in males than in females in the two species, being adapted to mobile flight for swarming of the males, while the lower speed of stroke enables the females to fly long distance. In particular, the larger wing size of *A. yushimai* females was considered to support their gentle flight for long-distance. The flying speed of *A. yushimai* was 22.9 m/min. The flying speed and the maximum continuous flying times indicated that the *A. yushimai* females carry the potential strong flight ability, i.e. continuously 13.7 km within 10 h. In addition, *A. yushimai* females have a habit to fly up to the sky after mating hence they are carried by wind for long distance. These results were reflected well in the different life history traits between *A. yushimai* and *A. aucubae*. The long-distance flight ability of *A. yushimai* supports the possibility of its host-searching behavior.

Further data of flying behavior and ability of gall midges should be accumulated to evaluate their dispersal beyond border because tiny gall midges may be able to move for a longer distance than our supposition.

7.8 Chemical Communication

Chemical communication plays an important role in reproductive activities of gall midges. According to Hall et al. (2012), the presence of female-produced sex pheromones has been known in 19 cecidomyiid species and components of the sex pheromones have been identified for 16 species among them (Table 7.1).

The chemical ecology of cecidomyiids was reviewed comprehensively by Harris and Foster (1999), thereafter by Wicker-Thomas (2007). More recently, Hall et al. (2012) reviewed progress on identification of sex pheromones, chemicals involved in location of host plants, the neurophysiology of reception of volatile chemicals, and application of semiochemicals to control pest cecidomyiids that occurred during the last decade. See Hall et al. (2012) for further references as to cecidomyiid chemicals.

7.9 Fecundity

From 40 to 400 ovarian eggs have been recorded for many gall-inducing cecidomyiids (Yukawa et al. 1976; Yukawa and Miyamoto 1979; Basedow and Schütte 1982; Sunose 1983; Miller and Borden 1984; Yukawa and Ohsaki 1988b; Harris and Rose 1989; Simbolon and Yukawa 1992; Kolesik 1993; Pivnick and Labbé 1993). Because ovarian eggs are normally mature at the time of emergence, we can easily examine the number of eggs by dissecting the female abdomen. Usually females of most species start ovipositing soon after mating but females of *Sitodiplosis mosellana* (Géhin), for example, do not start oviposition in the first night after mating, and the greatest mean daily fecundity occurs on the third night (Pivnick and Labbé 1993).

Only a few examples of realized fecundity have been recorded: 93% of 254 ovarian eggs were laid by *C. oregonensis* (Miller and Borden 1984), and 189 of 248, on average, by *P. neolitseae* (Yukawa et al. 1976). In addition, the presence of an oviposition deterrent has been suggested for *C. oregonensis* (Miller and Borden 1984).

7.10 Oviposition

Eggs are laid either outside or inside the tender plant tissue depending on the length and morphological characteristics of ovipositor. The outside plant tissue means not only the open surface of plant organs but also inside folded leaves or chinks between plant organs where eggs can be hidden. The gall midges with an aciculate ovipositor, such as species of *Asphondylia*, *Pseudasphondylia* and *Schizomyia*, lay their eggs into plant tissue. Some eggs are laid gregariously and others individually.

Table 7.1 Cecidomyiid species of which female sex pheromones have been found or identified

Species[a]	Host plants (Family)[b]	Reference (see Hall et al. 2012 for the details of literature cited here)
*Aphidoletes aphidimyza**		Choi et al. (2004)
*Contarinia nasturtii**	*Brassica* and other genera (Brassicaceae)	Hillbur et al. (2005), Boddum et al. (2009)
*Contarinia oregonensis**	*Pseudothuga menziesii* (Pinaceae)	Miller and Borden (1984), Gries et al. (2002)
*Contarinia pisi**	*Pisum* spp. (Fabaceae)	Wall et al. (1985), Hillbur et al. (1999, 2000)
*Contarinia pyrivora**	*Pyrus communis* (Rosaceae)	Amarawardana (2009)
*Dasineura gleditchiae**	*Gleditsia triacanthos* (Fabaceae)	Molner et al. (2009)
*Dasineura mali**	*Pyrus malus, P. sylvestris* (Rosaceae)	Harris et al. (1996), Heath et al. (1998, 2005), Cross and Hall (2007, 2009)
Dasineura napi	*Brassica napi, B. oleracea* (Brassicaceae)	Williams and Martin (1986)
*Dasineura plicatrix**	*Rubus caesius, Rubus* spp. (Rosaceae)	Hall et al. unpublished
*Dasineura pyri**	*Pyrus communis* (Rosaceae)	Amarawardana (2009)
*Dasineura tetensi**	*Ribes nigrum* (Saxifragaceae)	Garthwaite et al. (1986), Amarawardana (2009)
*Mayetiola destructor**	*Triticum, Hordeum* and other genera (Poaceae)	McKay and Hatchett (1984), Foster et al. (1991), Anderson et al. (2009)
*Mayetiola thujae**	*Thuja plicata* (Cupressaceae)	Gries et al. (2005)
*Orseolia oryzae**	*Oryza sativa, O. perrenis* (Poaceae)	Sain and Kalode (1985), Zao (1982)
*Resseliella theobaldi**	*Rubus* sp., *R. idaeus* (Rosaceae)	Hall et al. (2009)
*Rhopalomyia longicauda**	*Chrysanthemum* spp. (Asteraceae)	Liu et al. (2009)
*Sitodiplosis mosellana**	*Triticum* spp. (Poaceae)	Pivnick (1993), Gries et al. (2000)
Stenodiplosia sorghicola	*Sorghum bicolor, Sorghum* spp. (Poaceae)	Sharma and Vidyasagar (1992)
Thecodiplosis japonensis	*Pinus densiflora, P. thunbergii* (Pinaceae)	Lee and Lee (1985)

Mainly based on Hall et al. (2012)
[a]Asterisks indicate species of which female sex pheromones have been identified
[b]Based mainly on Gagné and Jaschhof (2017)

According to Kim et al. (2015), the females of *M. selenis* were observed ovipositing between 1:00 pm and 4:00 pm on half or fully opened new leaves of *Q. serrata* (Fig. 7.3c). They required about 30 s to lay one egg. One female laid four eggs successively within 120 s. It is remarkable that females of *M. selenis* lay their eggs both on the upper and under surfaces of the fresh leaves (Fig. 7.3d, e). This indicates that females may have no ability to distinguish chemical and physical signals from different surfaces of the leaf blade.

In contrast, oviposition site selection is clear for other leaf gall-inducing cecidomyiids. For examples, females of *Daphnephila machilicola* Yukawa, *M. pustulae*, *Lasioptera camelliae* Ohno & Yukawa, *H. faggalli* and its sibling species can discriminate upper or lower leaf surfaces at the time of oviposition (e.g. Yukawa 1974; Yukawa and Sunose 1976; Ohno and Yukawa 1984; Mishima et al. 2014). However, females of *M. pustulae* lay their eggs not only on fresh leaves of *Euonymus japonicus* Thunberg (Celastraceae) but also on 1- or 2-year old host leaves or on leaves of various understory plants, but galls have never developed on these leaves (Yukawa and Sunose 1976) (see Sect. 2.2, Mistaken oviposition).

In the sorghum midge *Stenodiplosis sorghicola* (Coquilett), ovipositional time was significantly greater for females on panicle of a midge-resistant than a midge-susceptible sorghum hybrid, indicating that, at the same gall midge density and time interval, fewer eggs are deposited in flowers of the resistant than the susceptible hybrid (Waquil et al. 1986).

In the case of *P. neolitseae* reported in Yukawa et al. (1976), the mated females remained on the under surface of the leaves for 1 h or so after mating, then, flew off between 10:30 am and 11:30 am to find suitable host buds for oviposition (22–38 mm in length). At 3:00 pm, the females reached a peak of oviposition. Once they found suitable buds on a host plant, they usually stayed at the plant throughout their oviposition period, which ended between 8 pm and 10 pm. In the census field, the females that have found suitable buds were usually fewer in number than those that had left the leaves after mating. The females moving in and out of the census filed, as well as those caught by predators such as spiders or empidid flies (Diptera: Empididae), are involved in a total dispersal loss during the course of the search for the buds.

7.11 Larval Behavior

Larvae that hatched from eggs outside plant tissue move to appropriate feeding sites. Larvae of some species penetrate the plant tissue without the help of a fungus. The first instars of *M. pustulae*, for example, enter the leaf blade of *E. japonicus* (Yukawa and Sunose 1976), and those of *D. machilicola* sink into the plant tissue of *Machilus thunbergii* Siebold & Zuccarini (Lauraceae) just beneath the eggshell (Yukawa 1974). A similar behavior is also evident in *Rabdophaga degeerii* (Bremi) on *Salix*. In contrast, larvae of *Lasioptera arundinis* (Schiner) penetrate the plant tissue of *Phragmites australis* (Cavanilles) Trinius ex Steudel (Poaceae) in association with a fungus (Skuhravá and Skuhravý 1992; Rohfritsch 1997). Such an association is common in ambrosia gall midges such as *Lasioptera* species.

All hatched larvae of *M. selenis*, even those derived from eggs on the under surface, crawl to the upper surface of the leaf margin (Fig. 7.3f), where they settle down (Kim et al. 2015). This means that hatched larvae can recognize appropriate places for settling down and gall induction (Fig. 7.3g). Crawling seems to be more dangerous for larvae from the under surface than those from the upper surface. Nevertheless, females did not develop safer ways of oviposition to benefit their progeny. Otherwise, laying eggs on both surfaces may contribute to spreading risks during the egg stage because various mortality factors, such as predators or rainfall, would operate differently on eggs on different surfaces.

Larvae of type IA species are more active than those of types IB, IIA and IIB. After they exit their galls, they drop to the ground and crawl their way into the soil. Larvae of *Contarinia* species spin cocoons in which they overwinter and pupate. In particular, larvae of *Contarinia* and *Thecodiplosis* can jump. Mature larvae of the Japanese cedar gall midge *Contarinia inouyei* Mani moved 62.8 ± 21.0 mm (mean ± s.d.) per random jump under uniform lighting conditions while moved 81.1 ± 42.5 mm (mean ± s.d.) per jump toward light source under unidirectional lighting conditions (Tokuhisa et al. 1979). This behavior appears to be an adaptation for exiting galls and finding a suitable site on the ground to construct a cocoon (Tokuhisa et al. 1979). In contrast, larvae of types IB, IIA and IIB are usually inactive and do not jump, probably because they do not exit their galls and do not search for cocooning sites on the ground.

References

Bai L, Yang D, Compton SG (2008) A gall midge inhabiting the figs of *Ficus benjamina* in Xishuangbanna, South-Western China. Symbiosis 45:149–152

Barnes HF (1930) On some factors governing the emergence of gall midges (Cecidomyiidae: Diptera). Proc Zool Soc London 2:381–393

Barnes HF (1931) The sex ratio at the time of emergence and the occurrence of unisexual families in the gall midges (Diptera, Cecidomyiidae). J Genet 24:225–234

Basedow T, Schütte F (1982) Die Populationsdynamik der Weizengallmüken *Contarinia tritici* (Kirby) und *Sitodiplosis mosellana* (Géhin) (Dipt., Cecidomyiidae) in zwei norddeutschen Weizenbaugebieten von 1969 bis 1976. *Zoologische Jahrbücher*. Abteilung für Systematic, Ökologie und Geographie der Tiere 109:33–82

Cloudsley-Thompson L (1955) The design of entomological aktograph apparatus. Entomologiste 88:153–161

Cloudsley-Thompson JL (1961) Rhythmic activity in animal physiology and behavior. Academic Press, New York and London

Dorchin N, Fredberg A (2004) Sex ratio in relation to season and host plant quality in a monogenous stem-galling midge (Diptera: Cecidomyiidae). Ecol Entomol 29:677–684

Dorchin N, Clarkin CE, Scott ER, Luongo MP, Abrahamson WG (2007) Taxonomy, life history, and population sex ratios of north American *Dasineura* (Diptera: Cecidomyiidae) on goldenrods (Asteraceae). Ann Entomol Soc Am 100:539–548

Fyodorova MV, Lapshin DN (1990) Responses of moths to ultrasound. In: Gribakin FG, Wiese K, Popov AV (eds) Sens Syst Commun Arthrop. Birkhauser Verlag, Basel, pp 178–181

Gallum RL, Hatchett JH (1969) Genetic evidence of elimination of chromosomes in the hessian fly. Ann Entomol Soc Am 63:1095–1101

Giron D, Huguet E, Stone GN, Body M (2016) Insect-induced effects on plants and possible effectors used by galling and leaf-mining insects to manipulate their host-plant. J Insect Physiol 84:70–89

Hall DR, Amarawardana L, Cross JV, Francke W, Boddum T, Hillbur Y (2012) The chemical ecology of cecidomyiid midges (Diptera: Cecidomyiidae). J Chem Ecol 38:2–22

Hao YN, Miao J, Wu YQ, Gong ZJ, Jiang YL, Duan Y, Li T, Cheng WN, Cui JX (2013) Flight performance of the orange wheat blossom midge (Diptera: Cecidomyiidae). J Econ Entomol 106:2043–2047

Harris MO, Foster SP (1999) Gall midges. In: Hardie J, Minks AK (eds) Pheromones of non-lepidopteran insects associated with agricultural plants. CABI Publishing, pp 27–49

Harris MO, Rose S (1989) Temporal changes in the egglaying behavior of the hessian fly, *Mayetiola destructor*. Entomol Exp Appl 53:17–29

Harris MO, Galanihe LD, Sandanayake M (1999) Adult emergence and reproductive behavior of the leafcuring midge *Dasineura mali* (Diptera: Cecidomyiidae). Ann Entomol Soc Am 92:748–757

Kanmiya K (1994) Flight analysis system using strain-gauge transducers to estimate power and pattern of flying insects. Jpn J Appl Entomol Zool 38:261–273. (In Japanese with English abstract)

Kanmiya K, Yukawa J (2020) Observation on swarming and mating behavior of *Campylomyza flavipes* Meigen (Diptera: Cecidomyiidae: Micromyiinae). Acta Dipterol 31:4–9

Kim W, Yukawa J, Harris KM, Minami T, Matsuo K, Skrzypczyńska M (2014) Description, host range and distribution of a new *Macrodiplosis* species (Diptera: Cecidomyiidae) that induces leaf-margin fold galls on deciduous *Quercus* (Fagaceae) with comparative notes on Palaearctic congeners. Zootaxa 3821:222–238

Kim W, Matsunaga K, Gyoutoku N, Matsuo K, Minami T, Yukawa J (2015) Life history strategy and adult and larval behavior of *Macrodiplosis selenis* (Diptera: Cecidomyiidae), a species that induces leaf-margin fold galls on deciduous *Quercus* (Fagaceae). Entomol Sci 18:470–478

Kolesik P (1993) Basic bionomics of the lentil gall midge (*Contarinia lentis* Aczel) (Dipt., Cecidomyiidae). J Appl Entomol 116:371–380

Laboulbéne A (1873) Métamorphoses de la cécidomyie du buis *Cecidomyia* (*Diplosis*) *buxi*. Ann Soc entomol Fr 3:313–326. pl. 9

Miao BG, Yang DR, Liu C et al (2011) The impact of a gall midge on the reproductive success of *Ficus benjamina*, a potentially invasive fig tree. Biol Control 59:228–233

Miller TA (1979) Insect neurophysiological techniques. Springer-Verlag, New York, Heiderberg, Berlin

Miller GE, Borden JH (1984) Reproductive behaviour of the Douglas-fir cone gall midge, *Contarinia oregonensis* (Diptera: Cecidomyiidae). Can Entomol 116:607–618

Mishima M, Sato S, Tsuda K, Yukawa J (2014) Sexual isolation between two known intraspecific populations of *Hartigiola* (Diptera: Cecidomyiidae) that induce leaf galls on upper and lower surfaces of *Fagus crenata* (Fagales: Fagaceae), indicating possible diversification into sibling species. Ann Entomol Soc Am 107:789–798

Ohno K, Yukawa J (1984) Description of a new gall midge (Diptera: Cecidomyiidae) causing leaf galls on *Camellia japonica* L., with notes on its bionomics. Kontyû 53:427–434

Pivnick KA, Labbé E (1993) Daily patterns of activity of females of the orange wheat blossom midge, *Sitodiplosis mosellana* (Géhin) (Diptera: Cecidomyiidae). Can Entomol 125:725–736

Pringle JWS (1957) Insect flight. Cambridge University Press, Cambridge, Massachusetts

Ribak G, Barkan S, Soroker V (2017) The aerodynamics of flight in an insect flight-mill. PLoS One 12. https://doi.org/10.1371/jounal.pone.0186441

Rohfritsch O (1997) Morphological and behavioral adaptations of the gall midge *Lasioptera arundinis* (Schiner) (Diptera, Cecidomyiidae) to collect and transport conidia of its fungal symbiont. Tijdschr Entomol 140:59–66

Roskam JC, Nadel H (1990) Redescription and immature stages of *Ficiomyia perarticulata* (Diptera: Cecidomyiidae), a gall midge inhabiting syconia of *Ficus citrifolia*. Proc Entomol Soc Wash 92:778–792
Simbolon H, Yukawa J (1992) Description of a new gall midge (Diptera, Cecidomyiidae) causing galls on *Lithocarpus edulis* (Fagaceae) in Japan. Jpn J Entomol 60:609–617
Skuhravá M, Skuhravý V (1992) Biology of gall midges on common reed in Czechoslovakia. In: Shorthouse JD, Rohfritsch O (eds) Biology of insect-induced galls. Oxford University Press, New York, pp 196–207
Skuhravá M, Skuhravý V, Dengler K (1998) Gall inducing and other gall midge species (Diptera: Cecidomyiidae) associated with oaks (*Quercus* spp.) (Fagaceae) in the Palaearctic region. In: Csóka G, Mattson W, Stone GN, Price PW (eds) The biology of gall-inducing arthropods, pp 1–11. USDA Forest Service, north central Research Station, general technical report NC-199
Skuhravý V, Skuhravá M, Brewer JW (1993) The saddle gall midge *Haplodiplosis marginata* (Diptera: Cecidomyiidae) in Czech Republic and Slovak Republic from 1971–1989. Acta Soc Zool Bohem 57:117–137
Summers CG (1975) Daily adult emergence in the sorghum midge, *Contarinia sorghicola*. Environ Entomol 4:495–498
Sunose T (1983) Studies on the population dynamics of the *Euonymus* gall midge *Masakimyia pustulae* Yukawa et Sunose (Diptera, Cecidomyiidae) under different environmental conditions. Kontyû 51:29–40
Tokuhisa E, Nagai S, Yukawa J (1979) Jumping behaviour of the larvae of the Japanese cedar gall midge, *Contarinia inouyei* Mani (Diptera: Cecidomyiidae). Kontyû 47:599–605. (In Japanese with English summary)
Uechi N, Kim W, Tokuda M, Fujii T, Kikuchi H, Kakizaki M, Iwasaki A, Paik JC, Yukawa J (2018) Genetic and ecological differences between *Asphondylia yushimai* and the ivy gall midge, *Asphondylia* sp. (Diptera: Cecidomyiidae) with a new distribution record of the former from Hokkaido and South Korea. Appl Entomol Zool 53:363–371
Waquil JM, Teetes GL, Peterson GC (1986) Sorghum midge (Diptera: Cecidomyiidae) adult ovipositional behavior on resistant and susceptible sorghum hybrids. J Econ Entomol 79:530–532
Wicker-Thomas C (2007) Pheromonal communication involved in courtship behavior in Diptera. J Insect Physiol 53:1089–1100
Yafuso M, Adaniya S, Yukawa J (2013) Host plant manipulation by fig gall midges (Diptera: Cecidomyiidae) that induce emergence projections on flower galls in the syconia of *Ficus microcarpa* (Moraceae). Entomol Sci 16:413–420
Yukawa J (1974) Descriptions of new Japanese gall midges (Diptera, Cecidomyiidae, Asphondyliidi) causing leaf galls on Lauraceae. Kontyû 42:293–304
Yukawa J (1983) Arthropod community centred upon the neolitsea leaf gall midge, *Pseudasphondylia neolitseae* Yukawa (Diptera, Cecidomyiidae) and its host plant, *Neolitsea sericea* (Blume) Koidz. (Lauraceae). Mem Fac Agric Kagoshima Univ 19:89–96
Yukawa J (1987) Life history strategies of univoltine gall-making Cecidomyiidae (Diptera) in Japan. Phytophaga 1:121–139
Yukawa J, Miyamoto K (1979) Redescription of *Asphondylia sphaera* Monzen (Diptera, Cecidomyiidae), with notes on its bionomics. Mem Fac Agric Kagoshima Univ 15:99–106
Yukawa J, Ohsaki N (1988a) Separation of the aucuba fruit midge, *Asphondylia aucubae* sp. nov. from the ampelopsis fruit midge, *Asphondylia baca* Monzen (Diptera, Cecidomyiidae). Kontyû 56:365–376
Yukawa J, Ohsaki N (1988b) Adult behaviour of the aucuba fruit midge, *Asphondylia aucubae* Yukawa & Ohsaki (Diptera: Cecidomyiidae). Kontyû 56:645–652
Yukawa J, Sunose T (1976) Description of a new gall midge (Diptera, Cecidomyiidae) on *Euonymus*, with notes on its bionomics. Kontyû 44:159–168
Yukawa J, Tsuda K (1987) A new gall midge (Diptera: Cecidomyiidae) causing conical leaf galls on *Celtis* (Ulmaceae) in Japan. Kontyû 55:123–131

Yukawa J, Takahashi K, Ohsaki N (1976) Population behaviour of the neolitsea leaf gall midge, *Pseudasphondylia neolitseae* Yukawa (Diptera, Cecidomyiidae). Kontyû 44:358–365

Yukawa J, Ohtani T, Yazawa Y (1983) Host-change experiments from wild plants to soybean in *Asphondylia* species (Diptera: Cecidomyiidae). Proc Assoc Plant Prot Kyushu 29:115–117. (In Japanese with English summary)

Yukawa J, Uechi N, Horikiri M, Tuda M (2003) Description of the soybean pod gall midge, *Asphondylia yushimai* sp. n. (Diptera: Cecidomyiidae), a major pest of soybean and findings of host alternation. Bull Entomol Res 93:73–86

Yukawa J, Abe J, Mizota K (2008) Improvement in the practical use of an aphidophagous gall midge, *Aphidoletes aphidimyza* (Diptera: Cecidomyiidae), in greenhouses. In: Mason PG, Gillespie DR, Vincent C (eds) Proceedings of ISBCA, vol 3, pp 77–78

Yukawa J, Sato S, Xu HL, Tokuda M (2011) Description of a new species of the genus *Resseliella* (Diptera: Cecidomyiidae), a pollinator of *Kadsura longipedunculata* (Schisandraceae) in China, with comments on its flower-visiting habit. Entomol Sci 14:297–303

Yukawa J, Nakagawa K, Saigou T, Awa T, Fukuda T, Higashi M (2013) Adult behavior of an ambrosia gall midge *Illiciomyia yukawai* (Diptera: Cecidomyiidae) and synchronization between its emergence and host plant phenology. Entomol Sci 16:400–412

Yukawa J, Ichinose M, Wanggyu K, Uechi N, Gyotoku N, Fujii T (2016) Lower development threshold temperatures and thermal constants for four species of *Asphondylia* (Diptera: Cecidomyiidae) in Japan and their larval developmental delay caused by heat stress. Appl Entomol Zool 57:71–80

Yukawa J, Moriya T, Kanmiya K (2019) Comparison in the flight ability between the soybean pod gall midge, *Asphondylia yushimai* and the aucuba fruit gall midge, *A. aucubae* (Diptera: Cecidomyiidae). Appl Entomol Zool 54:167–174

Chapter 8
Population Study

Junichi Yukawa

Abstract Population studies are essential to identify density-dependent and density-independent effects on population dynamics and to evaluate the relative contribution of bottom-up and top-down effects to population regulation. For these purposes, if possible, long-term life table data are needed. In the first section of this chapter, methods for field survey are explained by taking advantage of outstanding features and immobility of galls. The second section refers to the spatial and chronological distribution patterns of galls expressed mainly with relationships between mean gall density and mean crowding. In the third section, two examples of population dynamics based on long-term life table data are demonstrated. Biotic and abiotic effects on various gall midge populations are also mentioned by citing short-term field data.

Keywords Field survey · Key-factor · Life table · Mean crowding · Mortality factor · Population dynamics · Spatial distribution

8.1 Introduction

Galls provide life table data in a convenient way (e.g. Redfern and Cameron 1978). Galls can be easily counted continuously from early to final stage of the gall and the galler's development because of their outstanding features and immobility (e.g. Yukawa 2000, 2019). Therefore, population dynamics of gall-inducing cecidomyiids has been relatively intensively studied (e.g. Redfern and Cameron 1978, 1993; Keller and Schweizer 1994; Redfern and Hunter 2005; see also Table 10.1 in Price 2003 for other insect populations). In recent years, the method of key-factor/key-stage analysis (Yamamura 1999) has been used (Yamamura 2012; Yukawa et al. 2016a), instead of the conventional key-factor analysis (e.g. Morris

J. Yukawa (✉)
Entomological Laboratory, Faculty of Agriculture, Kyushu University, Fukuoka, Japan
e-mail: galling-cecidomyiids@kyudai.jp

J. Yukawa, M. Tokuda (eds.), *Biology of Gall Midges*, Entomology Monographs,
https://doi.org/10.1007/978-981-33-6534-6_8

1959; Varley and Gradwell 1960, 1968), to detect density-dependent and independent forces operating on gall midge populations at different developmental stages and to assess the relative strength of top-down and bottom-up effects.

The first section of this chapter explains methods of practical field survey of gall-inducing insects and analytical methods of field data, particularly for application of life table data to population studies. These methods have been practiced mainly for univoltine cecidomyiid species that induce leaf or fruit galls on evergreen broad-leaved trees (Yukawa and Akimoto 2006; Yukawa et al. 2006, 2013, 2016a, b, 2018) because the larvae do not leave the galls throughout their life and the adults emerge directly from the galls on the host plants (see Sect. 6.2 for type IIA or IIB life history strategy). In addition, we need to note that trees, if not too tall, and shrubs are more appropriate materials for field surveys than herbs that are frequently disturbed by artificial cutting.

Different insect species exhibit different spatial and chronological distribution patterns. Therefore, to grasp the distribution patterns of target insects is essential in population ecology. Using mean crowding (m^*) and mean density (m), Iwao (1968, 1970, 1977) proposed methods to analyze the spatial distribution pattern of animal populations, changes in the degree of contagiousness (m^*/m) in the action of mortality factors, and the degree of spatial and geographical association between two sympatric populations. In the second section, these methods are explained and examples of actual data are demonstrated.

The third section provides two known examples of long-term field surveys for population dynamics of gall-inducing cecidomyiids, and refers to biotic and abiotic effects on gall midge populations by citing short-term field data demonstrated for various gall midge species (Sunose and Yukawa 1979; Sunose 1983; Ehler 1985, 1986; Roskam 1986; Soné 1987; Manzoli-Palma and da Teles 1990; Axelsen 1994; Lampo 1994).

8.2 Methods of Field Survey for Population Study

8.2.1 Field Survey for Gall Midges with Type IIA or IIB Life History Strategy

8.2.1.1 Adult Stage

First, readers are requested to refer to Chap. 7 for the adult behavior of gall midges. Prior to field surveys, researchers need to gather information on adult daily activities, such as emergence, mating, swarming, oviposition and longevity, together with data of sex ratio and the number of ovarian eggs (Yukawa et al. 1976, 2013). Information on these behavioral traits can be obtained mostly by direct observation in the field and partly by laboratory experiments. For direct observation, researchers need to select 10–20 census host trees in the census field to obtain data sufficient for statistic analysis.

Hourly emergence pattern can be determined for diurnal gall midges by direct observation in the field and for nocturnal species by rearing experiments under natural lighting conditions in outdoor cages. Field surveys of daily emergence pattern are essential in referring to synchronization between gall midge emergence and host plant phenology. Because the daily emergence patterns fit well to the gamma distribution pattern (Yukawa and Akimoto 2006), the number of adults emerged can be surveyed at intervals of 3–4 days, instead of everyday observation, to determine the daily emergence pattern. The emergence of gall midges is distinguishable from that of parasitoids by the diameter of exit holes that are usually larger in gall midges and frequently retain their pupal exuviae, which should be taken out of the galls and kept in 75% ethanol for later morphological observation. Precise morphological observation of pupal exuviae sometimes gives us information to distinguish males and females, with which daily sex ratio can be determined. All galls with exit holes are marked with a felt pen to distinguish them from future emergence holes of gall midges and parasitoids.

Swarming and mating should be observed in the field. The number of females that were missed by swarming males can be evaluated. Because the observable number of females that are waiting for males on the host leaves or fruit is limited for one person in a day, several co-observers are needed to obtain ample data of the rate of females mated. If possible, researchers need to record the daily number of females that visit the host organs for oviposition. The daily number can be used to estimate the rate of females that survive factors operating on the adult stage, such as predation by spiders and dispersal loss, although identification of such factors is rather difficult.

Flight ability can be measured using the thread tethered flight method by modifying devices with a CCD camera and an apparatus for analyzing tethered flight sounds to measure the flight power of tiny gall midges (see Figs. 2 & 3 in Yukawa et al. 2019b for the devices and apparatus). Estimation of flight ability is needed to evaluate dispersal of gall midges; particularly those that exhibit host plant alternation.

Sometimes the sex ratio is female-biased and daily emergence patterns are different between sexes. Therefore, at least 50 individuals are needed to determine sex ratio. Because ovarian eggs are matured at the time of emergence, the eggs are countable under a binocular microscope by dissecting female abdomens with fine needles. Yukawa et al. (1976) recognized 248 ovarian eggs on average, and then they found 59 eggs on average left in the ovaries of females of *Pseudasphondylia neolitseae* Yukawa. The females fastened on the host buds in the evening and stopped ovipositing after fulfilling their duty. Thus, they calculated that 189 (248–59) of initial ovarian eggs were realized during oviposition on average per female. The number of adults that emerged, their sex ratio and the average number of ovarian eggs can be used to estimate the initial number of eggs of the following generation (E_n) by the equation described below:

$$E_n = A_{n-1} \times S \times O$$

$A_{n\text{-}1}$: The number of adults emerged from the census trees in the preceding generation.
S: Sex ratio (the ratio of females among adults emerged).
O: The number of ovarian eggs before oviposition. If possible, the number of realized eggs after oviposition is more appropriate.

8.2.1.2 Immature Stages

Usually life tables start from the egg stage of the first generation. Eggs can be directly counted when eggs are laid on the host leaf surface or scars of the oviposited position can be recognized after larval hatching. For example, such scars are made on the under surface of *Machilus* leaves by larvae of *Daphnephila* that hatched and sank into the leaf tissue (Fig. 9.2b). When eggs are laid inside plant tissue, the number of eggs of the first generation can be estimated using the aforementioned equation.

Immediately after galls become conspicuous, researchers need to record the number of early-stage galls containing first instars and the number of galled and ungalled host leaves to evaluate gall density per host organ. Then, monitoring of galls that surviving various biological and non-biological mortality factors is needed on the way to the maturity of galls. During this period, galls that could not mature because of larval death in the galls by unknown factors can be recognized. Later in the season (in October–November or in the following March in the case of type IIA and IIB univoltine gall midges), mature galls remaining on the host plants are also counted. During the growth of galls, researchers should collect some galls from the sampling trees occasionally (once or twice) and dissect them under a binocular microscope to know: (1) the developmental stadium of larvae, (2) presence or absence of ectoparasitoid larvae and their developmental stadium, and (3) presence or absence of inquilines and their developmental stadium, if any. About 2 weeks before emergence, mature galls are dissected again to record the aforementioned items. At this time, endoparasitoids, if any, pupate in the host larval skin. At every time of gall sampling, researchers should record the number of host organs remaining on the census trees. Thus, the number of larvae and pupae that survived age specific mortality factors can be recorded.

In the survey of fruit gall midges, special attention should be paid to the following two facts as shown in Yukawa et al. (2016a): (1) The first instars inhabiting flower buds or flowers drop to the ground and die because of flower abortion, which is tremendous in rate. However, the abortion of flowers is a normal event and is not caused by larval infestation. (2) Frequently galled fruit cannot be distinguished from normal fruit in appearance until late autumn. Therefore, occasional dissection of flower buds, flowers and young fruit is more important than direct field observation.

8.2.2 *Field Survey for Gall Midges with Type IA or IB Life History Strategy*

There are considerable difficulties in the life table study of gall midges with type IA life history strategy because mature larvae quit the galls and drop to the ground early in the season, and adults emerge from the ground in the following spring. Before maturity, usually many larvae gregariously live together in a chamber of galled organ. Density of a gall midge can be expressed as the rate of galled organs among all those targeted for oviposition during the season. For this purpose, the average number of first instars per galled organ should be investigated beforehand. Later, the average number can be compared with that of third instars to evaluate mortality during the period from first to third stadium larvae. Occasional dissection of galls is needed to find mortality factors operating in the galls. The life table data during the period from mature larvae to the pupal stage on the ground are hardly obtained.

Life tables covering all developmental stages from egg to adult are ideal for analysis, but partial life tables covering a part of the life history (only larval stage, for example) are available for analysis when there are some difficulties in the field survey. The partial life table data during the larval stage tell us about density-dependent and independent forces operating on the gall midge larvae at different developmental stadia. It is much better than nothing.

In contrast to the gall midges with type IA life history strategy, the life table study of gall midges with type IB strategy is relatively easy because mature larvae do not exit from galls, but galls or gall-bearing organs containing mature larvae drop to the ground and adults emerge from the galls on the ground (see Sect. 6.2). In addition, each gall usually contains one gall midge larva. Therefore, before they drop to the ground, life table data can be obtained from the galls on the host plants with the methods similar to those employed in the field survey for gall midges with type IIA or IIB life history strategy. Thereafter, if necessary, galls or gall-bearing organs can be collected from the ground for dissection to confirm the survival of larvae or pupae in the galls.

In the case of *Hartigiola faggalli* (Monzen) that induces leaf galls on *Fagus crenata* Blume (Fagaceae), for example, Mishima et al. (2014) randomly collected fallen leaves with galls from the forest floor under the host trees several days before the emergence season commenced. The leaves were put into dark cardboard boxes (H 25 cm, W 30 cm, D 45 cm), which were placed on the forest floor and covered with cheesecloth to exclude direct sunshine. A transparent plastic tube (2 cm × 15 cm) was inserted in the topside of the box and the top end of the tube was covered with a mesh bag. Adults were attracted towards the light and caught in the mesh bag. Thus, the daily or hourly number of adults emerged from the ground could be recorded during the emergence season. Mishima et al. (2014) also caught many swarming males by sweeping with an insect net over fallen leaves on the forest floor (about 30–60 cm high) and near lower branches of *F. crenata* with fresh leaves (about 2.5 m high). The adults collected were examined to determine sex ratio. Mating and oviposition were observed directly in the field.

8.3 Spatial and Chronological Distribution Patterns

8.3.1 *Mean Density and Mean Crowding*

Iwao (1968) proposed a new method to analyze the spatial distribution pattern of animal populations using the regression of mean crowding (m^*) (a parameter proposed by Lloyd 1967) on mean density (m) in a set of populations sampled. Iwao (1970) further demonstrated a method to analyze changes in the degree of contagiousness (mean crowding/mean density) in the action of mortality factors. In order to examine the degree of spatial and geographical association between two sympatric populations, Iwao (1977) proposed another equation and index.

Iwao (1968) summarized his method as follows: if mean crowding is plotted against mean density, the relation can be fitted to a linear regression in a wide variety of situations. The intercept of regression, α, would be called the "Index of Basic Contagion", which indicates that infinitesimal density an individual would be expected to live together with α other individuals in the same quadrat. Namely, α reflects the contagiousness inherent to the species, such as aggregation tendency stemed from mass deposition of eggs or from mutual attraction of individuals. The slope β may be termed the "Density-Contagiousness Coefficient", because it indicates the manner in which individuals or groups of individuals distribute themselves in their habitat with change in the mean density.

Briefly speaking, the spatial distribution pattern is determined as follows: even distribution when $-1 < \beta < 0$, random distribution when $\beta = 0$, and contagious distribution when $0 < \beta < 1$. See Iwao (1968) for the detailed relations of α and β to some important distribution models. Theoretically, both α and β can not be less than zero, but in some cases the negative value of α between 0 and -1 would be obtained by extrapolation of the regression, as seen in uniform distributions. Such a "negative" contagion would be expected when some repulsive interaction occurs between individuals. The analyses of insect counts verify the usefulness of α and β as the measures of dispersion pattern characteristic of the species. It is noteworthy that the two examples where the spatial distributions change from underdispersion to overdispersion and vice versa with increasing density also are fitted to single regression lines with $-1 < \alpha < 1$ and $\beta > 1$ and with $\alpha > 1$ and $\beta < 1$, respectively. Thus the aggregation pattern of animal populations involves two different aspects as indicated by α and β, which are independent from each other.

8.3.2 *Spatial Distribution Pattern of Galls of Various Gall Midge Species*

The spatial distribution pattern of galls can be examined using the following equations (Iwao 1968):

$$m* = \left(\sum x_j^2 / \sum x_j\right) - l \quad (8.1)$$

where $m*$ is mean crowding and x_j is the number of galls on leaf j.

On the basis of multiple data on mean and mean crowding, a regression line can be obtained as follows:

$$m* = \alpha + \beta m \quad (8.2)$$

where m is mean density, α is the index of basic contagion and β is density-contagiousness coefficient, both of which are consistent for each species (Iwao 1968, 1977).

Equation (8.2) can be transformed as follows:

$$m = (m* - \alpha)/\beta \quad (8.3)$$

If the correlation between the mean gall density and mean crowding expressed with the Eq. (8.2) is significantly supported, mean density (m) can be estimated from Eq. (8.3) without data for the number of ungalled leaves because α and β are fixed numbers, and mean crowding is calculated based only on the number of galls on gall-bearing leaves (Yukawa 1995).

Based on the field data, Kim et al. (2015) calculated mean gall density per leaf (m) and mean crowding ($m*$) of the leaf-margin folded galls induced by *Macrodiplosis selenis* Kim & Yukawa on *Quercus serrata* Thunberg ex Murray (Fagaceae) and obtained the results as follows: the mean gall density of *M. selenis* per leaf ranged from 0.002 to 0.087, varying with trees and localities. The degree of contagiousness (mean crowding/mean density) varied with trees and localities, ranging from 20.842 to 266.800. Thus, the spatial distribution pattern of galls was aggregated. Relationship between mean gall density per leaf and mean crowding was calculated as $m* = 57.56\ m + 0.449$ ($R^2 = 0.728$, $P = 0.003$).

Yukawa (1995) attempted similar calculations for *P. neolitseae* on *Neolitsea sericea* (Blume) Koidzumi (Lauraceae), *Contarinia* sp. on *Viburnum japonicum* (Thunberg ex Murray) Sprengel (Adoxaceae), *Daphnephila* sp. on *Machilus japonica* Siebold & Zuccarini ex Blume (Lauraceae), and *Lasioptera camelliae* Ohno & Yukawa on *Camellia japonica* Linnaeus (Theaceae). Correlation between mean gall density per leaf and mean crowding of each species varied from year to year even in the same census field and also with localities surveyed. Among the four species, the correlation was relatively high in 1992 ($r = 0.998$) for *P. neolitseae* and in 1994 ($r = 0.876$) for *Contarinia* sp. but significant relation was not supported when the multiple data in different years were summed. The correlation was relatively low in *Dasineura* sp. and *L. camelliae*.

Frequently, mean insect density per unit significantly correlates well with mean crowding (Iwao 1968). However, examples of such significant correlation are rather uncommon in gall-inducing insects when multiple data are summed. This is possibly because the number of galls is used in the calculation instead of the number of insect individuals. The timing of gall maturity differs with gall-inducing species. There is a more or less distinct time lag between the egg stage and the maturity of galls. For example, the leaf galls mature about one and half month later in *P. neolitseae* and 7 or 8 months later in *Daphnephila* sp. During this period, various mortality factors operate differently on eggs and young larvae of the gall midges. Therefore, characteristics of female egg-laying behavior may not be directly reflected in the spatial distribution pattern of mature galls. However, the spatial distribution pattern of mature galls is important when effects of mortality factors are evaluated in response to gall density and mean crowding.

Because of operation of mortality factors, the degree of contagiousness (mean crowding/mean density) changes from the egg stage to the time of gall maturity. The degree increases on one hand, or decreases on the other hand. This change can be calculated with the following equation (Iwao 1970).

$$\tau = (m^*/m_0^*) \cdot (m_0/m) \tag{8.4}$$

where m_0 and $m_0{}^*$ are initial mean density and initial mean crowding, respectively, and m and m^* are those after mortality factors operated on the population. When the mortality factors operate on the population randomly, $\tau = 1$; density dependently, $\tau < 1$; or inverse density dependently, $\tau > 1$.

Spatial distribution pattern of cecidomyiid leaf galls at the time of maturity is more or less contagious (e.g. Yukawa 1995) although female egg-laying behavior is not directly reflected. The index of basic contagion, "α", varies with species from year to year but a certain tendency can be recognized for each species. For example, the index was very high for *Dasineura* sp., varying from 26.08 in 1992 to 54.38 in 1994, and followed by *P. neolitseae* for which the index varied from 0.50 in 1994 to 10.30 in 1992 (Yukawa 1995). In contrast, the index was extremely low, less than 1.0 for *L. camelliae* and less than 1.0 or frequently less than −0.5 for *Contarinia* sp.

8.3.3 Chronological Distribution Pattern

Synchronization of the ovipositing season with host plant phenology is another important factor in determining the spatial distribution pattern (e.g. Yukawa 2000). The ovipositing season synchronizes well with the host available season in most gall-inducing cecidomyiids (see Sect. 10.2). Contagious distribution pattern in time and space is important for short-lived insects such as adult gall midges to meet with mating partners in the field. They have evolved to increase the rate of successful mating in this manner (see Chap. 7).

In calculating mean gall density per host organ and mean crowding, one important thing that should be taken into consideration is to what extent the host organs were targeted for oviposition. For example, *Pitydiplosis puerariae* Yukawa, Ikenaga & Sato is univoltine in Japan and South Korea and adults emerge in May, but its host plant, *Pueraria lobata* (Willdenow) Ohwi (Fabaceae), continues to produce fresh leaves from April to November (Yukawa et al. 2012). Therefore, many fresh leaves appeared in the seasons other than May were not targeted for oviposition. They should be excluded from the calculation of mean gall density per leaf. Otherwise, the index of basic contagion, "α", becomes tremendously high because of many ungalled leaves. In contrast, in the case of *P. neolitseae*, its host plant, *N. sericea*, extends new shoots, targeted organ for oviposition, within a relatively short period of time. Therefore, all leaves flushed in this period can be included in the calculation of the mean gall density per leaf. The calculation of mean gall density is possible for bi- or trivoltine gall midges if the border between generations is clear, while it is quite difficult for multivoltine species because of overlap of generations.

8.3.4 Degree of Spatial and Geographical Association between Two Populations

In order to examine if the two intra- or inter specific populations exhibit a sympatric distribution, the degree of spatial and geographical association between the two populations can be analyzed using the following equation and index (Iwao 1977):

$$\text{Degree of association}: \gamma = \frac{\sum XxjXyj}{\sqrt{\sum Xxj^2}\sqrt{Xyj^2}} \tag{8.5}$$

Xxj: The number of type A galls on leaf-*j*.
Xyj: The number of type B galls on leaf-*j*.
$\gamma = 1$: The two gall types are completely overlapping.
$\gamma = 0$: The two gall types are completely separated.

Mishima et al. (2014) studied reproductive isolation between intraspecific populations of *H. faggalli* that induce upper and lower type galls, respectively, on upper and lower leaf surfaces of *F. crenata* (see Sect. 2.6). The number of galls was recorded for each gall type on all leaves collected from the 25 localities. Gall density was expressed as the number of galls per 100 leaves. Relative abundance of the two gall-type populations was calculated based on the gall density. Then, the degree of spatial and geographical association between the two gall types was analyzed using the Eq. (8.5). The upper type galls were found in 23 of 25 localities surveyed and lower type galls in 21 localities. The degree of spatial association (γ) was 0.95963 ($n = 25$), indicating that the two gall types overlap in their geographical distribution. The degree of spatial association (γ) was 0.846 ($n = 130$), 0.588 ($n = 31$), and 0.751 ($n = 34$) in Takizawa, Ojiya and Tsunan, respectively, indicating that the two gall types have a sympatric rather than an allopatric distribution.

Because two arthropod species frequently induce different sorts of gall on the same host plant species, the Eq. (8.5) is useful to detect the degree of spatial and geographical association in relation to their competition for resource utilization.

8.4 Population Dynamics

8.4.1 Population Dynamics of Taxomyia taxi

The yew gall midge *Taxomyia taxi* (Inchbald) induces globe artichoke-shaped galls on the shoots of yew *Taxus baccata* Linnaeus (Taxaceae) (Barnes 1951). Some (< 10%) of *T. taxi* individuals are univoltine but most individuals require two years (rarely longer) to complete one generation (Redfern 1975) (see Sect. 6.4). This gall midge has been surveyed for ten years since 1966, and life table analysis indicated that the key-factor was failure to achieve maximum fecundity, although parasitism by *Mesopolobus diffinis* (Walker) (Hymenoptera: Pteromalidae) was equally important for univoltine individuals (Redfern and Cameron 1978). However, regulatory mechanisms could not be firmly established even with ten years of data, mainly because of the coexistence of 1-year and 2-years life cycles (Redfern and Cameron 1978). Thereafter, extension to a 24-years run revealed that mortality in the period between emergence of adult *T. taxi* and larval infestation of buds was density-dependent and accounted for the largest component of overall mortality (Redfern and Cameron 1978, 1993). In the 2-years life cycle, this mortality and that caused by *Torymus nigritarsus* (Walker) (Hymenoptera: Torymidae) were key-factors, and in the 1-year cycles mortality caused by *M. diffinis* was density-dependent and accounted for most within-gall losses. The 24 years field surveys also demonstrated long-term fluctuations in population densities, with corresponding changes in the balance of advantage between 1- and 2-year life cycles of *T. taxi*. Redfern and Hunter (2005) demonstrated, based on further population studies over 35 years in total that (1) 2-years type populations of *T. taxi* influenced the density of *M. diffinis* and, after a delay, that of *T. nigritarsus.* The current density of 2-years type populations depended on the density of its 2-years type parents that had been moderated by the negative effects of *T. nigritarsus* on these parents, (2) parasitism had no effect on the dynamics of univoltine *T. taxa*, and (3) tree growth has little effect on the yew gall midge.

These studies are model examples representing long-term field surveys of insect populations that consist of 1-year and 2-years type individuals. Particularly, the existence of 2-years type individual with a higher percentage than 1-year type would make field surveys relatively difficult and complicated. As shown in Table 6.2, various species with polymodal emergence patterns are awaiting for population studies.

8.4.2 Population Dynamics of Asphondylia sphaera

Yukawa et al. (2016a) studied population dynamics of *Asphondylia sphaera* Monzen (Diptera: Cecidomyiidae), a species that induces fruit galls on *Ligustrum japonicum* Thunberg (Oleaceae), from 1970 to 1996 in evergreen broad-leaved forests in Kagoshima, southern Japan. The numbers of fruit galls and emerging adults fluctuated greatly from year to year along with alternate year flowering of the host plant. To detect density-dependent and independent forces operating on the *A. sphaera* population and to assess the relative strength of top-down and bottom-up effects, they used the method of key-factor/key-stage analysis, which allowed them to avoid various problems (Royama 1996) of the conventional key-factor analysis (e.g. Varley and Gradwell 1960, 1968).

Yamamura (1999) proposed a key-factor/key-stage analysis by integrating the conventional key-factor analyses and ANOVA, emphasizing the importance of discriminating between the key-factor and the key-stage. This analysis can identify the key-factor, the key-stage, and the combination of factor and stage that is most influential in determining the fluctuation of total mortality. The effectiveness of the key-factor/key-stage analysis is demonstrated by using the life table data of *Pieris rapae crucivora* Boisduval (Lepidoptera: Pieridae) (Yamamura 1999). The key-factor/key-stage analysis was further extended to analyze longitudinal data in pharmaceutical experiments (Yamamura 2012).

In the life table data, five factors and seven stages were distinguished (Table 8.1). Table 8.2 shows the definition of stages and the calculation method for the survival rates based on a selected example of the 1978 life table data for *A. sphaera* (Table 8.1). Actual equations are indicated in Yukawa et al. (2016a) and an *'R'* function to perform key-factor/key-stage analyses is available from the following web site: (Yamamura 2015 personal information in Yukawa 2019):

http://cse.niaes.affrc.go.jp/yamamura/Key-factor_analysis_program.html

Key-factor/key-stage analyses indicated that the number of flower buds made the largest contribution to the annual changes of total survival rate by operating through the proportion of eggs and first instars that survived abortion of host flower buds and flowers, which contributed most to density dependency (see Tables 3 & 4 in Yukawa et al. 2016a). The population dynamics of *A. sphaera* has been strongly influenced by the bottom-up effects of both the abundance of flower buds and the abortion of flowers of *L. japonicum* (Fig. 8.1). A top-down effect of *B. asphondyliae* was weakly density-dependent and contributed less to the total survival rate.

Thus, this study emphasizes importance of using the method of key-factor/key-stage analyses for life table data to evaluate density-dependent and independent forces operating on insect populations and to assess the relative strength of top-down and bottom-up effects.

8.4.3 *Biotic Bottom-up Effects*

Various kinds of bottom-up effects of host plants have been known to operate on herbivore populations other than gall-inducing arthropods (see reference citation in Yukawa et al. 2016a). Therefore, many examples of bottom-up effects have been anticipated to operate also on gall-inducing cecidomyiids because they have close

Table 8.1 A selected example of life table for the 1978 generation of *Asphondylia sphaera* (data indicate mean/census tree at Shiroyama)

Developmental stage	Season	Mortality factor	*Lx*	*Dx*	% *Lx*
Eggs and first instars in flower buds and flowers	May		7555.00		100.00
		k1: Abortion of flower buds & flowers		7488.00	
First instars in young fruit galls	July		67.00		0.89
		k2: Galls fed by moth larvae		1.01	
First instars in young fruit galls	September		65.99		0.87
		k3: Galls decayed and dropped by unknown factors		14.06	
First and second instars in developed galls	March		51.93		0.69
		k4: Developmental delay, resulted in the death of larvae		0.33	
Third instars and pupae in fully developed galls	April		51.60		0.68
		k5: Larval death by unknown factors		19.10	
			32.50		0.43
		k6: *Eurytoma* sp.		3.35	
			19.15		0.39
		k7: *Bracon asphondyliae*		12.73	
Adults emerged	May		16.42		0.22
		(Sex ratio: 0.616)			
Females emerged	May		10.11		
Potential number of eggs for the following generation	May	(10.11 × 252 ovarian eggs/female)	2547.7		0.34[a]

Lx the number of living individuals, *Dx* the number of dead individuals caused by each mortality factor, *%Lx* survival rate from the initial number of *Lx* (7555.00)
See also Table 2 in Yukawa et al. 2016a for comparison with the life table in 1970
[a]Reproductive rate in comparison with the number of eggs in the previous generation (2547.7/7555.0)

association with their host plants (e.g. Yukawa et al. 2005). However, there are only a few examples indicating the bottom-up effects on populations of gall-inducing cecidomyiids. For example, the number of females of *P. neolitseae* that synchronized with the shoot extension phenology of its host plant *N. sericea* determined gall density in the following generation (Yukawa and Akimoto 2006). Upon the urban population of *Masakimyia pustulae* Yukawa & Sunose, earlier fall of heavily galled

Table 8.2 Stage specific mortality factors operating on the population of *Asphondylia sphaera* and the calculation of survival rates

The life-table of an imaginary gall midge is divided into four developmental stages. N_t is the potential number of eggs on the host leaf surface at the *t*th year. S_t is the rate of generation change of N_t, that is, N_{t+1}/N_t. Then, S_t can be given by the multiplication of the survival rate (or rate of change) at each developmental stage:
$S_t = s_{1t}s_{2t}s_{3t}s_{4t}s_{5t}s_{6t}s_{7t}s_{8t}$ (8.6)
$2547.7/7555 = 67/7555 \times 65.99/67 \times 51.93/65.99 \times 51.6/51.93 \times 32.5/51.6 \times 19.15/32.5 \times 16.42/19.15 \times 2547.7/16.42$ (actual data in 1978 from Table 9.1)
where s_{it} is the survival rate in the population entering the *i*th life stage at the *t*th year. The survival rate at each life stage is defined as follows:
s_{1t}: proportion of eggs and first instars that survived the abortion of flower buds and flowers,
s_{2t}: proportion of first instars that survived lepidopteran larval feeding on galls,
s_{3t}: proportion of first instars that survived the desiccation of galls,
s_{4t}: proportion of first and second instars that survived developmental delay,
s_{5t}: proportion of larvae that survived unknown factors,
s_{6t}: proportion of third instars that survived parasitism by *Eurytoma* sp.,
s_{7t}: proportion of third instars that survived parasitism by *Bracon asphondyliae*
s_{8t} is regarded as the rate of change caused by the oviposition of one female, which is calculated from the sex ratio and the mean number of ovarian eggs per female.
Equation 8.1 is expressed in common logarithms as follows:
$\log_e(S_t) = \sum_{i=1}^{8} \log_e(s_{it})$ (8.7)
(3.41–3.88) = (1.83–3.88) + (1.82–1.83) + (1.72–1.82) + (1.71–1.72) + (1.51–1.71) + (1.28–1.51) + (1.22–1.28) + (3.41–1.22)

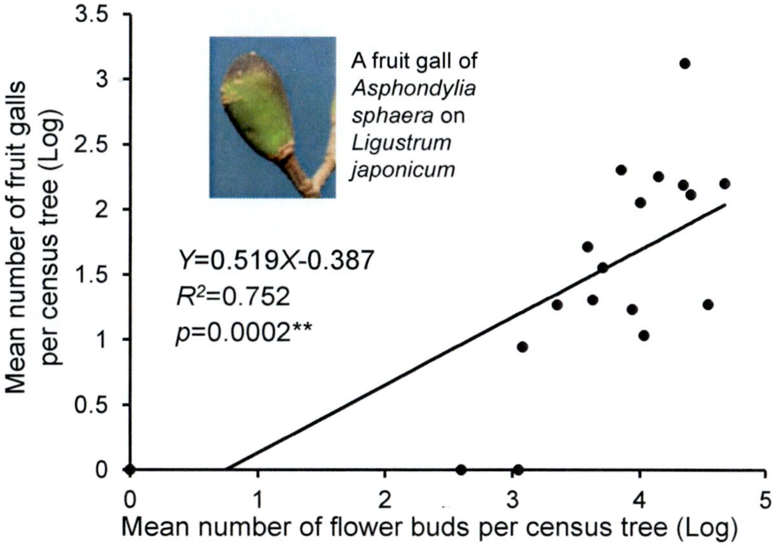

Fig. 8.1 Relationship between the mean number of flower buds per census tree and the mean number of fruit galls induced by *Asphondylia sphaera* per census tree of *Ligustrum japonicum*

leaves of *Euonymus japonicus* Thunberg (Celastraceae) operated as a density-dependent mortality factor (Sunose and Yukawa 1979). Future field surveys will provide further evidences of bottom-up effects on cecidomyiid populations.

8.4.4 Biotic Top-Down Effects

Life table data have demonstrated that top-down factors strongly affect the population dynamics of various gall midge species. In populations of the euonymus gall midge *M. pustulae*, larval mortality caused by a platygastrid was a key-factor (Sunose 1983). The populations of *M. pustulae* did not fluctuate greatly because of the high percentage of parasitism but increased to an outbreak situation following a decline in parasitism (Sunose 1985). Another example of parasitoid influence on host gall midge populations has been known for the pea gall midge, *Contarinia pisi* (Loew), following a 10-years investigation (Keller and Schweizer 1994). In addition to the aforementioned long-term studies, effects of various parasitoid wasps on the density of host gall midges have been evaluated in short-term studies: *Platygaster* (Hymenoptera: Platygastridae) on *Semudobia* species (Roskam 1986), *Aprostocetus diplosidis* Crawford (Hymenoptera: Eulophidae) on *Stenodiplosis sorghicola* (Coquillett) (Manzoli-Palma and da Teles 1990; Lampo 1994), *Platygaster oebalus* Walker (Hymenoptera: Platygastridae) on *Dasineura napi* (Loew) (=*D. brassicae*) (Axelsen 1994), and a guild of five interacting parasitoids on *Rhopalomyia californica* Felt (Ehler 1985, 1986). These examples demonstrate that top-down factors strongly affect the population dynamics of gall-inducing cecidomyiids. Exceptionally, a population of the Japanese pine-needle gall midge, *Thecodiplosis japonensis* Uchida & Inouye, was usually maintained at a low level by abiotic factors rather than biotic, and failure to achieve the potential fecundity was a key-factor (Soné 1987).

8.4.5 Meteorological Factors

Among meteorological factors, effects of temperature, precipitation and humidity on arthropod populations and communities have been frequently evaluated particularly in recent years in relation to climate change (e.g. Masters et al. 1998; Kiritani and Yukawa 2010; Kiritani 2011; Jamieson et al. 2012; Karuppaiah and Sujayanad 2012; Pearce-Higgins 2013). Based on 47- and 65-year field surveys, Whittaker and Tribe (1998) demonstrated that annual changes in the density of *Neophilaenus lineatus* (Linnaeus) (Hemiptera: Auchenorrhyncha) were density-independent and correlated with mean minimum temperature in September. In a 32-year study, Price and Hunter (2015) identified strong bottom-up effects of precipitation and plant growth on population dynamics of a gall-inducing sawfly, *Euura lasiolepis* Smith (Hymenoptera: Tenthredinidae).

Typhoons, hurricanes and cyclones are catastrophic events for organisms and their habitats but the effects of strong wind and heavy rainfall on ecosystems have been relatively roughly evaluated on a large scale (e.g. Tanner et al. 1991; Mabry et al. 1998; Xi et al. 2012). In contrast, direct and indirect effects of strong wind on particular plant species and their associated herbivores have been seldom evaluated intensively possibly because of difficulties in obtaining field data just before and after a catastrophic event.

Yukawa et al. (2019a) evaluated effects of typhoons on populations of a gall midge *P. neolitseae* and its associated arthropods through leaf fall of the host plant, *N. sericea*. The very strong typhoon No. 13 ("8513-Pat") attacked Kagoshima, southern Kyushu, Japan in August 1985 with the maximum instantaneous wind velocity, 56 m/s. It caused heavy leaf fall particularly on forest edge trees, resulted in the larval death of gall midge and its endoparasitoid *Gastrancistrus* sp. (Hymenoptera: Pteromalidae). The heavy leaf fall promoted the production of lammas shoots (= late season shoots) in autumn as a compensatory reaction, resulted in the reduction of number of the following spring shoots. The shortage of spring shoots was considered to be influential to the gall midge females in searching oviposition sites. Thus, strong wind affects insect populations directly and indirectly through various ways.

As a recent trend, population fluctuation has frequently been easily and simply attributed to global warming without necessary field data. However, as has been demonstrated in this section, increase and decrease of population numbers are determined by the combination of various biotic and abiotic factors. Therefore, sufficient field data are essential to determine important factors.

References

Axelsen JA (1994) Host-parasitoid interactions in an agricultural ecosystem: a computer simulation. Ecol Model 73:189–203

Barnes HF (1951) Gall midges of economic importance volume. V. Trees. Crosby Lockwood & Son Ltd., London

Ehler LE (1985) Species-dependent mortality in a parasite guild and its relevance to biological control. Environ Entomol 14:1–6

Ehler LE (1986) Distribution of progeny in two ineffective parasites of a gall midge (Diptera: Cecidomyiidae). Environ Entomol 15:1268–1271

Iwao S (1968) A new regression method for analyzing the aggregation pattern of animal populations. Res Popul Ecol 10:1–20

Iwao S (1970) Analysis of contagiousness in the action of mortality factors on the western tent caterpillar population by using $m^*- m$ relationship. Popul Ecol 12:100–110

Iwao S (1977) Analysis of spatial association between two species based on interspecies mean crowding. Res Popul Ecol 18:243–260

Jamieson MA, Trowbridge AM, Raffa KF, Lindroth RL (2012) Consequences of climate warming and altered precipitation patterns for plant-insect and multitrophic interactions. Plant Physiol 160:1719–1727

Karuppaiah V, Sujayanad GK (2012) Impact of climate change on population dynamics of insect pests. World J Agric Sci 8:240–246

Keller S, Schweizer C (1994) Populationsdynamische Untersuchungen an der Erbsengallmücke *Contarinia pisi* Winn. (Dipt., Cecidomyiidae) und ihrer Parasitoide. J Appl Entomol 118:281–299

Kim W, Matsunaga K, Gyoutoku N, Matsuo L, Minami T, Yukawa J (2015) Life history strategy and adult and larval behavior of *Macrodiplosis selenis* (Diptera: Cecidomyiidae), a species that induces leaf-margin fold galls on deciduous *Quercus* (Fagaceae). Entomol Sci 18:470–478

Kiritani K (2011) Impacts of global warming on *Nezara viridula* and its native congeneric species. J Asia-Pac Entomol 14:221–226

Kiritani K, Yukawa J (2010) Effects of global warming on insects. Zenkoku Nôson Kyôiku Kyôkai, Tokyo. (In Japanese)

Lampo M (1994) The importance of refuges in the interaction between *Contarinia sorghicola* and its parasitic wasp *Aprostocetus diplosidis*. J Anim Ecol 63:176–186

Lloyd M (1967) Mean crowding. J Anim Ecol 36:1–30

Mabry CM, Hamburg SP, Lin TC, Horng FW, King HB, Hsia YJ (1998) Typhoon disturbance and stand-level damage patterns at a subtropical forest in Taiwan. Biotropica 30:238–250

Manzoli-Palma MF, da Teles MC (1990) Ocorrência de inimigos naturais de *Contarinia sorghicola* (Coq., 1898) (Diptera: Cecidomyiidae) na regiao de Ribeirao Preto. Rev Bras Entomol 34:589–594

Masters GJ, Brown VK, Clarke IP, Whittaker JB, Hollier JA (1998) Direct and indirect effects of climate change on insect herbivores: Auchenorrhyncha (Homoptera). Ecol Entomol 23:45–52

Mishima M, Sato S, Tsuda K, Yukawa J (2014) Sexual isolation between two known intraspecific populations of *Hartigiola* (Diptera: Cecidomyiidae) that induce leaf galls on upper and lower surfaces of *Fagus crenata* (Fagales: Fagaceae), indicating possible diversification into sibling species. Ann Entomol Soc Am 107:789–798

Morris RF (1959) Single factor analysis in population dynamics. Ecology 40:580–588

Pearce-Higgins JW (2013) Evidence of climate change impacts on populations using long-term datasets. Terrestrial biodiversity climate change impacts report card technical papers 7

Price PW (2003) Macroevolutionary theory on macroecological patterns. Cambridge University Press, Cambridge

Price PW, Hunter MD (2015) Population dynamics of an insect herbivore over 32 years are driven by precipitation and host-plant effects: testing model predictions. Environ Entomol 44:463–473

Redfern M (1975) The life history and morphology of the early stage of the yew gall midge *Taxomyia taxi* (Inchbald) (Diptera: Cecidomyiidae). J Nat Hist 9:513–533

Redfern M, Cameron RAD (1978) Population dynamics of the yew gall midge *Taxomyia taxi* (Inchbald) (Diptera: Cecidomyiidae). Ecol Entomol 3:251–263

Redfern M, Cameron RAD (1993) Population dynamics of the yew gall midge *Taxomyia taxi* and its chalcid parasitoids: a 24-year study. Ecol Entomol 18:365–378

Redfern M, Hunter M (2005) Time tells: long-term patterns in the population dynamics of the yew gall midge, *Taxomyia taxi* (Cecidomyiidae), over 35 years. Ecol Entomol 30:86–95

Roskam JC (1986) Biosystematics of insects living in female birch catkins. IV. egg-larval parasitoids of the genera *Platygaster* Latreille and *Metaclisis* Förster (Hymenoptera, Platygastridae). Tijdschr Entomol 129:125–140

Royama T (1996) A fundamental problem in key factor analysis. Ecology 77:87–93

Soné K (1987) Population dynamics of the pine needle gall midge, *Thecodiplosis japonensis* Uchida et Inouye (Diptera, Cecidomyiidae). Z Angew Entomol 103:386–402

Sunose T (1983) Studies on the population dynamics of the *Euonymus* gall midge *Masakimyia pustulae* Yukawa et Sunose (Diptera, Cecidomyiidae) under different environmental conditions. Kontyû 51:29–40

Sunose T (1985) Population regulation of the euonymus gall midge *Masakimyia pustulae* Yukawa and Sunose (Diptera: Cecidomyiidae) by hymenopterous parasitoids. Res Popul Ecol 27:287–300

Sunose T, Yukawa J (1979) Interrelationship between the leaf longevity of the evergreen spindle tree, *Euonymus japonicus* Thunb. and the euonymus gall midge, *Masakimyia pustulae* Yukawa and Sunose (Diptera, Cecidomyiidae) in different environments. Jpn J Ecol 29:29–34

Tanner V, Kapos V, Healey J (1991) Hurricane effects on forest ecosystems in the Caribbean. Biotropica 23:513–521

Varley GC, Gradwell GR (1960) Key factors in population studies. J Anim Ecol 29:399–401

Varley GC, Gradwell GR (1968) Population models for the winter moth. In: Southwood TRE (ed) Insect abundance. Blackwell, Oxford, pp 132–142

Whittaker JB, Tribe NP (1998) Predicting numbers of an insect (*Neophilaenus lineatus*: Homoptera) in a changing climate. J Anim Ecol 67:987–991

Xi W, Chen SHV, Chu YC (2012) The synergistic effects of typhoon and earthquake disturbances on Forest ecosystem: lessons from Taiwan for ecological restoration and sustainable management. Tree and Forestry Science and Biotechnology, Global Science Books, London

Yamamura K (1999) Key-factor/key-stage analysis for life table data. Ecology 80:533–537

Yamamura K (2012) Extended key-factor/key-stage analysis for longitudinal data. J Biopharm Stat 22:1–15

Yukawa J (1995) Estimation of population density of gall-inducing insects using relationships between mean and mean crowding. Nat Kagoshima 21:15–18. (In Japanese)

Yukawa J (2000) Synchronization of gallers with host plant phenology. Popul Ecol 42:105–113

Yukawa J (2019) Galls provide us with good information for ecological studies - methods of practical field survey and data analysis. Form Entomol 38:25–32

Yukawa J, Akimoto K (2006) Influence of synchronization between adult emergence and host plant phenology on the population density of *Pseudasphondylia neolitseae* (Diptera: Cecidomyiidae) inducing leaf galls on *Neolitseae sericea* (Lauraceae). Popul Ecol 48:13–21

Yukawa J, Fujimoto K, Ganaha-Kikumura T, Wanggyu K (2019a) Effects of typhoons on an arthropod community centered upon a leaf gall midge, *Pseudasphondylia neolitseae* (Diptera: Cecidomyiidae) and its host plant, *Neolitsea sericea* (Lauraceae) through leaf fall and late-season shoot production. Ecol Res 35:252–264

Yukawa J, Haitsuka S, Miyaji K, Kamikado T (2006) Influence of the population dynamics of a gall-inducing cecidomyiid and its parasitoids on the abundance of a successor, *Lasioptera yadokariae* (Diptera: Cecidomyiidae). In: Ozaki K, Yukawa J, Ohgushi T, Price PW (eds) Galling arthropods and their associates: ecology and evolution. Springer-Verlag, Tokyo, pp 241–249

Yukawa J, Ikenaga H, Sato S, Tokuda M, Ganaha-Kikumura T, Uechi N, Matsuo K, Mishima M, Tung GH, Paik JC, Ren BQ, Don XO (2012) Description and ecological traits of a new species of *Pitydiplosis* (Diptera: Cecidomyiidae) that induces leaf galls on *Pueraria* (Fabaceae) in East Asia, with a possible diversification scenario of intraspecific groups. Entomol Sci 15:81–98

Yukawa J, Miyamoto K, Yamaguchi T, Takesaki K, Uechi N, Matsuo K (2016a) Key factor/key stage analysis of long-term life table data for a fruit gall midge, *Asphondylia sphaera* (Diptera: Cecidomyiidae). Ecol Entomol 41:516–526

Yukawa J, Moriya T, Kanmiya K (2019b) Comparison in the flight ability between the soybean pod gall midge, *Asphondylia yushimai* and the aucuba fruit gall midge, *A. aucubae* (Diptera: Cecidomyiidae). Appl Entomol Zool 54:167–174

Yukawa J, Nakagawa K, Saigou T, Awa T, Fukuda T, Higashi M (2013) Adult behavior of an ambrosia gall midge *Illiciomyia yukawai* (Diptera: Cecidomyiidae) and synchronization between its emergence and host plant phenology. Entomol Sci 16:400–412

Yukawa J, Nakagawa K, Kohno A, Tokuda M, Kiritani K, Matsuo K, Mitsui H, Fujii T (2016b) Geographical and annual variations in the proportion of extended diapausing individuals of *Illiciomyia yukawai* (Diptera: Cecidomyiidae) with reference to an adaptive significance of its bimodal emergence pattern. Entomol Sci 19:275–289

Yukawa J, Takahashi K, Ohsaki N (1976) Population behaviour of the neolitsea leaf gall midge, *Pseudasphondylia neolitseae* Yukawa (Diptera, Cecidomyiidae). Kontyû 44:358–365

Yukawa J, Uechi N, Tokuda M, Sato S (2005) Radiation of gall midges (Diptera: Cecidomyiidae) in Japan. Basic Appl Ecol 5:453–461

Yukawa J, Yamamura K, Fujimoto K, Tokuda M, Muroya S, Maki T, Sadoshima T, Fukuda T (2018) Key-factors and key-stages that determine the leaf longevity of an evergreen broad-leaved tree, *Neolitsea sericea* (Lauraceae) at different growing sites in southern Japan. Ecol Res 33:175–190

Part V
Community Centered upon Galls and Gall Inducers

Chapter 9
Arthropod Community

Junichi Yukawa

Abstract The arthropod community centered upon gall midges and their host plants including various additional organisms that rely on the existence of galls is treated in this chapter. First, a new food web connecting two communities centered upon *Pseudasphondylia neolitseae* on *Neolitsea sericea* and *Daphnephila machilicola* on *Machilus thunbergii* is demonstrated. Effects of predators and herbivores on the community members are explained. In particular, direct and indirect effects of a stem borer, *Oberea hebescens*, on *N. sericea* and its associated insects are highlighted. Parasitoids are excluded from the explanation because they are mentioned in Chap. 11. Then, examples of inquiline, successor, and cecidophagy are reviewed. Cynipid inquilines are briefly introduced for comparison with cecidomyiid inquilines. Detailed explanation is given for ecological traits of a successor, *Lasioptera yadokariae*, based on a long-term field survey. The last section elaborates on various types of cecidophagy including gall-sucking behavior.

Keywords Food web · Community · Predator · Herbivore · Inquiline · Successor · Cecidophagy

9.1 Food Web

9.1.1 Introduction

Galls provide extremely favorable breeding conditions to a great variety of organism with its localized concentration of highly nutritive substances and its marked succulence (Mani 1964). The arthropod community centered upon gall midges and their host plants includes various additional organisms that rely on the existence of galls, such as inquilines, successors, and cecidophagous species, hence it is much

J. Yukawa (✉)
Entomological Laboratory, Faculty of Agriculture, Kyushu University, Fukuoka, Japan
e-mail: galling-cecidomyiids@kyudai.jp

J. Yukawa, M. Tokuda (eds.), *Biology of Gall Midges*, Entomology Monographs,
https://doi.org/10.1007/978-981-33-6534-6_9

more complex than common communities of external herbivores and their host plants. Interactions among tritrophic levels have been studied from various aspects, such as population ecology, chemical ecology, biological control, etc. (e.g. Price et al. 1980; Price 1986; Turlings and Erb 2018). Thus, the survey of arthropod community is fundamentally necessary for better understanding the inter-relations of these diverse organisms and for promoting further studies of the gall-inducing arthropods and their associated inhabitants (Yukawa 1983). In addition, we need to note that community ecology has been facilitating studies of a geographic mosaic of coevolution among members belonging to tritrophic levels (e.g. Craig et al. 2020; see Sect. 2.2).

9.1.2 A Food Web Consisting of Two Jointed Arthropod Communities

Previously Yukawa (1983) demonstrated a food web of the arthropod community centered upon *Pseudasphondylia neolitseae* Yukawa on *Neolitsea sericea* (Blume) Koidzumi. In this chapter, a new food web is provided in Fig. 9.1 by joining another community centered upon *Daphnephila machilicola* Yukawa on *Machilus thunbergii* Siebold & Zuccarini to the previous one. These two communities are associated with lauraceous host plants and co-existing in the same evergreen broad-leaved forest in southern Kyushu, Japan. Some minor members of the communities are excluded from the food web because their densities are very low, or they do not

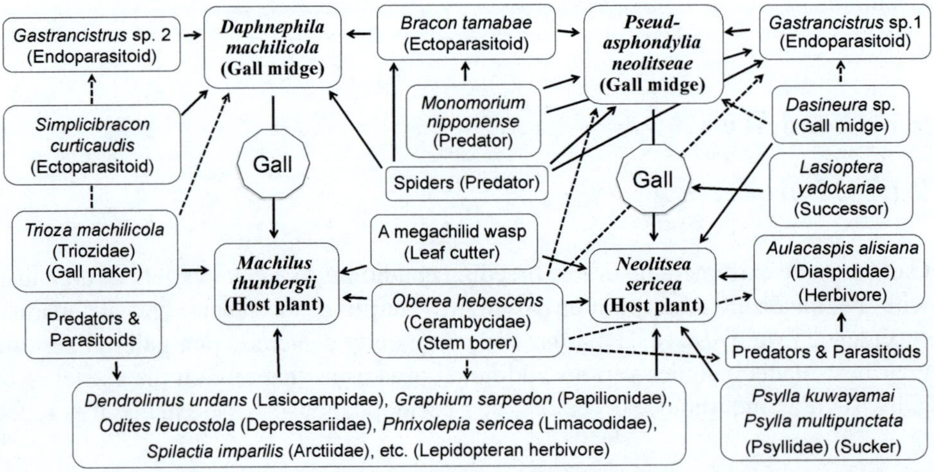

Fig. 9.1 A food web of two jointed arthropod communities centered upon *Pseudasphondylia neolitseae* on *Neolitsea sericea* and *Daphnephila machilicola* on *Machilus thunbergii*. The two communities are associated with lauraceous host plants and co-existing in the same evergreen broad-leaved forest in southern Kyushu, Japan. Allows point toward food resources. Dashed lines indicate indirect effects on organisms pointed

always appear in the communities. Naturally, the new food web becomes more complex than the web of single community, and some community members are related to both communities.

9.1.3 Parasitoids and Predators

Among parasitoids, the two species of *Gastrancistrus* (Pteromalidae) (endoparasitic early attacker) (Fig. 9.2c) are naturally monophagous, while *Bracon tamabae* Maeto (Braconidae) (ectoparasitic late attacker) (Fig. 9.2d) attacks both *P. neolitseae* and *D. machilicola* (Maetô 1991) (see Chapter 11 for details of the parasitoids). No further information on host range is available as to *Simplicibracon curticaudis* Maetô (Braconidae).

In March and April, workers of the ant, *Monomorium intrudens* Smith (= *nipponense*) (Formicidae) were observed to remove mature larvae and pupae of *P. neolitseae* and its parasitoids from the galls by opening the operculum on the hypophyllous surface of the galls (Yukawa 1983). Evidences of the predation were observed almost every year on a few restricted host trees, where the local gall midge populations were largely and concentratedly destroyed.

The ant can easily open the operculum because the mature larva prepares an exit before pupation by making layers beneath the exit very thin. Pupae push open the operculum with their apical spines and stretch themselves halfway out of galls for emergence (Yukawa et al. 1976). In contrast, such predation by ants has never been observed until today for the galls of *D. machilicola* on *M. thunbergii* possibly because the larval cavity is plugged with a thick subglobular lid, which falls off when adult emerges (Yukawa 1974).

At least 11 species of spiders were observed to feed on adults of *P. neolitseae* at the time of emergence or of oviposition (Yukawa 1983). Under the high population density of *P. neolitseae*, heavily galled leaves of *N. sericea* increase and are frequently folded outward, providing favorable webbing conditions to the spiders. The adults of *P. neolitseae* are not only killed by direct predation but also caught in the webs and then attacked by the spiders.

Unlike the leaves of *N. sericea*, those of *M. thunbergii* are thick and tough, and heavily galled leaves are not folded outward but rather inward. In addition, the barrel-shaped galls on the lower leaf surface are about 7 mm tall. Thus, the galled leaves of *M. thunbergii* do not provide favorable conditions to the spiders for webbing between or over the galls, resulted in the low mortality of *D. machilicola* at the time of emergence.

The spiders are sometimes observed to catch females of *P. neolitseae* or *D. machilicola* after their arrival at newly extending host buds for oviposition. Even if the spiders fail to catch the females, their oviposition action is disturbed and they must fly off the host buds to find the others, which occasionally promotes mistaken oviposition (see Sect. 2.1).

Fig. 9.2 Galls and some of community members in the food web shown in Fig. 9.1. (**a**), leaf galls induced by *Pseudasphondylia neolitseae* on *Neolitsea sericea*; (**b**), leaf galls induced by *Daphnephila machilicola* on *Machilus thunbergii* (arrows point toward scars made by the larvae that hatched and sank into the leaf tissue but did not induce galls in the first year, postponed to the second year due to long-term diapause: see Sect. 6.3); (**c**), *Gastrancistrus* sp. 1; (**d**), *Bracon tamabae*; (**e**), *Oberea hebescens* ♂; (**f**), ibid. ♀; (**g**), pit galls induced by *Trioza machilicola* on a leaf of *Machilus thunbergii*; (**h**), leaf galls induced by *P. neolitseae* on *N. sericea* (showing lower

9.1.4 *A Longicorn Beetle,* Oberea hebescens

Oberea hebescens Bates (Coleoptera: Cerambycidae) (Fig. 9.2e, f) is a stem borer of trees of Lauraceae, such as *N. sericea*, *Cinnamomum tenuifolium* (Makino) Sugimoto ex Hara and *Lindera praecox* (Siebold & Zuccarini) Blume (Ohbayashi and Niisato 2007). In May, females of *O. hebescens* lay their eggs one by one in the distal portion of newly extended shoots. Through mid-May to October, hatched larvae bore tunnels downward in the shoot toward the basal end of current twig and subsequently toward the 1-year old twigs. Current leaves attached to the infested shoot dry out because of the original infestation. When the cerambycid larvae reach the 1-year old twig, leaves attached to non-oviposited current shoots extending from the same 1-year old shoot dry out (Yukawa 1977). At the same time, galls on the current leaves of *N. sericea* dry out and the larvae of *P. neolitseae* and *Gastrancistrus* sp. (Pteromalidae) inhabiting the galls die (Yukawa and Akimoto 2006). Herbivores feeding on these leaves also lose their food resources.

Larvae of the cerambycid have never been observed boring the shoots and twigs of *M. thunbergii* that grows nearby *N. sericea*. The main shoot of *M. thunbergii*, as well as *Machilus japonica* Siebold & Zuccarini ex Blume, is quite short and the side shoots are not thick enough for the cerambycid larvae to bore (Yukawa et al. 2018). Adults of *O. hebescens* feed on the midribs of lauraceous plants, but this infestation was negligible in contrast to the prominent damage of shoots and twigs by larvae.

9.1.5 *Herbivores*

There are various species of herbivores associated with *N. sericea* or *M. thunbergii*. A few of them are monophagous, some are oligophagous on Lauraceae, and the others are polyphagous. Some of them feed on galls facultatively or rarely obligatorily. Cecidophagy will be mentioned in Sect. 9.4.

Dasineura sp. (Cecidomyiidae) is a monophagous and univoltine species. According to Yukawa (1983), in April, the females lay their eggs inside the leaf buds, which remain closed and wither gradually. Later, the leaf buds containing the mature larvae fall to the ground, where they spend summer, autumn and winter. At the same time, the eggs or the first instars of *P. neolitseae*, if any, are killed by the fall and the successive decay of the infested leaf buds.

Auracaspis alisiana Takagi (Hemiptera: Diaspididae) is an invasive pest, being host-specific to *N. sericea*, and has three overlapped generations in Fukuoka, Japan. The nymphs settle on the lower surface of *N. sericea* leaves, but they do not attack

Fig. 9.2 (continued) surface of the leaf, arrows point towards vacated galls plugged with fungal mycelium in which a larva of successor *Lasioptera yadokariae* lives); (**i**), scars of *Machilus thunbergii* leaves cut by an unidentified megachilid wasp newly emerged leaves although they generally prefer younger leaves (Mediga and Takasu 2018). Therefore, the larvae of *P. neolitseae* in the galls on the current leaves would not suffer bad influence by the scale insect.

Two species of Psyllidae (Heteroptera), *Psylla kuwayamaii* (Grawford) and *Psylla multipunctata* Miyatake are frequently found on the current leaves of *N. sericea*. They do not induce galls, but cause fading on the affected parts by sucking, resulted in yellowish flecks on the leaves. There is, however, no evidence that the galls or the gall midge larvae are ill influences by the psyllids on such leaves (Yukawa 1983). *Trioza machilicola* Miyatake (Triozidae) is oligophagous and univoltine, and many nymphs settle on the lower surface of current leaves of *M. thunbergii* or *M. japonica* (Fig. 9.2g) (Yukawa and Masuda 1996). They induce as many as 500 pit-hole galls on upper surface of the leaves. No barrel-shaped galls of *D. machilicola* can appear on the lower surface of current leaves under such highly dense conditions of the psyllid galls.

Several herbivorous species of Lepidoptera can be seen occasionally on the leaves of lauraceous plants including *N. sericea* and *M. thunbergii* (Fig. 9.1). Since all these species are oligo- or polyphagous species, they are not tightly associated with *N. sericea* and *M. thunbergii*. In addition, larvae of most herbivorous species, such as *Graphium sarpedon* Linnaeus (Papilionidae), avoid galled parts when they feed on the host leaf blade. Therefore, these herbivores do not cause damages to both the host trees and the gall midge populations.

Adults of an unidentified megachilid (Hymenoptera) sometimes cut off a part of *N. sericea* leaves in summer (Fig. 9.2i), but they always avoid collecting the galled portion of the leaves.

Each of these herbivorous species is attacked by a wide variety of parasitoids or predators, hence the number of community members increase accordingly. Of course, every member cannot play the lead in community ecology. However, unlike the study of population dynamics, multiple members should be regarded as leading characters in the study of "community dynamics" to clarify long-term interrelations between members of the community.

9.2 Inquiline

9.2.1 Introduction

An inquiline is an animal that lives commensally in the nest, gall, burrow, or dwelling place of an animal of another species. Unlike parasitoids or predators, inquilines do not have a deleterious effect on the host species. Among insects, the most commonly known inquilines are found in the nests of social insects, especially ants and termites. Arthropods other than ants and termites inhabit their nest are called myrmecophiles and termitophiles, respectively.

Inquilines inhabiting arthropod galls are known among cecidomyiids (e.g. Yukawa and Rohfritsch 2005), cynipids (Askew 1984; Ronquist 1994), aphids

(Akimoto 1988; Miller and Crespi 2003), thrips (Raman and Ananthakrishnan 1984), tanaostigmatids (LaSalle 1997), and psyllids (Yang et al. 2001). Among them, both cecidomyiids and cynipids are distinctly abundant in the Holarctic Region, while Maia et al. (2014) reported that cecidomyiids are the most frequent gall-inhabiting inquilines in the Brazilian costal vegetation.

Gall-inhabiting inquilines are treated in this section separated from other arthropods associated with gall inducers because they are important members constituting arthropod communities centered upon gall-inducing species and their host plants.

9.2.2 Cynipid Inquilines

The family Cynipidae includes about 1400 species belonging to 12 tribes (Ronquist et al. 2015), many of which are gall-inducing herbivores (e.g. Stone et al. 2002; Abe et al. 2007). Most cynipid inquilines belong to Synergini, which comprise about 180 species of seven genera including *Synergus* (Ide et al. 2018). *Synergus* is the most species-rich genus including at least 110 species in the world (Schwéger et al. 2015). Cynipid inquilines usually live in cynipid galls, but rarely in galls of other arthropods. For example, *Synergus yukawai* (Wachi et al. 2011) inhabits cecidomyiid stem galls on *Quercus acutissima* Blume (Fagaceae) (Wachi et al. 2011) and *Rhoophilus loewi* Mayr is an inquiline in galls induced by species of *Cecidoses* and *Eucecidoses* (Lepidoptera: Cecidosidae) on *Searsia* (previously identified as *Rhus*) and *Schinus* species (Anacardiaceae) (Van Noort et al. 2007).

Until recently, all species of Synergini were thought to be inquilines and not to be able to initiate gall formation although they can modify the galls of other species. However, Abe et al. (2011) showed that *Synergus itoensis* Abe, Ide & Wachi induces galls by itself inside the acorns of *Quercus glauca* Thunberg (Fagaceae). Then, Abe et al. (2011) speculated that an ability of gall induction might have evolved within Synergini. Subsequently, *Saphonecrus hupingshanensis* Liu, Yang & Zhu and *Synergus castaneus* Pujade-Villar, Bernardo & Viggiani were reported as possible gall inducers in Synergini (Liu et al. 2012; Bernardo et al. 2013). Based on a comprehensive phylogenetic analysis of Synergini, Ide et al. (2018) concluded that *S. itoensis* together with its two unidentified congeners evolved from inquilines and recently acquired their ability to induce galls independently from other cynipid gall inducers. If the two species mentioned by Liu et al. (2012) and Bernardo et al. (2013) are gall inducers, gall-inducing ability is considered to have evolved from inquilines independently at least three times in Synergini (Ide et al. 2018).

9.2.3 Tanaostigmatid Inquilines Inhabiting Cecidomyiid Galls

Tanaostigmatidae are predominantly phytophagous. According to LaSalle (1997), they are mainly associated with galls on fabaceous plants and, in general, have been presumed to be the gall formers although detailed biological information is very few for members. Many records of tanaostigmatids as gall-formers are undoubtedly correct, but inquilinism may be more common than previously suspected.

Protanaostigma kyushuana Masi (Hymenoptera: Tanaostigmatidae) was described, without any biological information, based on a single female collected in 1938 from Kyushu, Japan (Masi 1940). Later, Kamijo (2000) confirmed that it is an inquilinous wasp inhabiting lenticular cecidomyiid galls on the leaves of *Pueraria lobata* (Willdenow) Ohwi (Fabaceae) (Fig. 2.10a, b). Thereafter, Yukawa et al. (2012) described a new species *Pitydiplosis puerariae* Yukawa, Ikenaga & Sato that is responsible for the lenticular leaf galls on *P. lobata* and *Pueraria montana* (Loureiro) Merrill in Japan, mainland China, Taiwan, and South Korea. Until today, many larvae, pupae and adults of the inquilinous species have been obtained both from the galls of *P. puerariae* and from subglobular galls induced by an unidentified gall midge species on the under-leaf surface of *P. lobata* and *P. montana*. Therefore, redescription and an exact identification of *P. kyushuana* are needed based on these specimens in addition to the single female type specimen.

Tanaostigmodes puerariae Yang & Pitts (Hymenoptera: Tanaostigmatidae) was described from mainland China as an inducer of the lenticular gall on *Pueraria* (Yang et al. 2004). However, Yukawa et al. (2012) pointed out that it is apparently not a gall-inducing species but an inquilinous wasp inhabiting the cecidomyiid galls.

From Brazil, Fernandes et al. (1987) reported one inquiline species in a gall of *Anadiplosis* sp. on *Machaerium aculeatum* Raddi (Fabaceae). LaSalle (1987) subsequently described this species as *Tanaostigmodes ferbandesi* LaSalle (Hymenoptera: Tanaostigmatidae). The galls containing the inquiline were larger than normal cecidomyiid galls, and gall shape was different between them. Because the cecidomyiid larvae were never found in the galls inhabited by *T. ferbandesi*, they were possibly killed and/or consumed by the inquiline larvae. This information suggests that the same thing may also happen in *P. puerariae* galls inhabited by inquilines, *P. kyushuana* and *T. puerariae.*

9.2.4 Cecidomyiid Inquilines

According to Gagné and Jaschhof (2017), at least 110 species of Cecidomyiinae have been regarded as inquilines or possibly inquilines inhabiting galls induced by other cecidomyiid species. (Table 9.1). They are known from three genera of Cecidomyiidi and four of Lasiopteridi. This means that the inquilinous habit has evolved more than once in the subfamily Cecidomyiinae.

Table 9.1 The number of cecidomyiid inquiline species, their supertribal position, distribution and host cecidomyiids (picked up from Gagné and Jaschhof 2017, based on a key word "inquiline")

Supertribe	Genus of inquilines	Distribution and the number of species	Larval habit	Host cecidomyiids
Cecidomyiidi	*Anabremia*	Palearctic; 6 spp.	All inquilines or possibly inquilines	*Dasineura* spp. on Fabaceae
Cecidomyiidi	*Tricholaba*	Palearctic; 5 spp.	Inquilines or possibly inquilines	*Dasineura* spp. on Fabaceae
Cecidomyiidi	*Youngomyia*	Nearctic, neotropical, oriental; 6 spp.	Inquilines or possibly predaceous	Other cecidomyiines
Lasiopteridi	*Camptoneuromyia*	Neotropical, nearctic; 15 spp.	Mostly inquilines	Other Cecidomyiinae
Lasiopteridi	*Domolasioptera*	Neotropical; 7 spp.	All inquilines	Other Cecidomyiinae
Lasiopteridi	*Macrolabis*	Palearctic; 61 spp., nearctic 1sp.	Most species are inquilines	Other cecidomyiids
Lasiopteridi	*Trotteria*	Cosmopolitan; 23 spp.	All inquilines	Chiefly Asphondyliini

Among the seven genera listed in Table 9.1, the genus *Macrolabis* contains the most abundant inquilinous species, 61 from the Palaearctic Region and one from the Nearctic Region. For example, the larvae of *Macrolabis mali* Anfora were found to live as inquilines in leaf galls of the apple leaf curling gall midge *Dasineura mali* (Kieffer) on *Malus domestica* Borkhausen (Rosaceae) in Italy (Anfora et al. 2005). Each gall usually contains 20–50 larvae of *D. mali* and 20–40 larvae of *M. mali*. The larvae quit the galls to drop to the ground, where they pupate. Both species are multivoltine, five to six generations per year. At the end of the season, the number of *M. mali* larvae in a gall becomes larger than that of *D, mali*, resulted in the death of host larvae (Anfora et al. 2005).

Dorchin et al. (2006) newly discovered an inquilinous gall midge of the genus *Macrolabis* in the Nearctic Region. The larvae of *Macrolabis americana* Dorchin inhabit galls induced by *Dasineura folliculi* Felt on *Solidago rugosa* Miller and *Solidago gigantea* Aiton (Asteraceae). Unlike known *Macrolabis* species in the Palaearctic Region, *M. americana* is multivoltine and is found in the galls successively through the generations. The larvae of both species leave the galls to pupate in the soil. Emergence occurred between dusk and dawn and adults lived 1–2 days. Sex ratio was female biased, 1:2 to 1:4 (Dorchin et al. 2006).

In addition to the seven genera in Table 9.1, the genus *Dasineura* also contains some inquilinous species. For example, the larvae of *Dasineura balsamicola* (Lintner) inhabit bud galls induced by the balsam gall midge *Paradiplosis tumifex* Gagné on balsam fir *Abies balsamea* (Linnaeus) Miller (Pinaceae) in Canada and USA. Of 376 incipient galls of *P. tumifex* examined, 75 contained one, 20 contained two, and seven contained more than two larvae of *D. balsamicola* (Akar and Osgood 1987). Carleton et al. (2015) observed daily activities of both *P. tumifex* and *D.*

balsamicola. Both species were most active during 17:00–22:00 h. Male gall inducer activity was largely confined to the space below the crown. Calling and mating by the gall inducer occurred at ground level and were followed by dispersal of females to the vegetative crown. Peak oviposition occurred between 20:00 and 21:00 h. Female inquilines displayed similar activity patterns, although no calling or mating behavior was observed. Inquiline flight and search for oviposition sites were more active than the gall inducer, with shorter rest periods and more buds visited than the gall inducer.

Because *P. tumifex* is a major pest for the Christmas tree industry and the inquiline *D. balsamicola* is involved in the population dynamics of the balsam gall midge, Guay et al. (2016) surveyed their spring phenology in the field, as well as their survival during winter diapause under simulated climatic scenarios in the laboratory. They observed that spring asynchrony between fir host trees and *P. tumifex* in the first year of study had no impact on summer abundance, but they clarified that overwintering habitats and strategies differ between the galler and its inquiline. Experimental overwintering data suggested that diapausing conditions affect these species differentially.

Dasineura hybanthi Kolesik & Skuhravá was found from galls caused by an unknown gall midge on *Hybanthus floribundus* (Lindley) F. Mueller (Violaceae) in Australia but its detailed life history traits have never been studied (Kolesik and Skuhravá 1997).

Although most inquiline species inhabit galls induced by other cecidomyiid species, they are rarely found living in the galls induced by arthropods other than gall midges. For example, the larvae of *Resseliella* (= *Profeltiella*) *dizygomyzae* (Barnes) live gregariously as inquilines in mines of *Dizygomyza barnesi* Hendel (Diptera: Agromyzidae) on *Salix* spp. (Salicaceae) (Barnes 1933).

Thus, inquilines often affect their host populations by crowding or other forms of competition. Therefore, the existence of inquilines in arthropod communities cannot be ignored, but intensive and long-term population dynamics of inquilines have not yet been studied.

9.3 A Successor, *Lasioptera yadokariae*, Using Vacated Galls

9.3.1 Introduction

Successors inhabit galls after the gall-inducing organisms and their associates, parasitoids, predators, and inquilines, have departed (Mani 1964). As the galls usually do not decay sometime after the departure of these organisms, the vacated galls provide shelters to a variety of arthropods and breeding ground for fungi. Some of the successors utilize galls as occasional shelters or nests, but others are associated with particular sorts of galls and have become wholly specialized for living in vacated galls. Many species of arthropods, such as ants, aphids, thrips, psocids, and mites, are known to inhabit empty galls (e.g. Walsh 1864/1866; Ping 1920; Mani 1964, Yukawa 1983; Patton 1987; Yukawa and Rohfritsch 2005).

In the subfamily Cecidomyiinae, *Lasioptera cerasiphera* Stelter, *Lasioptera querciflorae* Felt, *Lasioptera tibialis* Felt, and *Lasioptera yadokariae* Yukawa & Haitsuka have been known as successor species that live in vacated galls (Gagné 1989; Yukawa and Haitsuka 1994; Gagné and Jaschhof 2017). Among these species, *L. yadokariae* has been most intensively studied by Yukawa and Haitsuka (1994).

9.3.2 *Galls Utilized by* Lasioptera yadokariae *and Associated Fungus*

Lasioptera yadokariae utilizes the following three sorts of gall (Yukawa and Haitsuka 1994): brownish swellings induced by *P. neolitseae* on the leaves of *N. sericea*; angular, pear- or club-shaped galls of *Daphnephila* sp. on the under leaf surface of *M. japonica*; and thick type suboval leaf blisters (see Sect. 2.3) induced by *Masakimyia pustulae* Yukawa & Sunose on *Euonymus japonicus* Thunberg (Celastraceae). *Daphnephila machilicola* also produces barrel-shape galls on the under surface of the leaves of a related host plant, *M. thunbergii*, but *L. yadokariae* has never been found on this plant, possibly because the larval cavity and the exit hole are too large for the successors to inhabit.

Galls inhabited by larvae of *L. yadokariae* contain fungal mycelium on which the larvae feed. During the first larval stadium, mycelium of the host fungus is not distinct in the gall. When the larvae become the second instars, the exit hole of the gall is plugged with a thick growth of the mycelium (Fig. 9.2h). After mature larvae quit the galls to drop to the ground, the mycelium is weakened and disappears. The fungus obtained from the vacated gall on *N. sericea* was identified as *Pestalotia* sp. (Fungi Imperfecti: Melanconiales: Melanconiaceae) (Yukawa and Haitsuka 1994).

9.3.3 *Life History*

In southern Kyushu, Japan, the aforementioned three gall-inducing host midges are normally univoltine, and adults emerge in April and May (Yukawa 1974, 1983; Yukawa and Sunose 1976; Yukawa and Haitsuka 1994). Therefore, vacated galls are available for *L. yadokariae* in this season. Exit holes of the gall midge and its parasitoids are made under leaf side of the galls. The emergence of *L. yadokariae* from the ground usually begins in early April and continues until early May.

The duration of the egg stage is not precisely known. According to Yukawa and Haitsuka (1994), 84% of the larvae found in the vacated galls on *N. sericea* were first instars in mid-April. The dissection of galls indicated that two or three first instars inhabited a gall more frequently than what expected from random distribution. The first instars rapidly developed into second, and then 48% of the larvae became third instars in early May. At this time, there was no case where two or more third instars coexist in a gall. This suggests that there was an intraspecific competition among the first or second instars. There was no sign of attack by any parasitoid species. All third instars quitted the galls to drop to the ground by mid-May. Thereafter, neither successor larvae nor fungus mycelia were found in the empty galls in the field.

The larvae possibly spin cocoons on the ground and pass through the summer, autumn, and winter, and pupate in the following spring. In addition, 33 out of the 129 larvae collected in April 1987 became adults after being exposed for one cycle of the four artificial seasons in an incubator for a relatively short period of time (see Table 1 in Yukawa and Haitsuka 1994). These results suggest that the successor is univoltine.

It should be remarked here that two of the 129 larvae became adults after being exposed for two cycles of the four artificial seasons. This result indicates the possibility that some individuals of *L. yadokariae* have a 2-year life cycle as has been noted for other gall midge species such as *Sitodiplosis mosellana* Géhin and *Contarinia tritici* Kirby, some of whose larvae diapause in the ground for more than 1 year (e.g. Barnes 1952, 1958; see Sect. 6.3). No adult successor emerged after three cycles of the four artificial seasons, which means that the remainder of the 129 larvae died during the incubation (Yukawa and Haitsuka 1994).

9.3.4 Emergence, Sex Ratio, Life Span, and Oviposition

Laboratory experiments indicated that the emergence of both males and females occurred mostly in the morning between 5:00 and 12:00. A few adults emerged in the afternoon. Sex ratio at the time of emergence was nearly 1:1. The average life span of unmated males and females was 6.67 ± 2.16 h ($n = 6$) and 6.67 ± 2.74 h ($n = 12$), respectively (Yukawa and Haitsuka 1994).

After mating, the females start to fly actively under the host leaves in search of vacated galls and lay their eggs inside the vacated galls one by one. The average time required for one egg laying was 14.2 ± 6.9 s ($n = 21$). The females of *L. yadokariae* did not show any preference between the galls vacated by the gall midge or its parasitoids, *Gastrancistrus* sp. and *B. tamabae*. In addition, the females did not distinguish unused empty galls from those already utilized by themselves or other conspecific females. We observed sometimes in the field that a female came back to the gall where she had oviposited before several minutes. There was a case where two females oviposited in a gall at the same time.

9.3.5 Population Fluctuation

The density of gall-inducing species and their parasitoids, which fluctuates independently of the density of the successor, is considered to influence the abundance and survival of *L. yadokariae*. In addition, synchronization of the emergence season of *L. yadokariae* adults with that of gall-inducing cecidomyiids and their parasitoids is an important factor in determining the abundance of the successor because its adult life span is very short and the females choose fresh vacant galls for oviposition (Yukawa and Haitsuka 1994; Yukawa et al. 2006).

Figure 9.3 shows the annual changes in the number of vacated galls that were unused and used by *L. yadokariae*. The vacated galls include those made by *P. neolitseae* and its parasitoids, *Gastrancistrus* sp. and *B. tamabae*. The vacated galls used by *L. yadokariae* fluctuated from 9 in 1979 up to 108 in 1989. The proportion of vacated galls utilized by *L. yadokariae* fluctuated from year to year between 3.1% in 1985 and 36.0% in 1987 (Yukawa et al. 2006). The vacated galls of *B. tamabae* were most frequently used by *L. yadokariae*, followed by those of *P. neolitseae*, while those of *Gastrancistrus* sp. were less frequently used.

Yukawa et al. (2006) concluded that when the density of *P. neolitseae* or *B. tamabae* is high, *L. yadokariae* increases its population number by utilizing the abundant vacated galls made by their emergence prior to the emergence of *L. yadokariae*, whereas the high density of *Gastrancistrus* sp. does not fully contribute to an increase of the successor because its emergence season occur a little after the emergence of *L. yadokariae*. The high density of *B. tamabae*, however, would decrease the number of *L. yadokariae* in the following generation, because the high percentage parasitism will reduce the density of *P. neolitseae*, resulting in a shortage of vacated galls.

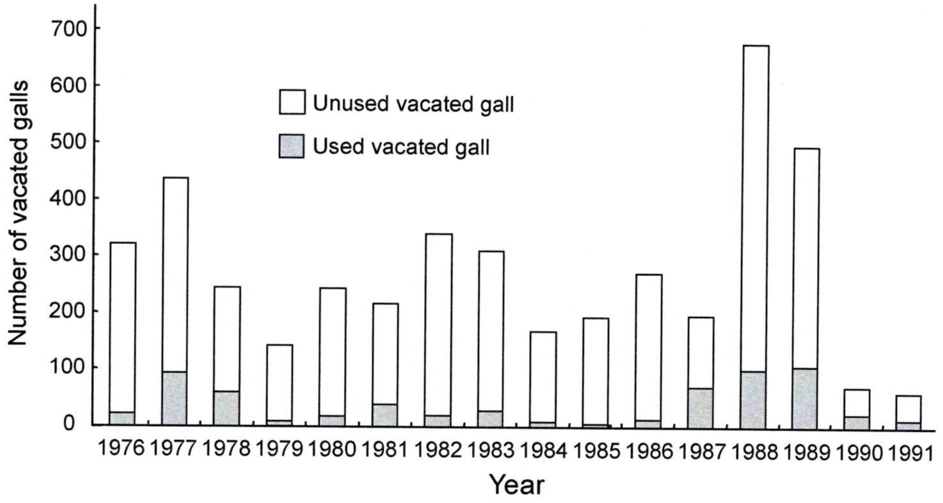

Fig. 9.3 Annual changes in the number of vacated galls that were unused and used by *Lasioptera yadokariae* (modified from Fig. 4 in Yukawa et al. 2006)

9.4 Cecidophagy

9.4.1 Introduction

Cecidophagy is a habit of feeding on the inside and outside of galls either obligatorily or facultatively, and frequently causes the death of gall inducers by destroying galls and leaving feces. In comparison with accidental feeding on galls, facultative cecidophagy is considered to be rather common in various cecidophagous species, although examples of obligatory cecidophagy are relatively rare. Some of facultative cecidophagous species prefer galled tissues or organs to ungalled parts.

9.4.2 Cecidomyiid Galls Fed by Dipteran, Lepidopteran and Coleopteran Insects

Miyatake et al. (2000) found *Bactrocera scutellata* (Hendel) (Diptera: Tephritidae) emerging from stem galls induced by a gall midge, *Lasioptera* sp., on wild gourds, *Diplocyclos palmatus* (Linnaeus) C. Jeffrey, *Trichosanthes cucumeroides* (Seringe) Maximowicz, and *Melothria liukiuensis* Nakai (Cucurbitaceae), on Okinawa Island, Japan. Because *B. scutellata* preferably feeds on the stem galls to ungalled parts of the gourds, it is regard as a cecidophagous fruit fly.

Sugiura and Yamazaki (2006) observed larvae of *Chyliza splendida* Iwasa (Diptera: Psilidae) feeding on leaf bud galls induced by *Asphondylia baca* Monzen on its winter host, *Weigela floribunda* (Siebold & Zuccarini) K. Koch (Caprifoliaceae) (Fig. 6.4e). The mortality rate of *A. baca* was extremely high, 98.0% ($n = 51$) and 98.2% ($n = 58$), in the galls with and without *C. splendida* larvae, respectively, but there was no significant difference between the rates. Therefore, the effect of *C. splendida* larvae on the survivorship of *A. baca* was considered to be minimal.

Yamazaki and Sugiura (2003a) observed larvae of *Amaya stellate* Butler (Lepidoptera: Noctuidae) feeding on the epidermis of stem galls induced by *Lasioptera achyranthii* Shinji on *Achyranthes bidentata* Blume (Amaranthaceae) and larvae of *Nippoptilia vitis* (Sasaki) (Lepidoptera: Pterophoridae) feeding inside fruit galls induced by *A. baca* on its summer host, *Ampelopsis glandulosa* var. *heterophylla* (Thunberg) Momiyama (Vitaceae). *Amaya stellate* little affected *A. baca* larvae, while *N. vitis* frequently caused the death of the gall midge larvae by destructing the fruit galls.

Two tortricid moth species emerged from axillary bud galls of *Schizomyia sasakii* (Monzen) on *Ilex crenata* Thunberg (Fig. 2.10c) (Aquifoliaceae), and from those of *Schizomyia soyogo* Kikuti on *Ilex chinensis* Sims, *Ilex integra* (Thunberg) (Fig. 2.10d) and *Ilex pedunculosa* Miquel (Sugiura et al. 2006) (see Elsayed et al. 2018 and Sect. 2.7 for taxonomic treatments and host ranges of the two *Schizomyia* species). Larvae of *Rhopobota ustomaculana* (Curtis) (Lepidoptera: Tortricidae) bored into the galls of *S. sasakii* on *I. crenata*, and *S. soyogo* on *I. pedunculosa*, respectively. *Rhopobota ustomaculana* (Curtis) larvae fed on leaves as well as gall tissues of *I. pedunculosa*, suggesting that it is a facultative cecidophage. Comparison of chemical properties between the gall and normal plant tissues of *I. pedunculosa* indicated that the axillary bud galls seem to be relatively favorable food resources for *R. ustomaculana* larvae because of higher water content and lower carbon and polyphenol contents in spite of poorer nitrogen content.

The infestation of young fruit of *Ligustrum japonicum* Thunberg (Oleaceae) by larvae of *Lemyra imparilis* (Butler) (Arctiidae) caused the death of first instars of *Asphondylia sphaera* Monzen by desiccation of internal tissues of the fruit, but Key-factor and key-stage analysis indicated that this infestation less contributed than bottom up mortality factors to the annual changes of total survival rate of the gall midge (Yukawa et al. 2016).

Fujii et al. (2012) obtained adults of *Darumazo distinctus* Morimoto & Miyakawa (Coleoptera: Curculionidae) from axillary bud galls induced by *S. sasakii* on *Ilex crenata* Thunberg var. *hachijoensis* Nakai (Aquifoliaceae). Later, Kojima (2013) found larvae of *D. distinctus* in the galls of *S. soyogo* (as *S. sasakii*, misidentification) on *I. integra* and regarded it as a cecidophagous insect. Further investigations are needed to confirm if its feeding habit is facultative or obligatory.

9.4.3 *Cecidophagy on the Galls Other Than Those of Cecidomyiids*

9.4.3.1 By Lepidopteran Species

Larvae of *Nola innocua* Butler (Lepidoptera: Nolidae) are obligatorily cecidophagous, feeding on the internal tissue of aphid galls on the leaves of *Distylium racemosum* Siebold & Zuccarini (Hamamelidaceae) (Itô and Hattori 1982, 1983). *Batrachedra striolata* Zeller (Lepidoptera: Batrachedridae) larvae fed on gall tissue within the gall of *Euura lasiolepis* Smith (Hymenoptera: Tenthredinidae) but killed only 1% of the sawfly larvae (Price and Craig 1984).

Yamazaki and Sugiura (2003a) reared four lepidopteran species, *Anatrachyntis japonica* Kuroko (Cosmopterygidae), *Hypatima* sp. (Gelechiidae), *Andrioplecta pulverula* Meyrick (Tortricidae), and *Characoma ruficirra* Hampson (Noctuidae) from twig galls of *Trichagalma serratae* (Ashmead) (Hymenoptera: Cynipidae) on *Q. acutissima*. Larvae of *A. japonica* fed on the lignified gall tissues and destroyed the larval cells, while cecidophagy by the larvae of *Hypatima* sp. could not be confirmed (Yamazaki and Sugiura 2003a). Both *A. pulverula* and *C. ruficirra* are facultatively cecidophagous species and their larvae fed more on unisexual galls of *T. serratae* than on leaves of *Q. acutissima* (Abe 1995).

The larvae of *Stathmopoda auriferella* (Walker) (Lepidoptera: Stathmopodidae) were observed feeding on the stem galls induced by *Eucosma metzneriana*

(Treitschke) (Lepidoptera: Tortricidae) on *Artemisia indica* Willdenow var. *maximowiczii* (Nakai) H. Hara (Asteraceae) (Yamazaki and Sugiura 2003b).

9.4.3.2 By Coleopteran Species

Larvae of *Orchestes hustachei* (Klima) (Coleoptera: Curculionidae) were known to feed on the inner wall of leaf galls induced by *Paracolopha morrisoni* (Baker) (Hemiptera: Aphididae) on *Zelkova serrata* (Thunberg) Makino (Ulmaceae) (Yamazaki and Sugiura 2001) but it is now regarded as an aphidophagous rather than a cecidophagous species (Tomisawa and Akimoto 2004). Females of *Demotina fasciculata* Baly (Coleoptera: Chrysomelidae) feed on tissues of bisexual generation galls induced by *Andricus moriokae* Monzen (Hymenoptera: Cynipidae) on the leaves of *Quercus serrata* Thunberg ex Murray (Fagaceae) (Katsuda and Yukawa 2010).

9.4.4 Gall-Sucking Behavior

There are various observations of gall-sucking behavior by hemipteran species (e.g. Pepper and Tissot 1973; Wachi 2010; Yang et al. 2001; Miller and Crespi 2003). Yamazaki and Sugiura (2005) observed in Osaka and Kyoto Prefectures, Japan seven hemipteran species sucking on six sorts of gall induced by six species of insects on six plant species belonging to six families (see Table 1 in Yamazaki and Sugiura 2005). Among them, there was only one example in which cecidomyiid galls were sucked by a pentatomid. Adults of *Carbula humerigera* (Uhler) (Hemiptera: Pentatomidae) were found sucking on oval or subglobular stem galls induced by *L. achyranthii* on *A. bidentata*. Because *C. humerigera* is known to suck from stems and flowers of various asteraceous herbs and rosaceous shrubs, it is regarded as a facultative gall-sacking species.

9.4.5 Gall Wall Thickness

Generally, parasitoids are restricted to attacking larvae inhabiting galls within reach of their ovipositor. Therefore, a thick gall-wall is considered to serve in defense of gallers from natural enemy attacks (e.g. Abrahamson and Weis 1997; Craig et al. 1990; Jones 1983; Stone et al. 2002; Katsuda and Yukawa 2010). In contrast, thick-walled galls may attract cecidophagous insects because many works have demonstrated a high concentration of potentially nutritive compounds in galls (e.g. McNeill and Southwood 1977; Rohfritsch and Shorthouse 1982; Price et al. 1987). However, there are some examples in which nutrient concentrations are lower in galls than in ungalled tissues (e.g. Hartley 1998; St. John and Shorthouse 2000) and some chemicals that could be deleterious to herbivores accumulate in galls (Cornell

1983; Rohfritsch and Shorthouse 1982; Price et al. 1987). Nevertheless, various insects have overcome some of such difficulties, if any, and evolved to be able to feed on gall tissues either obligatorily or facultatively, as shown in this section.

References

Abe Y (1995) Relationships between the gall wasp, *Trichagalma serratae* (Ashmead) (Hymenoptera: Cynipidae), and two moth species, *Andrioplecta pulverula* (Meyrick) (Lepidoptera: Tortricidae) and *Characoma ruficirra* (Hampson) (Lepidoptera: Noctuidae). Appl Entomol Zool 30:83–89

Abe Y, Melika G, Stone GN (2007) The diversity and phylogeography of cynipid gall wasps (Hymenoptera: Cynipidae) of the oriental and eastern Palearctic regions, and their associated communities. Orient Insects 41:169–212

Abe Y, Ide T, Wachi N (2011) Discovery of a new gall-inducing species in the inquiline tribe Synergini (Hymenoptera: Cynipidae): inconsistent implications from biology and morphology. Ann Entomol Soc Am 104:115–120

Abrahamson WG, Weis AE (1997) Evolutionary ecology across three trophic levels: goldenrods, Gallmakers, and natural enemies, monographs in population biology no 29. Princeton University Press, Princeton, NJ

Akar H, Osgood EA (1987) The method by which larvae of *Dasineura balsamicola* (Lintner) (Diptera: Cecidomyiidae) gain access to the interior of galls induced by *Paradiplosis tumifex* Gagné (Diptera: Cecidomyiidae). Can Entomol 119:395–396

Akimoto S (1988) The evolution of gall parasitism accompanied by a host shift in the gall aphid, *Eriosoma yangi* (Homoptera: Aphidoidea). Biol J Linn Soc 35:297–312

Anfora G, Isidoro N, De Cristofaro A, Ioriatti C (2005) Description of *Macrolabis mali* sp. nov. (Diptera Cecidomyiidae), a new inquiline gall midge species from galls of *Dasineura mali* on apple in Italy. Bull Insectol 58:95–99

Askew RR (1984) The biology of gall wasps. In: Ananthakrishnan TN (ed) Biology of gall insects. Oxford and IBH Publishing Co., New Delhi, pp 223–271

Barnes HF (1933) A cambium miner of basket willows (Agromyzidae) and its inquiline gall-midge (Cecidomyidae). Ann Appl Biol 20:498–519

Barnes HF (1952) Studies of fluctuations in insect populations. XII further evidence of prolonged larval life in the wheat-blossom midges. Ann Appl Biol 39:370–373

Barnes HF (1958) Wheat blossom midges on Broadbalk, Rothamstead experiment station, 1927–56. Proceedings of 10th International Congress of Entomology, Montreal, vol. 3, pp 367–374

Bernardo U, Gebiola M, Xiao Z, Zhu CD, Pujade-Villar J, Viggiani G (2013) Description of *Synergus castaneus* n. sp. (Hymenoptera: Cynipidae: Synergini) associated with an unknown gall on *Castanea* spp. (Fagaceae) in China. Ann Entomol Soc Am 106:437–446

Carleton RD, Johns RC, Edwards BV, Eveleigh ES, Silk PJ (2015) Diel activity patterns of *Paradiplosis tumifex* (Diptera: Cecidomyiidae) and its inquiline, *Dasineura balsamicola* (Diptera: Cecidomyiidae), within young balsam fir. Can Entomol 147:181–192

Cornell HV (1983) The secondary chemistry and complex morphology of galls formed by the Cynipinae (Hymenoptera): why and how? Am Midl Nat 110:220–234

Craig TP, Itami JK, Price PW (1990) The window of vulnerability of a shoot-galling sawfly to attack by a parasitoid. Ecology 71:1471–1482

Craig TP, Livingston-Anderson A, Itami JK (2020) A small-tiled geographic mosaic of coevolution between *Eurosta solidaginis* and its natural enemies and host plant. Ecosphere 11:e03182. https://doi.org/10.1002/ecs2.3182

Dorchin N, Scott ER, Abrahamson WG (2006) First record of *Macrolabis* (Diptera: Cecidomyiidae) in America: a new inquiline species from *Dasineura folliculi* galls on goldenrods. Ann Entomol Soc Am 99:656–661

Elsayed AK, Yukawa J, Tokuda M (2018) A taxonomic revision and molecular phylogeny of the eastern Palaearctic species of the genera *Schizomyia* Kieffer and *Asteralobia* Kovalev (Diptera: Cecidomyiidae: Asphondyliini), with descriptions of five new species of *Schizomyia* from Japan. ZooKeys 808:123–160

Fernandes GW, Martins RP, Neto ET (1987) Food web relationships involving *Anadiplosis* sp. galls (Diptera: Cecidomyiidae) on *Machaerium aculeatum* (Leguminosae). Rev Bras Bot 10:117–123

Fujii T, Yoshitake H, Matsuo K, Tokuda M (2012) Collection records of *Darumazo distinctus* (Coleoptera: Curculionidae) from galls induced by *Asteralobia sasakii* (Diptera: Cecidomyiidae) on the Izu Islands, Japan. Jpn J Syst Entomol 18:253–256

Gagné RJ (1989) The plant–feeding gall midges of North America. Cornell University Press, Ithaca, NY

Gagné RJ, Jaschhof M (2017) A catalog of the Cecidomyiidae (Diptera) of the World, 4th edn. Digital. [Cited 10 May 2020] Available from URL: https://www.ars.usda.gov/ARSUserFiles/80420580/Gagne_2017_World_Cat_4th_ed.pdf

Guay JF, Bulot D, Béland JM, Cloutier C (2016) Gall-maker *Paradiplosis tumifex* (Diptera: Cecidomyiidae) and its inquiline *Dasineura balsamicola* (Diptera: Cecidomyiidae): an update on epidemic episodes and seasonal ecology in Québec, Canada. Can Entomol 148:452–465

Hartley SE (1998) The chemical composition of plant galls: are levels of nutrients and secondary compounds controlled by the gall-former? Oecologia 113:492–501

Ide T, Kusumi J, Miura K, Abe Y (2018) Gall inducers arose from inquilines: phylogenetic position of a gall-inducing species and its relatives in the inquiline tribe Synergini (Hymenoptera: Cynipidae). Ann Entomol Soc Am 111:6–12

Itô Y, Hattori I (1982) A kleptoparasitic moth, *Nola innocua* Butler, attacking aphid galls. Ecol Entomol 7:475–478

Itô Y, Hattori I (1983) Relationship between *Nola innocua* Butler (Lepidoptera: Nolidae), a kleptoparasite, and aphids which cause gall on *Distylium racemosum* trees. Appl Entomol Zool 18:361–370

Jones D (1983) The influence of host density and gall shape on the survivorship of *Diastrophus kinkaidii* gill. (Hymenoptera: Cynipidae). Can J Zool 61:2138–2142

Kamijo K (2000) Chalcidoidea of the Imperial palace, Tokyo. Mem Natl Sci Mus 36:325–333. (In Japanese with English summary)

Katsuda T, Yukawa J (2010) Cecidophagy in adults of *Demotina fasciculata* (Coleoptera: Chrysomelidae) and its effects on the survival of *Andricus moriokae* (Hymenoptera: Cynipidae) inhabiting leaf galls on *Quercus serrata* (Fagaceae). Entomol Sci 13:8–16

Kojima H (2013) A new host record for a cecidophagous molytine weevil, *Darumazo distinctus* (Coleoptera: Curculionidae). Elytra 3:185–186

Kolesik P, Skuhravá M (1997) *Dasineura hybanthi* spec. nov., a new inquiline species of Cecidomyiidae (Diptera) from galls on *Hybanthus floribundus* (Violaceae) in Australia. Stud Dipterol 4:239–245

LaSalle J (1987) New world Tanaostigmatidae (Hymenoptera: Chalcidoidea). Contrib Am Entomol Inst 23:1–181

LaSalle J (1997) Chapter 19. Tanaostigmatidae. In: Gibson GAP, Huber JT, Woolley JB (eds) Annotated keys to the genera of Nearctic Chalcidoidea (Hymenoptera). NRC Research Press, Ottawa, pp 700–704

Liu Z, Yang XH, Zhu DH, He YY (2012) A new species of *Saphonecrus* (Hymenoptera, Cynipoidea) associated with plant galls on *Castanopsis* (Fagaceae) in China. Ann Entomol Soc Am 105:555–561

Maetô K (1991) Braconid parasitoids (Hymenoptera) of the gall-making Cecidomyiidae (Diptera) in Japan. Jpn J Entomol 59:295–313

Maia VC, Carvalho-Fernandes SP, Rodrigues AR, Ascendino S (2014) Galls in the Brazilian costal vegetation. In: Fernandes GW, Santos JC (eds) Neotropical insect galls. Springer, Dordrecht/Heidelberg/New York/London, pp 295–362

Mani MS (1964) Ecology of plant galls. Dr. W. Junk, The Hague

Masi L (1940) Description of a new tanaostigmine from Japan (Hymen. Chalcidoidea). Mushi 3:29–32

McNeill S, Southwood TRE (1977) The role of nitrogen in the development of insect/plant relationships. In: Harborne JB (ed) Biochemical aspects of plant and animal coevolution. Academic Press, New York, pp 77–98

Mediga CAO, Takasu K (2018) Life history strategies of the armored scale, *Aulacaspis alisiana* (Hemiptera: Coccoidea: Diaspididae) on the Japanese silver tree *Neolitsea sericea* (Bl.) Koidz. (Lauraceae) in Fukuoka, Japan. J Plant Sci Phytopathol 2:55–58

Miller DG, Crespi B (2003) The evolution of inquilinism, host-plant use and mitochondrial substation rates in *Tamalia* gall aphids. J Evol Biol 16:731–743

Miyatake T, Kuba H, Yukawa J (2000) Seasonal occurrence of *Bactrocera acutellata* (Diptera: Tephritidae) a cecidophage of stem galls produced by *Lasioptera* sp. (Diptera: Cecidomyiidae) on wild gourds (Cucurbitaceae). Ann Entomol Soc Am 93:1274–1279

Ohbayashi N, Niisato T (2007) Longicorn beetles of Japan. Tokai University Press, Kanagawa. (In Japanese)

Patton WH (1987) A gall inhabiting ant. Am Nat l:26–l27

Pepper JO, Tissot AN (1973) Pine-feeding Species of Cinara in the Eastern United States (Homoptera: Aphididae). Florida agricultural experiment stations monograph series no. 3, Florida, USA

Ping C (1920) Some inhabitants of the round gall of golden-rod. J Entomol Zool 7:161–177

Price PW (1986) Ecological aspects of host plant resistance and biological control: interactions among three trophic levels. In: Boethel DJ, Eikenbary RD (eds) Interactions of plant resistance and parasitoids and predators of insects. Wiley & Sons, Chichester, pp 11–30

Price PW, Craig TP (1984) Life history, phenology, and survivorship of a stem-galling sawfly, *Euura lasiolepis* (Hymenoptera: Tenthredinidae), on the arroyo willow, *Salix lasiolepis*, in northern Arizona. Ann Entomol Soc Am 77:712–719

Price PW, Bouton CE, McPherson BA, Thompson JN, Weis AE (1980) Interactions among three trophic levels: influences of plants on interactions between insect herbivores and natural enemies. Annu Rev Ecol Syst 11:41–65

Price PW, Fernandes GW, Waring GL (1987) Adaptive nature of insect galls. Environ Entomol 16:15–24

Raman A, Ananthakrishnan TN (1984) Biology of gall thrips (Thysanoptera: Insecta). In: Ananthakrishnan TN (ed) Biology of gall insects. Oxford and IBH Publishing Co., New Delhi, pp 107–128

Rohfritsch O, Shorthouse JD (1982) Insect galls. In: Kahl G, Schell JS (eds) Molecular biology of plant tumors. Academic, New York, pp 131–152

Ronquist F (1994) Evolution of parasitism among closely related species: phylogenetic relationships and the origin of inquilinism in gall wasps (Hymenoptera, Cynipidae). Evolution 48:241–266

Ronquist F, Nieves-Aldrey JL, Buffington ML, Liu Z, Liljeblad J, Nylander JAA (2015) Phylogeny, evolution and classification of gall wasps: the plot thickens. PLoS One 10:e0123301

Schwéger S, Melika G, Tang CT, Bihari P, Bozsó M, Stone GN, Nicholls JA, Pénzes Z (2015) New species of cynipid inquilines of the genus *Synergus* (Hymenoptera: Cynipidae: Synergini) from the eastern Palaearctic. Zootaxa 3999:451–497

St. John MG, Shorthouse JD (2000) Allocation patterns of organic nitrogen and mineral nutrients within stem galls of *Diplolepis spinosa* and *Diplolepis triforma* (Hymenoptera: Cynipidae) on wild roses (Rosaceae). Can Entomol 132:635–648

Stone GN, Schonrogge K, Atkinson RJ, Bellido D, Pujade-Villar J (2002) The population biology of oak gall wasps (Hymenoptera: Cynipidae). Annu Rev Entomol 47:633–668

Sugiura S, Yamazaki K (2006) First record of a psilid fly feeding on cecidomyiid galls. Can Entomol 138:235–237

Sugiura S, Yamazaki K, Osono T (2006) Consequences of gall tissues as a food resource for a tortricid moth attacking cecidomyiid galls. Can Entomol 138:390–398

Tomisawa R, Akimoto S (2004) Host range and host preference of a flea weevil, *Orchestes hustachei*, parasitizing aphid galls. Entomol Sci 7:21–30

Turlings TCJ, Erb M (2018) Tritrophic interactions mediated by herbivore-induced plant volatiles: mechanisms, ecological relevance, and application potential. Annu Rev Entomol 63:433–452

Van Noort S, Stone GN, Whithead VB, Nieves-Aldrey JL (2007) Biology of *Rhoophilus loewi* (Hymenoptera: Cynipoidea: Cynipidae), with implications for the evolution of inquilinism in gall wasps. Biol J Linn Soc 90:153–172

Wachi N (2010) Gall-feeding behavior by *Dinorhynchus dybowskyi* (Hemiptera: Pentatomidae). Pulex 89:551–553. (In Japanese)

Wachi N, Ide T, Abe Y (2011) A new inquiline species of *Saphonecrus* (Hymenoptera: Cynipidae: Synergini) associated with cecidomyiid galls on oak trees in Japan. Ann Entomol Soc Am 104:369–373

Walsh BD (1864/1866) On the insects, coleopterous, hymenopterous and dipterous, inhabiting the galls of certain species of willows. Proc Entomol Soc Philadelphia 3:543–641; 6:223–288

Yamazaki K, Sugiura S (2001) Bionomics of the gall-parasitic flea weevil *Rhynchaenus hustachei* (Coleoptera: Curculionidae). Entomol Sci 4:239–242

Yamazaki K, Sugiura S (2003a) Gall-feeding habits in Lepidoptera of Japan. I: three different types of galls. Trans Lepidopterol Soc Jpn 54:31–39

Yamazaki K, Sugiura S (2003b) Gall-feeding habits in Lepidoptera of Japan. II: a cecidophagous stathmopodid moth attacking the gall of a tortricid moth on the Japanese mugwort. Trans Lepidopterol Soc Jpn 54:143–146

Yamazaki K, Sugiura S (2005) Hemiptera as cecidophages. Entomol News 116:121–126

Yang MM, Mitter C, Miller DR (2001) First incidence of inquilinism in gall-forming psyllids, with a description of the new inquiline species (Insecta, Hemiptera, Psylloidea, Psyllidae, Scondyliaspidinae). Zool Scr 30:97–113

Yang ZQ, Sun JH, Oitts JP (2004) First discovery of the family Tanaostigmatidae (Hymenoptera: Chalcidoidea) from China with a description of a new gall-making species utilizing kudzu leaves. J Entomol Sci 39:275–280

Yukawa J (1974) Descriptions of new Japanese gall midges (Diptera, Cecidomyiidae, Asphondyliidi) causing leaf galls on Lauraceae. Kontyû 42:293–304

Yukawa J (1977) Life of *Oberea hebescens* and its relation to *Pseudasphondylia neolitseae*. Insectarium 14:152–155. (In Japanese)

Yukawa J (1983) Arthropod community centred upon the neolitsea leaf gall midge, *Pseudasphondylia neolitseae* Yukawa (Diptera, Cecidomyiidae) and its host plant, *Neolitsea sericea* (Blume) Koidz. (Lauraceae). Mem Fac Agric Kagoshima Univ 19:89–96

Yukawa J, Akimoto K (2006) Influence of synchronization between adult emergence and host plant phenology on the population density of *Pseudasphondylia neolitseae* (Diptera: Cecidomyiidae) inducing leaf galls on *Neolitsea sericea* (Lauraceae). Popul Ecol 48:13–21

Yukawa J, Haitsuka S (1994) A new cecidomyiid successor (Diptera) inhabiting empty midge galls. Jpn J Entomol 62:709–718

Yukawa J, Masuda H (1996) Insect and mite galls of Japan in colors. Zenkoku Nôson Kyôiku Kyôkai, Tokyo. (In Japanese with English explanations for color plates)

Yukawa J, Rohfritsch O (2005) Biology and ecology of gall–inducing Cecidomyiidae (Diptera). In: Raman A, Schaefer CW, Withers TM (eds) Biology, ecology, and evolution of gall–inducing arthropods, vol 1. Science Publications, Enfield, pp 273–304

Yukawa J, Sunose T (1976) Description of a new gall midge (Diptera, Cecidomyiidae) on *Euonymus*, with notes on its bionomics. Kontyû 44:159–168

Yukawa J, Haitsuka S, Miyaji K, Kamikado T (2006) Influence of the population dynamics of a gall-inducing cecidomyiid and its parasitoids on the abundance of a successor, *Lasioptera yadokariae* (Diptera: Cecidomyiidae). In: Ozaki K, Yukawa J, Ohgushi T, Price PW (eds) Galling arthropods and their associates: ecology and evolution. Springer-Verlag, Tokyo, pp 241–249

Yukawa J, Ikenaga H, Sato S, Tokuda M, Ganaha-Kikumura T, Uechi N, Matsuo K, Mishima M, Tung GH, Paik JC, Ren BQ, Don XO (2012) Description and ecological traits of a new species of *Pitydiplosis* (Diptera: Cecidomyiidae) that induces leaf galls on *Pueraria* (Fabaceae) in East Asia, with a possible diversification scenario of intraspecific groups. Entomol Sci 15:81–98

Yukawa J, Miyamoto K, Yamaguchi T, Takesaki K, Uechi N, Matsuo K (2016) Key factor/key stage analysis of long-term life table data for a fruit gall midge, *Asphondylia sphaera* (Diptera: Cecidomyiidae). Ecol Entomol 41:516–526

Yukawa J, Yamamura K, Fujimoto K, Tokuda M, Muroya S, Maki T, Sadoshima T, Fukuda T (2018) Key-factors and key-stages that determine the leaf longevity of an evergreen broad-leaved tree, *Neolitsea sericea* (Lauraceae) at different growing sites in southern Japan. Ecol Res 33:175–190

Yukawa J, Takahashi K, Ohsaki N (1976) Population behaviour of the neolitsea leaf gall midge, *Pseudasphondylia neolitseae* Yukawa (Diptera, Cecidomyiidae). Kontyû 44:358–365

Chapter 10
Association with Host Plants

Junichi Yukawa and Makoto Tokuda

Abstract Gall-inducing cecidomyiids and their host plants interact closely with each other in various ways. Interaction includes, for example, synchronization of herbivores with host plant phenology, negative and positive effects of gall induction on host plants, manipulation of host plants, and critical relation of gall inducers with phytohormones. In addition, recent studies have elucidated molecular basis of interactions between gall midges and their host plants. In this chapter, we summarize various types of interaction between gall midges and their host plants.

Keywords Asynchrony · Galled organ · Interaction · Phenology · Phytohormone · Synchronization

10.1 Introduction

Because gall induction is a delicate event during the course of cell division, tissue differentiation, shoot extension, and other processes of plant growth (e.g. Yukawa and Uechi 1999), gall-inducing organisms have established particular relationships with their host plants and exhibit various modes of interaction with them. The synchronization of herbivores with their host plant phenology is one of such interactions and is a critical event, in particular for such short-lived insects as gall midges (Yukawa 2000, 2010; Yukawa and Akimoto 2006; Yukawa et al. 2013). In addition to various bottom-up effects, a time lag in the synchronization determines quality and quantity of food resources and affects the survival rates and abundance of herbivores (e.g. Kerslake and Hartley 1997; Yukawa 2000). In other cases, galls may

J. Yukawa (✉)
Entomological Laboratory, Faculty of Agriculture, Kyushu University, Fukuoka, Japan
e-mail: galling-cecidomyiids@kyudai.jp

M. Tokuda
Laboratory of Systems Ecology, Faculty of Agriculture, Saga University, Saga, Japan
e-mail: tokudam@cc.saga-u.ac.jp

J. Yukawa, M. Tokuda (eds.), *Biology of Gall Midges*, Entomology Monographs,
https://doi.org/10.1007/978-981-33-6534-6_10

have a deleterious effect on host plants by changing plant architecture, reducing leaf longevity, reducing photosynthesis, seed production, and biomass. Such deleterious effects become severe when gall density is high, resulting in the death of gall midge larvae accompanied by the modification or drop of plant organs and in the shortage of galling sites for the succeeding generations. In contrast, gall-inducing organisms may manipulate host plants and positively affect the existence of galled organs, which in turn directly or indirectly improve their survival. This chapter first demonstrates examples of synchronization between gall midge emergence and their host plant phenology, and then refers to negative and positive effects of gall induction on galled host plant organs. Secondarily, examples are presented for two cecidomyiid species that exhibit critical relation with phytohormones, such as auxin (indole-3-acetic acid, IAA) and cytokinins, in their gall induction (Tooker and De Moraes 2011; Zhu et al. 2011; Tanaka et al. 2013).

Studies of two major gall midge pests associated with cereal crops, the Hessian fly *Mayetiola destructor* (Say) on wheat *Triticum aestivum* Linnaeus (Poaceae) and the Asian rice gall midge *Orseolia oryzae* (Wood-Mason) on rice *Oryza sativa* Linnaeus (Poaceae), have revealed that, as in plant-pathogen interactions, gene-for-gene concept can be applicable between host resistant genes and gall midge biotypes (Harris et al. 2003; Stuart et al. 2012; Bentur et al. 2016). Recent progress in genomic and transcriptomic analyses and several novel aspects of interactions between these gall midges and their host plants are summarized in the last section of this chapter.

10.2 Synchronization Between Gall Midge Emergence and Host Plant Phenology

10.2.1 Synchronization Patterns

The combination of period of gall midge emergence and period of host available season gives rise to the four patterns of synchronization between gall midge emergence and host plant phenology (Table 10.1) (Yukawa 2000). In pattern I, the synchronization is quite difficult for gall midges because even a slight speed or

Table 10.1 Four synchronization patterns in the combination of period of gall midge emergence season and period of host availability

	Pattern I	Pattern II	Pattern III	Pattern IV
Period of emergence season	Short	Short	Long	Long
Period of host availability	Short	Long	Short	Long
	No example	Common in univoltine species	No example	Common in multivoltine species

delay in emergence may cause loss of the chance to encounter a suitable oviposition site. To avoid such a risk, the gall midge must adapt by having (1) an oligophagous habit, by which they can utilize an alternative host, or (2) a particular way to receive information about physiological changes related to host plant phenology. There are, however, no clear evidences to support the two possibilities.

Pattern II is quite common in temperate evergreen broad-leaved forests, being represented by such examples as *Illiciomyia yukawai* Tokuda on *Illicium anisatum* Linnaeus (Schisandraceae) (Yukawa et al. 2013) and *Pseudasphondylia neolitseae* Yukawa on *Neolitsea sericea* (Blume) Koidzumi (Lauraceae) (Yukawa and Akimoto 2006). Details of their patterns will be demonstrated later in this section. In this pattern, the period of host available season is longer than the period of gall midge emergence, but the duration of host available period varies with plant species. In most tree species, it does not exceed 3 weeks. An extreme example of this pattern can be seen in the univoltine population of *Pitydiplosis puerariae* Yukawa, Ikenaga & Sato on a vine, *Pueraria lobata* (Willdenow) Ohwi (Fabaceae). Although host available period lasts from March to October, the gall midge utilizes for galling leaves that opened in April to June, while Taiwanese populations of this gall midge are multivoltine on *Pueraria montana* (Loureiro) Merrill (Yukawa et al. 2012) (see also Sect. 2.7).

In pattern III, only a small part of gall midge population can synchronize with host plant phenology while the remaining large majority cannot. Such a waste way of life history strategy would not evolve. Pattern IV can be seen in multivoltine gall midges but not in uni- or bivoltine species.

10.2.2 Asphondylia aucubae *and* Aucuba japonica

Emergence of *Asphondylia aucubae* Yukawa & Ohsaki occurs within a considerably short period of time (14–20 days) and the young fruit of *Aucuba japonica* Thunberg (Aucubaceae) persist more than a month from late April to early June (Yukawa and Ohsaki 1988b). Therefore, host available period was thought to be long enough for *A. aucubae* as shown in the synchronization pattern 3. However, Imai and Ohsaki (2004a, b) and Imai (2006) revealed that the young fruit becomes available for oviposition only within a restricted period of the season and found that the synchronization with developmental conditions of *A. japonica* fruit was important for successful oviposition by *A. aucubae*. The hard endocarp of *A. aucuba* provided effective mechanical protection of the young developing fruit from the gall midge, which deposits eggs by inserting its aciculae ovipositor into the fruit. The protection was retained only while the fruit was susceptible to gall induction. The gall midge succeeds in oviposition only when the ovipositor is inserted through cracks in the endocarp otherwise oviposition is almost completely unsuccessful. The endocarp, consequently, inhibited 33.2% (1211/3652) of observed ovipositor thrusts (Imai and Ohsaki 2004a). This example demonstrates that the mechanical traits of host plant affect the oviposition efficiency of a gall-inducing insect even within a period when emergence and host plant phenology are synchronizing.

10.2.3 Illiciomyia yukawai *on* Illicium anisatum

Illiciomyia yukawai is distributed in Honshu, Shikoku and Kyushu, Japan and induces hemispherical galls usually on the upper leaf surface of *I. anisatum* (Yukawa and Masuda 1996) (Fig. 10.1a). This gall midge is fundamentally univoltine with summer diapause as first instars and overwinters as third instars in the galls on the host trees (Yukawa et al. 2013). Some individuals require two years to complete one generation (Yukawa et al. 2016c) (see also Sect. 6.4). In this section, adult emergence data of 1- and 2-year populations are summed because they emerge almost simultaneously in May. Females alight on host shoots that are about to open and lay their eggs in the shoots with the aciculate ovipositor.

Yukawa et al. (2013) demonstrated that the lower development threshold temperature for overwintered larvae of *I. yukawai* was 14 °C, while the thermal constant for emergence differed with individuals. Thermal totals above 14 °C up to the 50%

Fig. 10.1 Galls mentioned in this chapter. (**a**), leaf galls of *Illiciomyia yukawai* on *Illicium anisatum*; (**b**), leaf galls of *Contarinia* sp. on *Quercus glauca*; (**c**), fruit galls of *Kiefferia ezoensis* on *Angelica ursina*; (**d**), leaf galls of *Rhopalomyia yomogicola* on *Artemisia indica*

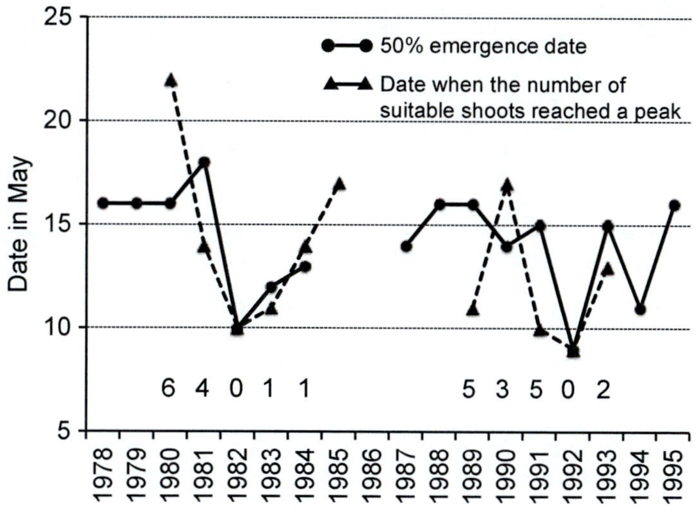

Fig. 10.2 Annual changes in the 50% emergence date of *Illiciomyia yukawai* and the 50% host shoot available date for oviposition. Numerals above years indicate time lag in days between the two dates (modified from Yukawa et al. 2013)

emergence date varied yearly from 33.1 to 68.7 degree-days. The 50% emergence date varied from 9 to 18 May (Fig. 10.2). The thermal totals significantly correlated with the 50% emergence date but did not correlate with the date when 50% of shoots became suitable for oviposition. However, the maximum time lag between the 50% emergence date and the 50% shoot available date was 6 days, which is shorter than 10 days in the case of *P. neolitseae* (Yukawa and Akimoto 2006).

10.2.4 **Pseudasphondylia neolitseae** *and* **Neolitsea sericea**

According to Yukawa et al. (1976), *P. neolitseae* is fundamentally univoltine and a monophagous gall midge inducing leaf galls on *N. sericea* (Fig. 2.6c, d). Larvae pass through summer as the first instars, molt into the second instars in October, and overwinter as full-grown larvae in the leaf galls on the host plant. Pupation takes place in February or March. Adults emerge in late March or April and females deposit their eggs in the terminal leaf buds of *N. sericea*. Leaf buds suitable for oviposition are those with a length ranging from 22 to 38 mm.

Based on the long-term study of population dynamics of *P. neolitseae* and leaf bud extension data, Yukawa and Akimoto (2006) first examined time lag in the synchronization between the peak emergence date and the peak date in the number of suitable buds for oviposition. The time lag varied from one in 1974 to ten days in 1977, indicating that *P. neolitseae* has been synchronizing well with the host plant phenology.

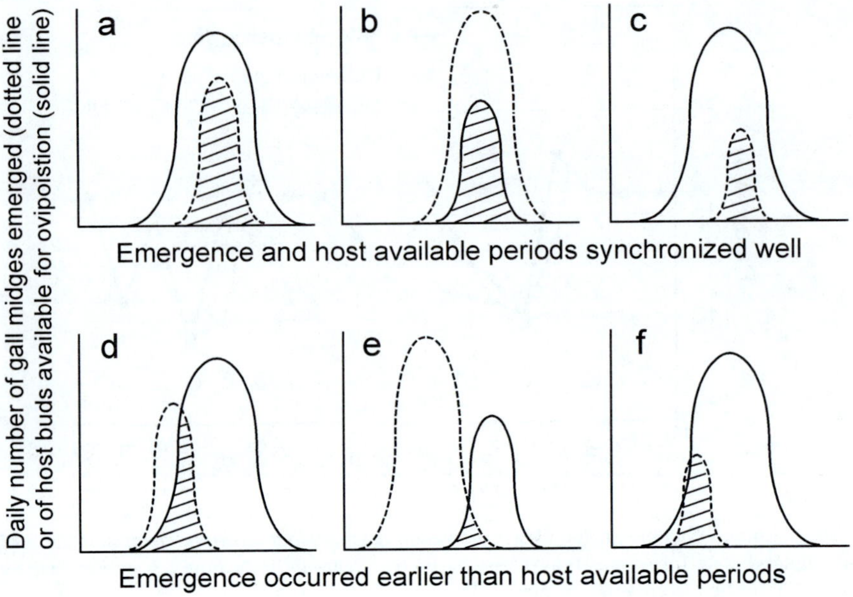

Fig. 10.3 Six patterns of combination expressed by superposing the gall midge emergence curve (dashed line) on the host bud available curve (solid line). (**a**, **b**, and **c**) indicate patterns in which the two curves synchronized well but (**d**, **e**, and **f**) indicate patterns of asynchrony between the two curves. The height of curves indicates the number of gall midges emerged (gall midge density) or the number of host buds available for oviposition (amount of food resources). Oblique lines were drawn in the overlapped area of the two curves, indicating a portion of gall midge population that could synchronize with the host bud available period

Secondly, Yukawa and Akimoto (2006) evaluated the influence of synchronization on the gall midge population density in the following year. In previous studies, the degree of synchronization has been expressed with only time lag between the dates of emergence and host available seasons. However, Yukawa and Akimoto (2006) considered that the time lag did not always reflect the genuine degree of synchronization when the amount of food resources was not included in the evaluation of synchronization. They successfully applied the gamma distribution model to the emergence curve of *P. neolitseae* and the normal distribution model to the daily changes in the number of buds suitable for oviposition. By superposing the two models and calculating overlapped area, both the time lag and the amount of host resources were combined, and the degree of synchronization could be evaluated more accurately than the previous methods (Fig. 10.3).

When the amount of host buds was plenty enough, almost all females could synchronize with the host buds (Fig. 10.3a, c), even when the time lag was distinct (Fig. 10.3f). In contrast, the proportion of females that had synchronized with the suitable host buds became very low under the relative shortage of host buds, even though the peak emergence and host phenology synchronized almost completely (Fig. 10.3b). Under such conditions, the contest-type competition would oblige the losers to leave suitable buds, or to fly from tree to tree in search of the suitable buds, resulted in mistaken oviposition (see Sect. 2.1). In the case when gall midges emerged distinctly earlier than usual, most females could not synchronize with the host bud available period (Fig. 10.3d, e).

10.2.5 *Effects of Global Warming on the Synchronization*

Climate changes affect the synchronization between herbivores and host plants (Dewar and Watt 1992; Harrington et al. 1999; Watt and Woiwod 1999; Yukawa 2000). Global warming possibly causes asynchrony between herbivores and host plants more seriously than before because herbivores and host plants respond to thermal effects differently (Yukawa 2010). The emergence season of gall midges can

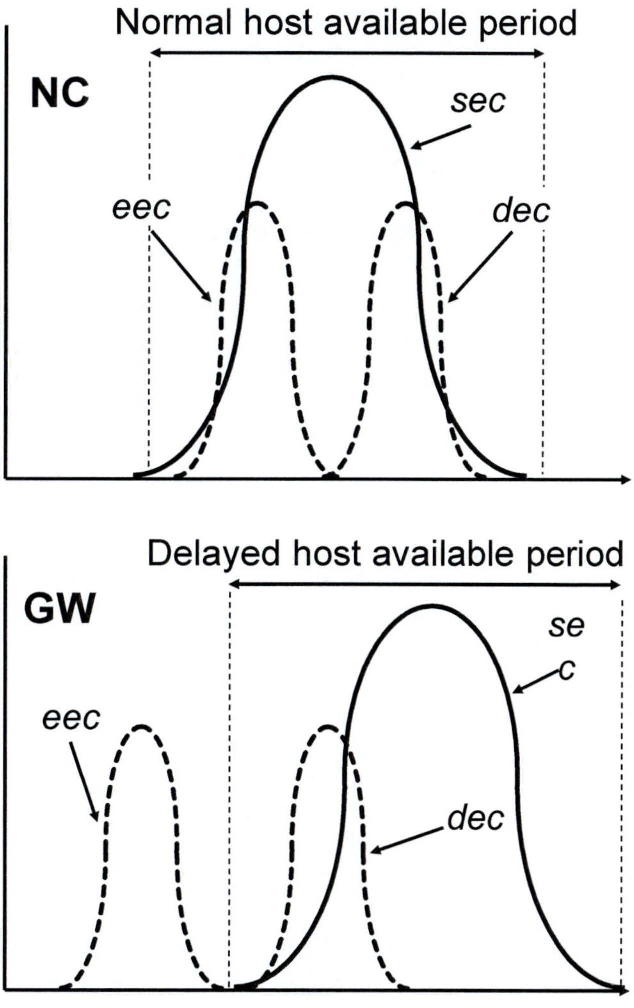

Fig. 10.4 Schematic representation of synchronization between gall midge emergence and host available period under normal conditions (NC) and of asynchrony under global warming conditions (GW). *sec* shoot extension curve, *eec* early emergence curve in relatively warmer years, *dec* delayed emergence curve in relatively cooler years

be forecasted by the accumulation of temperature above the lower development zero point (Yukawa 2000), but phenological events of plant cannot be forecasted simply by the accumulation of temperature, because a period of chilling followed by warming was required for plant growth (e.g. Murray et al. 1989). Therefore, a warm spring may promote the earlier emergence of gall midge but may delay the budburst of host plant (Yukawa 2000) (Fig. 10.4).

As demonstrated in this section, *A. aucubae*, *I. yukawai* and *P. neolitseae* have been synchronizing well with the host plant phenology. However, there was a tendency for *P. neolitseae* to emerge earlier than the bud extension in recent years (Yukawa and Akimoto 2006). Gall midges have been adapting to avoid asynchrony with host plant phenology to a certain extent by retaining individual variations as to thermal constant required for development, but rapid global warming may cause serious asynchrony between herbivores and host plants beyond the degree of adaptation by gall midges. This possibly causes an extinction of local populations, otherwise promotes shifts of host plant or distribution ranges.

10.3 Survival of Galled Organs in Relation to Gall Induction

10.3.1 Earlier Fall of Galled Leaves

Mean longevity of heavily galled leaves of *Euonymus japonicus* Thunberg (Celastraceae) (Fig. 2.6e, f) was considerably shortened, whereas ungalled and lightly galled leaves were seldom shed within a year (see Table 2 in Sunose and Yukawa 1979). Such an earlier fall occurred more severely in the urban environments in association with anthracnose disease caused by *Gloeosporium euonymicolum* Hemmi (Fungi: Ascomycota: Dermateaceae). The earlier drop of galled leaves operates as a density-dependent mortality factor on the population of *Masakimyia pustulae* Yukawa & Sunose.

On the fresh leaves of an evergreen broad-leaved tree *Quercus glauca* Thunberg (Fagaceae), *Contarinia* sp. induces subglobular galls along the midrib, and the galled leaves remain folded upward (Fig. 10.1b). Gall induction by this gall midges also caused earlier drop of galled leaves in April and May, whereas ungalled leaves survived more than 26 months on an average (Yukawa and Tsuda 1986). However, earlier drop of fresh leaves did not kill the larvae because they had exited the galls and dropped to the ground before defoliation.

Usually, the survival rate of *N. sericea* leaves galled by *P. neolitseae* was a little lower than that of ungalled leaves, but the former was remarkably and significantly lower than the latter in 1985 when strong typhoons attacked the forest in Kagoshima, southern Japan (Yukawa et al. 2019). In particular, the survival rate of galled leaves on a tree growing along forest edge was significantly lower than the rates on the other census trees. In addition, heavily galled leaves on *N. sericea* are folded outward to various extents, providing favorable webbing sites for spiders (Yukawa 1983) (see also Sect. 9.1). Because adults of *P. neolitseae* emerged from the lower surface of the gall, they were either trapped in the web or by spiders directly. This mortality is density-dependent.

10.3.2 Survival of Fruit Galls

Asphondylia aucubae induces fruit galls on *A. japonica* (Yukawa and Masuda 1996). Both ungalled and galled fruits cannot be distinguished with size and color until August. Thereafter, the ungalled fruit rapidly grows and reddens at maturity in December (Fig. 2.4a). Some of them remain on the trees in wintertime but the remainders drop to the ground. Frequently, beak-marks can be seen on the surface of fruit on the ground. In contrast, galled fruit slowly grows and remains without beak-marks on the host tree until the following May or June when adults emerge from the galls (Yukawa and Ohsaki 1988a). The galled fruit is distinctly smaller than ungalled fruit and is only partly tinged with red but not the whole of galled fruit. It is highly probable that gall midge larvae manipulate host fruit not to mature normally and to disturb formation of stratum disjunction for keeping galls on the trees.

A similar example has been known for ungalled and galled fruit induced by *Asphondylia sphaera* Monzen on *Ligustrum* spp. (Oleaceae) although abortion of flower buds and young fruit occur greatly regardless of presence or absence of gall midge larvae in the organs (Yukawa and Miyamoto 1979; Yukawa et al. 2016b). Flower bud and fruit gall midges of the genus *Asphondylia* that have type IIB life history strategy (see Sect. 6.2) exhibit more or less similar patterns in gall maturity.

In contrast, fruit galls caused by *Kiefferia* species (Fig. 10.1c) drop to the ground earlier in the season (Kim et al. 2019) even though they belong to tribe Asphondyliini together with *Asphondylia*. Comparison of fruit manipulation system among various fruit gall midge taxa would be one of the interesting study subjects in association of gall midges with the host plant species.

10.4 Involvement of Phytohormones in Gall Induction

Phytohormones, especially auxin (indole-3-acetic acid, IAA) and cytokinins are critically related to gall inductions by various insects (Tokuda et al. 2013; Tooker and Helms 2014; Kai et al. 2017). Although whether insects have an ability to synthesize phytohormones by themselves or not were remained unclear, Yamaguchi et al. (2012) demonstrated that a gall-inducing sawfly *Pontania* sp. (Hymenoptera: Tenthredinidae) possesses the synthetic ability of auxin, and probably also cytokinins, in their body.

In Cecidomyiidae, larvae of the Hessian fly *M. destructor* that fed on susceptible wheat variety elevated the IAA level but did not increase phytohormones such as jasmonic (JA) and salicylic (SA) acids that are responsible for plant defense, which suggests that the gall midge larvae nutritionally modify plant tissue and manipulate it to avoid plant defense (Tooker and De Moraes 2011). In contrast, Zhu et al. (2011) revealed that, in a resistant wheat variety, both JA and SA levels increased following the infestation by Hessian fly larvae.

Rhopalomyia yomogicola (Matsumura) induces leaf galls on some species of *Artemisia* (Asteraceae) (Table 4.1; Fig. 10.1d). Tanaka et al. (2013) found that larvae of *R. yomogicola* possessed high levels of IAA and cytokinins, and had an ability to synthesize IAA as in *Pontania* sp. Based on these findings, they suggested that these phytohormones might have functions to maintain the feeding part as fresh nutritive tissue in the larval chamber of gall.

10.5 Plant Resistance to Pest Gall Midges

Plant resistance to various pest gall midges has been studied. *Dasineura* and *Rabdophaga* species on *Salix* were studied from various aspects, such as oviposition choice and larval survival (Larsson and Strong 1992; Larsson et al. 1995), heritability of host plant resistance (Strong et al. 1993), interspecific hybridization of plants and resistance to herbivores (Fritz et al. 1994), gall initiation success and fecundity (Glynn and Larsson 1994), hypersensitive response associated with resistance (Ollerstam et al. 2002), and preference-performance relationship (Nyman et al. 2011).

The sorghum midge *Stenodiplosis sorghicola* (Coquilett) is the most important pest of grain sorghum worldwide. Franzmann and Butler (1993), Sharma (1993), Sharma et al. (1993), and Teetes et al. (1994) studied plant resistance to the sorghum midge. Local losses in tropical Africa and Asia may exceed 50%. The gall midge resistant sorghum DSV 3 reduced the damage by 74% over untreated control. Furthermore, it proved superior to all the chemical control treatment by recording 41% lower damage than mean of insecticide control (Kamatar et al. 2009).

The resistance of *Pinus densiflora* Siebold & Zuccarini, *Pinus thunbergii* Parlatore and other species of the genus *Pinus* (Pinaceae) against the Japanese pine-needle gall midge *Thecodiplosis japonensis* Uchida & Inouye was studied (Lee et al. 1993). The resistant trees were characterized by higher contents (compared with susceptible trees) of limonene and lower contents of â-pinene in needles. The resistant trees also had higher contents of terpinolene and á-pinene (Kim et al. 1975; Kim et al. 1977). Females laid eggs on all the pine species in the study, though the gall formation rate was 0% in *Pinus virginiana* Miller, 9% in *P. thunbergii* × *P. virginiana* and 22% in *P. thunbergii*. Substances present in needles of *P. virginiana* killed the young larvae. It was suggested that the phenolic compounds contained in needles of *P. virginiana*, i.e. salicylic acid and chitinase, may be substances that confer resistance to *T. japonensis* (Son et al. 1996a, b).

10.6 Molecular Basis Interactions of Gall Midges with Their Host Plants

10.6.1 Genomic and Transcriptomic Analyses of the Hessian Fly Mayetiola destructor

Molecular and genetic studies focusing on interactions between the Hessian fly *M. destructor* and wheat have been reviewed in Harris et al. (2003), Stuart et al. (2012) and Stuart (2015). After these, Chen et al. (2016a) compared gene expressions of *M. destructor* larvae between resistant and susceptive wheat cultivars and detected that larvae on the resistant cultivar expressed lower abundance of genes related to nutrient metabolism, RNA synthesis, and protein synthesis. So, the resistant cultivar is suggested to inhibit nutrient metabolism and protein production in *M. destructor* larvae. In addition, *M. destructor* larvae on the resistant cultivar expressed more cytochrome P450 genes, which have many functions including detoxification of toxins, at higher abundance than those on the susceptible cultivar, suggesting that toxic chemicals from the resistant cultivar are responsible for the larval death of *M. destructor* on the resistant cultivar (Chen et al. 2016a). Furthermore, Chen et al. (2016b) analyzed gene expressions in different aged *M. destructor* larvae as well as pupae and adults.

Secreted salivary gland proteins (SSGPs) are considered to have critical roles for the gall induction by *M. destructor* (Chen et al. 2008). Several candidates of effector proteins were detected by transcriptomic analyses of SSGPs for the first instar of *M. destructor* (Al-Jbory et al. 2018a) as well as for the barley midge *Mayetiola hordei* Kieffer and the oat midge *Mayetiola avenae* Marchal (Al-Jbory et al. 2018b). Crane et al. (2020) analyzed salivary proteins in different biotypes of *M. destructor* but any variations among biotypes were associated with the virulence to any wheat resistance genes.

Zhao et al. (2016) mapped an avirulence gene (*vH24*) in *M. destructor* genome and proposed that the gene encodes a protein containing a type-2 serine/threonine protein phosphatase (PP2C) domain. This protein is suggested to act as an effector in wheat. Wang et al. (2018) analyzed proteins secreted by *M. destructor* larvae and suggested that nucleoside diphosphate kinase (NDK) as an effector protein responsible for host manipulation by the gall midge.

10.6.2 Interactions Between Rice and the Asian Rice Gall Midge

Relationships between rice resistant genes and biotypes of the Asian rice gall midge *O. oryzae* were reviewed in Bentur et al. (2016) and Sinha et al. (2017). Sinha et al. (2015) identified apoptotic cell death gene (*OoDAD1*) expressed by the feeding of *O. orizae* and suggested that the gene expression induces hypersensitive response in resistant rice against infestation by *O. oryzae*. Agarrwal et al. (2016) analyzed transcriptomic and metabolic changes in resistant rice variety infested by an avirulent biotype of *O. oryzae* and proposed resistant mechanism of rice against *O. oryzae* through the occurrence of hypersensitive response at the feeding site. In addition, Atray et al. (2016) determined complete mitochondrial genome of *O. oryzae* including several unique features such as rearrangement in the order of tRNAs and protein coding genes and reported sequential differences in the tandem repeat region among different biotypes of *O. oryzae*.

References

Agarrwal R, Padmakumari AP, Bentur JS, Nair S (2016) Metabolic and transcriptomic changes induced in host during hypersensitive response mediated resistance in rice against the Asian rice gall midge. Rice 9:5. https://doi.org/10.1186/s12284-016-0077-6

Al-Jbory Z, Anderson KM, Harris MO, Mittapalli O, Whitworth RJ, Cheng MS (2018a) Transcriptomic analyses of secreted proteins from the salivary glands of wheat midge larvae. J Insect Sci 18:17. https://doi.org/10.1093/jisesa/iey009

Al-Jbory Z, El Bouhssini M, Chen MS (2018b) Conserved and unique putative effectors expressed in the salivary glands of three related gall midge species. J Insect Sci 18:15. https://doi.org/10.1093/jisesa/iey094

Atray I, Bentur JS, Nair S (2016) The Asian rice gall midge (*Orseolia oryzae*) mitogenome has evolved novel gene boundaries and tandem repeats that distinguish its biotypes. PLoS One 10: e0134625. https://doi.org/10.1371/journal.pone.0134625

Bentur JS, Rawat N, Divya D, Sinha DK, Agarrwal R, Atray I, Nair S (2016) Rice–gall midge interactions: battle for survival. J Insect Physiol 84:40–49

Chen M-S, Zhao H-X, Zhu YC, Scheffler B, Liu X, Liu X, Hulbert S, Stuart JJ (2008) Analysis of transcripts and proteins expressed in the salivary glands of hessian fly (*Mayetiola destructor*) larvae. J Insect Physiol 54:1–16

Chen M-S, Liu S, Wang H, Cheng X, El Bouhssini M, Whitworth RJ (2016a) Genes expressed differentially in hessian fly larvae feeding in resistant and susceptible plants. Int J Mol Sci 17:1324. https://doi.org/10.3390/ijms17081324

Chen M-S, Liu S, Wang H, Cheng X, El Bouhssini M, Whitworth RJ (2016b) Massive shift in gene expression during transitions between developmental stages of the gall midge, *Mayetiola destructor*. PLoS One 11:e0155616. https://doi.org/10.1371/journal.pone.0155616

Crane YM, Crane CF, Schemerhorn BJ (2020) Relationship of secreted salivary protein variants to virulence in hessian fly (*Mayetiola destructor* (say)). Adv Entomol 8:15–33

Dewar RC, Watt AD (1992) Predicted changes in the synchrony of larval emergence and budburst under climatic warming. Oecologia 89:557–559

Franzmann BA, Butler DG (1993) Compensation for reduction in seed set due to sorghum midge (Contarinia sorghicola) in midge-susceptible and midge-resistant sorghum (Sorghum bicolor). Aust J Exp Agric 33:193–196

Fritz RS, Nichols-Orians CM, Brunsfeld SJ (1994) Interspecific hybridization of plants and resistance to herbivores: hypotheses, genetics, and variable responses in a diverse herbivore community. Oecologia 97:106–117

Glynn C, Larsson S (1994) Gall initiation success and fecundity of *Dasineura marginemtorquens* on variable *Salix viminalis* host plants. Entomol Exp Appl 73:11–17

Harrington R, Woiwod I, Sparks T (1999) Climate change and trophic interactions. Trends Ecol Evol 14:146–150

Harris MO, Stuart JJ, Mohan M, Nair S, Lamb RJ, Rohfritsch O (2003) Grasses and gall midges: plant defense and insect adaptation. Annu Rev Entomol 48:549–577

Imai (2006) A prospective mechanism in the host plant, *Aucuba*, against oviposition by the fruit gall midge, *Asphondylia aucubae* (Diptera: Cecidomyiidae). In: Ozaki K, Yukawa J, Ohgushi T, Price PW (eds) Galling arthropods and their associates: ecology and evolution. Springer-Verlag, Tokyo, pp 169–176

Imai K, Ohsaki N (2004a) Internal structure of developing aucuba fruit as a defense increasing oviposition costs of its gall midge *Asphondylia aucubae*. Ecol Entomol 29:420–428

Imai K, Ohsaki N (2004b) Oviposition site of and gall formation by the fruit gall midge *Asphondylia aucubae* (Diptera: Cecidomyiidae) in relation to internal fruit structure. Entomol Sci 7:133–137

Kai S, Kumashiro S, Adachi S, Suzuki Y, Shiomi Y, Matsunaga K, Gyoutoku N, Asami T, Tokuda M (2017) Life history of *Stenopsylla nigricornis* (Hemiptera: Psylloidea: Triozidae) and phytohormones involved in its gall induction. Arthropod Plant Interact 11:99–108

Kamatar MY, Yadwad A, Shekharappa BSG (2009) Host plant resistance vs. insecticidal control of sorghum midge. Plant Arch 9:831–833

Kerslake JE, Hartley SE (1997) Phenology of winter moth feeding on common heather: effects on source population and experimental manipulation of hatch dates. J Anim Ecol 66:373–385

Kim CS, Hong SH, Choi C, Kim JW, Ryu JB, Park MH (1975) Breeding of varieties of pines resistant to pine gall midges (*Thecodiplosis japonensis*). I degree of gall formation and monoterpene composition of resistant trees. Korean J Breed 7:135–142

Kim JW, Kim CS, Ryu JB, Kim JS, Park CK (1977) Studies on the monoterpene composition in the oleoresin of *Pinus thunbergii* Parl and the resistance to pine gall midge. J Korean Agric Chem Soc 20:81–87. 30 refs

Kim W, Minami T, Tokuda M, Matsuo K, Harris KM, Yukawa J (2019) Detection of two new cryptic species of *Kiefferia* (Diptera: Cecidomyiidae) by means of morphological, molecular and ecological studies. Entomol Sci 22:450–462

Larsson S, Strong DR (1992) Oviposition choice and larval survival of *Dasineura marginemtorquens* (Diptera: Cecidomyiidae) on resistant and susceptible *Salix viminalis*. Ecol Entomol 17:227–232

Larsson S, Glynn C, Hoglund S (1995) High oviposition rate of *Dasineura marginemtorquens* on *Salix viminalis* genotypes unsuitable for offspring survival. Entomol Exp Appl 77:263–270

Lee DK, Kim KS, Kwon KW, Hoh DS (1993) Comparing resistance to the pine gall midge among seedlings produced from controlled pollination and grafting, (pp 153–156). In: Price PW, Mattson WJ, Baranchikov YM (eds) The ecology and evolution of gall-forming insects. General technical report NC-174. USDA Forest Service, north central Forest Experiment Station, North Carolina, USA

Murray MB, Cannell MGR, Smith RI (1989) Date of budburst of 15 tree species in Britain following climatic warming. J Appl Ecol 26:693–700

Nyman T, Paajanen R, Heiska S, Julkunen-Tiitto R (2011) Preference–performance relationship in the gall midge *Rabdophaga rosaria*: insights from a common-garden experiment with nine willow clones. Ecol Entomol 36:200–211

Ollerstam O, Rohfritsch O, Höglund S, Larsson S (2002) A rapid hypersensitive response associated with resistance in the willow Salix viminalis against the gall midge Dasineura marginemtorquens. Entomol Exp Appl 102:153–162

Sharma HC (1993) Host-plant resistance to insects in sorghum and its role in integrated pest management. Crop Prot 12:11–34

Sharma HC, Vidyasagar P, Nwanze KF (1993) Effect of host-plant resistance on economic injury levels for the sorghum midge, Contarinia sorghicola. Int J Pest Manag 39:435–444

Sinha DK, Atray I, Bentur JS, Nair S (2015) Feeding on resistant rice leads to enhanced expression of defender against apoptotic cell death (*OoDAD1*) in the Asian rice gall midge. BMC Plant Biol 15:235. https://doi.org/10.1186/s12870-015-0618-y

Sinha DK, Atray I, Agarrwal R, Bentur JS, Nair S (2017) Genomics of the Asian rice gall midge and its interactions with rice. Curr Opinion Insect Sci 19:76–81

Son DS, Eom TJ, Seo JD, Lee SR (1996a) A study of resistant substance to pine needle gall midge among phenolic compounds in pine needles. J Korean Forest Soc 85:372–380

Son DS, Eom TJ, Seo SD, Lee SR (1996b) Potential resistance factors in pine needles to pine gall midge. J Korean Forest Soc 85:244–250

Strong DR, Larsson S, Gullberg U (1993) Heritability of host plant resistance to herbivory changes with gall midge density during an outbreak on willow. Evolution 47:291–300

Stuart JJ, Chen MS, Shukle R, Harris MO (2012) Gall midges (hessian flies) as plant pathogens. Annu Rev Phytopathol 50:339–357

Stuart JJ (2015) Insect effectors and gene-for-gene interactions with host plants. Curr Opinion Insect Sci 9:56–61

Sunose T, Yukawa J (1979) Interrelationship between the leaf longevity of the evergreen spindle tree, *Euonymus japonicus* Thunb. and the euonymus gall midge, *Masakimyia pustulae* Yukawa and Sunose (Diptera, Cecidomyiidae) in different environments. Jpn J Ecol 29:29–34

Tanaka Y, Okada K, Asami T, Suzuki Y (2013) Phytohormones in Japanese mugwort gall induction by a gall-inducing gall midge. Biosci Biotechnol Biochem 77:1942–1948

Teetes GL, Anderson RM, Peterson GC (1994) Exploitation of sorghum nonpreference resistance to sorghum midge (Diptera: Cecidomyiidae) using mixed plantings of resistant and susceptible sorghum hybrids. J Econ Entomol 87:826–831

Tokuda M, Jikumaru Y, Matsukura K, Takebayashi Y, Kumashiro S, Matsumura M Kamiya Y (2013) Phytohormones related to host plant manipulation by gall-inducing leafhopper. PLoS One 8:e62350. https://doi.org/10.1371/journal.pone.0062350

Tooker JF, De Moraes CM (2011) Feeding by hessian fly (*Mayetiola destructor* [say]) larvae on wheat increases levels of fatty acids and indole-3-acetic acid but not hormones involved in plant-defense signaling. J Plant Growth Regul 30:158–165

Tooker JF, Helms AM (2014) Phytohormone dynamics associated with gall insects, and their potential role in the evolution of the gall-inducing habit. J Chem Ecol 40:742–753

Wang Z, Ge J-Q, Chen H, Cheng X, Yang Y, Li J, Whiworth RJ, Chen M-S (2018) An insect nucleoside diphosphate kinase (NDK) functions as an effector protein in wheat - hessian fly interactions. Insect Biochem Mol Biol 100:30–38

Watt AD, Woiwod IP (1999) The effect of phenological asynchrony on population dynamics: analysis of fluctuations of British macrolepidoptera. Oikos 87:411–416

Yamaguchi H, Tanaka H, Hasegawa M, Tokuda M, Asami T, Suzuki Y (2012) Phytohormones and willow gall induction by a gall-inducing sawfly. New Phytol 196:586–595

Yukawa J (1983) Arthropod community centred upon the neolitsea leaf gall midge, *Pseudasphondylia neolitseae* Yukawa (Diptera, Cecidomyiidae) and its host plant, *Neolitsea sericea* (Blume) Koidz. (Lauraceae). Mem Fac Agric Kagoshima Univ 19:89–96

Yukawa J (2000) Synchronization of gallers with host plant phenology. Popul Ecol 42:105–113

Yukawa J (2010) Synchronization between insects and their host plants. In: Kiritani K, Yukawa J (eds) Effects of global warming on insects. Zenkoku Nôson Kyôiku Kyôkai, Tokyo, pp 121–139. (In Japanese)

Yukawa J, Akimoto K (2006) Influence of synchronization between adult emergence and host plant phenology on the population density of *Pseudasphondylia neolitseae* (Diptera: Cecidomyiidae) inducing leaf galls on *Neolitseae sericea* (Lauraceae). Popul Ecol 48:13–21

Yukawa J, Masuda H (1996) Insect and mite galls of Japan in colors. Zenkoku Nôson Kyôiku Kyôkai, Tokyo. (In Japanese with English explanation for color plates)

Yukawa J, Miyamoto K (1979) Redescription of *Asphondylia sphaera* Monzen (Diptera, Cecidomyiidae), with notes on its bionomics. Mem Fac Agric Kagoshima University 15:99–106

Yukawa J, Ohsaki N (1988a) Separation of the aucuba fruit midge, *Asphondylia aucubae* sp. nov. from the *ampelopsis* fruit midge, *Asphondylia baca* Monzen (Diptera: Cecidomyiidae). Kontyû 56:365–376

Yukawa J, Ohsaki N (1988b) Adult behaviour of the aucuba fruit midge, *Asphondylia aucubae* Yukawa & Ohsaki (Diptera: Cecidomyiidae). Kontyû 56:645–652

Yukawa J, Tsuda K (1986) Leaf longevity of *Quercus glauca* Thunb., with reference to the influence of gall formation by *Contarinia* sp. (Diptera: Cecidomyiidae) on the early mortality of fresh leaves. Mem Fac Agric Kagoshima Univ 22:73–77

Yukawa J, Uechi N (1999) Can Gallers expand the host range to alien plants within a short period of time? Esakia 39:1–7

Yukawa J, Takahashi K, Ohsaki N (1976) Population behaviour of the neolitsea leaf gall midge, *Pseudasphondylia neolitseae* Yukawa (Diptera, Cecidomyiidae). Kontyû 44:358–365

Yukawa J, Ikenaga H, Sato S, Tokuda M, Ganaha-Kikumura T, Uechi N, Matsuo K, Mishima M, Tung GH, Paik JC, Ren BQ, Don XO (2012) Description and ecological traits of a new species of *Pitydiplosis* (Diptera: Cecidomyiidae) that induces leaf galls on *Pueraria* (Fabaceae) in East Asia, with a possible diversification scenario of intraspecific groups. Entomol Sci 15:81–98

Yukawa J, Nakagawa K, Saigou T, Awa T, Fukuda T, Higashi M (2013) Population behavior of an ambrosia gall midge, *Illiciomyia yukawai* (Diptera: Cecidomyiidae) and synchronization between its emergence and host plant phenology. Entomol Sci 16:400–412

Yukawa J, Miyamoto K, Yamaguchi T, Takesaki K, Uechi N, Matsuo K (2016b) Key factor/key stage analysis of long-term life table data for a fruit gall midge, *Asphondylia sphaera* (Diptera: Cecidomyiidae). Ecol Entomol 41:516–526

Yukawa J, Nakagawa K, Kohno A, Tokuda M, Kiritani K, Matsuo K, Mitsui H, Fujii T (2016c) Geographical and annual variations in the proportion of extended diapausing individuals of *Illiciomyia yukawai* (Diptera: Cecidomyiidae) with reference to an adaptive significance of its bimodal emergence patterns. Entomol Sci 19:275–289

Yukawa J, Fujimoto K, Ganaha-Kikumura T, Wanggyu K (2019) Effects of typhoons on an arthropod community centered upon a leaf gall midge, *Pseudasphondylia neolitseae* (Diptera: Cecidomyiidae) and its host plant, *Neolitsea sericea* (Lauraceae) through leaf fall and late-season shoot production. Ecol Res 35:252–264

Zhao C, Shukle R, Navarro-Escalante L, Chen M, Richards S, Stuart JJ (2016) *Avirulence* gene mapping in the hessian fly (*Mayetiola destructor*) reveals a protein phosphatase 2C effector gene family. J Insect Physiol 84:22–31

Zhu L, Chen MS, Liu X (2011) Changes in phytohormones and fatty acids in wheat and rice seedlings in response to hessian fly (Diptera: Cecidomyiidae) infestation. J Econ Entomol 104:1384–1392

Chapter 11
Natural Enemies with Special Reference to Parasitic Wasps

Check for updates

Junichi Yukawa, Kazunori Matsuo, and Tomohisa Fujii

Abstract Parasitic hymenopterans are most predominant among arthropod community members centered upon gall inducers and their host plants. They frequently play an important role as biotic mortality factors operating on gall midge populations. To evaluate their impact on gall midges, this Chapter treats necessary information as to their taxonomic status and various behavioral and ecological traits such as parasitic strategies, host ranges, and interaction with host gall midges. We also emphasize necessity of expressing the abundance of parasitoids as population numbers instead of percentage parasitism.

Keywords Biotic mortality factor · Early parasitoid · Endoparasitism · Ectoparasitism · Host manipulation · Host range · Late parasitoid · Parasitic strategies · Population dynamics

11.1 Introduction

Based on current evidence, Sharkey et al. (2012) mentioned that parasitism arose once in the common ancestor of the Orussoidea and Apocrita within the order Hymenoptera and many subsequent elaborations of the parasitic mode of life (e.g. endoparasitism, secondary phytophagy, etc.) apparently evolved independently

J. Yukawa (✉)
Entomological Laboratory, Faculty of Agriculture, Kyushu University, Fukuoka, Japan
e-mail: galling-cecidomyiids@kyudai.jp

K. Matsuo
Biosystematics Laboratory, Faculty of Social and Cultural Studies, Kyushu University, Fukuoka, Japan
e-mail: matsuosudachi@scs.kyushu-u.ac.jp

T. Fujii
Kyushu Okinawa Agricultural Research Center, National Agriculture and Food Research Organization, Kumamoto, Japan
e-mail: tomofuji@affrc.go.jp

J. Yukawa, M. Tokuda (eds.), *Biology of Gall Midges*, Entomology Monographs,
https://doi.org/10.1007/978-981-33-6534-6_11

more than once. Therefore, many hymenopteran taxa have been known as parasitoids of various arthropod families. Among them, Platygastridae, Ceraphronidae, Eulophidae, Eurytomidae, Torymidae, Pteromalidae and Braconidae are predominantly associated with gall-inducing cecidomyiid species and exhibit various modes of parasitic life.

Parasitic hymenopterans are most predominant both in the numbers of species and individuals among arthropod community members centered upon gall inducers and their host plants. They frequently play an important role as density dependent or independent biotic mortality factors operating on gall midge populations. To evaluate their impact on gall midges, it is essential to understand their taxonomic status, parasitic strategies, host ranges, interaction with host gall midges, and population dynamics. This Chapter treats the necessary information on parasitic hymenopterans.

11.2 Parasitic Strategies

11.2.1 Parasitoids of Oak Gall Wasps

Askew (1975) divided parasitoids of oak gall wasps (Hymenoptera: Cynipidae) into two groups based on their different parasitic strategies. The first group includes early parasitoids that attack small (young) hosts and are usually univoltine, monophagous and endoparasitic (r-strategist). Adult life span of these parasitoids is relatively short, and they spend much of their energy to produce more progeny. The second group consists of late parasitoids that attack large (older) hosts. These are usually multivoltine, polyphagous and ectoparasitic (K-strategist). Their adult life span is relatively long, and they devote most of their energy to searching for suitable hosts. Such contrasted strategies are quite understandable and useful in studying behavioral and ecological traits of parasitic wasps that attack gall-inducing arthropods.

11.2.2 Parasitoids of Gall Midges

A similar contrast in the parasitic strategy has also been noted for the parasitoids that attack gall midges of the genus *Asphondylia* (Yukawa et al. 1981; Yukawa 1987). Species of Platygastridae attack eggs and larvae of gall midges. They are usually monophagous and endoparasitic, synchronizing their emergence season with the egg stage or first larval stadium of their hosts. Their adult life span is short, and they try to lay their eggs as many as possible within a short period of life span, of which behavior is naturally supported by the fact that survival curve indicates that host individuals are most abundant in this period. Both early endoparasitoids and late ectoparasitoids have been known in Pteromalidae, Eulophidae, and Eurytomidae.

In contrast, species of Torymidae and Braconidae attack third instars or pupae of gall midges, being usually multivoltine, polyphagous, and ectoparasitic. Their adult life span inevitably becomes long because their generation time is relatively short, and they have to search for suitable host larvae or pupae successively in the season. In addition, the individual number of older hosts is much less abundant than young hosts as has been demonstrated in survival curves for various r-strategists like insects, of which conditions make ectoparasitoids difficult in finding host targets.

Thereafter, the two contrasting parasitic strategies have been referred to parasitoids of various gall midge species (e.g. Maeto 1991; Tabuchi and Amano 2003; Tokuda et al. 2004; Matsuo and Yukawa 2009b; Tokuda 2012; Yukawa et al. 2012; Matsuo et al. 2013; Kim et al. 2015; Matsuo et al. 2016b; Matsuo et al. 2018). The two parasitic strategies are taken into consideration in describing behavioral and ecological traits of parasitoids also throughout this book.

11.3 Diversity of Parasitoids and their Host Range

11.3.1 Platygastridae

Platygastrids are all endoparasitoids that attack the eggs (rarely young larvae) of other arthropods. Unlike Scelionidae of which development is completed in the host egg, Platygastridae do not emerge until the host becomes a full-grown larva (Gauld and Bolton 1988). Thus, species of *Platygaster* that parasitize gall-inducing cecidomyiids are known as typical early attackers (Askew 1975).

The genus *Platygaster* Latreille (Hymenoptera: Platygastridae) (Fig. 11.1a) contains more than 600 species, of which approximately 300 species have been described from various biogeographic regions during the last two decades (e.g. Buhl 1994a, b, Buhl 2006a, b). Eleven species of *Platygaster* are now known to occur in Japan, of which nine parasitize gall midge species (Ashmead 1904a; Ishii 1953; Yoshida and Hirashima 1979; Yamagishi 1980; Buhl and Duso 2008; Vlug 1995; Matsuo et al. 2018).

Host specificity of *Platygaster* has been paid special attention in behavioral and ecological studies, particularly host-parasitoid interactions. For example, Stireman III et al. (2006) demonstrated host-associated genetic differentiation in *Platygaster variabilis* Fouts that attacks *Rhopalomyia solidaginis* (Loew) (Diptera: Cecidomyiidae). Yamagishi (1980) reported that larvae of *Rabdophaga rosaeformis* Kovalev (Diptera: Cecidomyiidae) parasitized by *Platygaster stimulator* Yamagishi mature in summer whereas those unparasitized pass through the summer as first instars. In terms of reproductive strategy, some species including *Platygaster robiniae* Buhl & Duso are known to be gregarious parasitoids (Kim et al. 2011). In addition, polyembryony, the production of genetically identical embryos from a single egg through clonal division, has been found in several species, such as *Platygaster feltii* Fouts (Segoli et al. 2010) and *Platygaster vernalis* (Myers) (Leiby and Hill 1924).

Fig. 11.1 Females of parasitoid wasp. (**a**), *Platygaster ingeniosus* (Platygastridae); (**b**), *Aphanogmus flavigastris* (Ceraphronidae); (**c**), *Sigmophora tricolor* (Eulophidae); (**d**), *Eurytoma erythrinae* (Eurytomidae); (**e**), *Torymus wanggyui* (Torymidae); (**f**), *Gastrancistrus fuscicornis* (Pteromalidae); (**g**), *Simplicibracon curticaudis* (Braconidae); (**h**), *Testudobracon pleuralis* (Braconidae)

11.3.2 Ceraphronidae

Up to the present, Ceraphronidae are known to consist of about 300 described species in 15 genera worldwide (Johnson and Musetti 2004; Mikó and Deans 2009) but many species have been remaining undescribed.

The genus *Aphanogmus* Thomson (Hymenoptera: Ceraphronidae) contains at least 100 species (Johnson and Musetti 2004; Evans et al. 2005; Buhl et al. 2010; Matsuo et al. 2016a). About 20% of them have been known as parasitoids of various insects including Cecidomyiidae (Diptera), Bethylidae, Ichneumonidae (Hymenoptera) and Cybocephalidae (Coleoptera) (Oatman 1985; Gilkeson et al. 1993; Polaszak and Dessart 1996; Evans et al. 2005). Host information for the remaining 80% has not been provided. Three species, *Aphanogmus floridanus* Ashmead, *Aphanogmus fulmeki* Szelényi (=*A. parvulus* Roberti), and *Aphanogmus flavigastris* Matsuo (Fig. 11.1b) have been known to parasitize predacious or mycophagous gall midges. *Aphanogmus floridanus* is an endoparasitoid of *Feltiella acarivora* (Zehntner) in the Holarctic Region (Oatman 1985; Johnson and Musetti 2004) and has been regarded as a negative force in controlling *Tetranychus urticae* Koch (Acarina: Tetranychidae) on strawberry in California due to its high parasitism rate (Oatman 1985). *Aphanogmus fulmeki*, a Palaearctic species, attacks *Aphidoletes aphidimyza* (Rondani), *Feltiella acarisuga* (Vallot), *F. acarivora*, (Rondani) and *Mycodiplosis* sp. (Dessart 1992) (see Chap. 13 for the details of *Aphidoletes* and *Feltiella*). *Aphanogmus flavigastris* was reared from cocoons that had been made possibly by *F. acarisuga* and *F. acarivora* in Okinawa, Japan (Abe et al. 2011; Ganaha-Kikumura et al. 2012; Matsuo et al. 2016a).

11.3.3 Eulophidae

Eulophidae is one of the largest families of parasitoid wasps with about 300 genera and 4000 species placed in five subfamilies (Noyes 2019). The majority of species are parasitoids but a few are phytophagous or predatory species (Gauthier et al. 2000). Parasitic modes of life are various, being endo- or ectoparasitoids, koino- or idiobionts, solitary or gregarious, specialists or generalists, and primary parasitoids, hyperparasitoids or facultative hyperparasitoids.

Sigmophora contains 18 species that have been recorded from the Afrotropical, Australasia, Indomalaya, and Palearctic Regions (Ikeda 1999; Matsuo et al. 2013; Noyes 2019). Host information is not available for 12 out of the 18 species. The remaining species are parasitoids of Cecidomyiidae or Lyonetiidae (Lepidoptera) (Ikeda 1999; Yefremova and Yegorenkova 2009; Matsuo et al. 2013), except *Sigmophora brevicornis* (Panzer) whose host range covers at least four insect orders, Diptera, Hymenoptera, Coleoptera, and Lepidoptera (Ikeda 1999; Noyes 2019).

In Japan, Ashmead (1904b) described *Tetrastichus tricolor* Ashmead (Fig. 11.1c), without host information. Later, *Tetrastichus sayatamabae* Ishii was described as a parasitoid of the soybean pod gall midge *Asphondylia yushimai* Yukawa & Uechi (Ishii 1950). These two species were transferred to *Sigmophora* and were synonymized with *S. brevicornis* (Ikeda 1999; Noyes 2019). However, Matsuo et al. (2013) resurrected *Sigmophora tricolor* (Ashmead) and *Sigmophora sayatamabae* (Ishii) from *S. brevicornis* based on morphological differences. At the same time, *S. sayatamabae* was synonymized with *S. tricolor* and an additional new species, *Sigmophora longiseta* Matsuo was described (Matsuo et al. 2013). *Sigmophora tricolor* is a solitary, multivoltine ectoparasitoid and has been reared from *A. yushimai* on *Glycine max* (Linnaeus) Merrill and *Lespedeza bicolor* Turczaninov (Fabaceae) and from *Pseudasphondylia neolitseae* Yukawa on *Neolitsea sericea* (Blume) Koidzumi (Lauraceae) in Japan. *Sigmophora longiseta* is an ectoparasitoid of *Asphondylia morivorella* (Naito) on *Morus alba* Linnaeus (Moraceae) and an unidentified gall midge that induces fruit galls on *Hydrangea serrata* (Thunberg) Seringe (Hydrangeaceae).

11.3.4 Eurytomidae

The family Eurytomidae currently includes more than 80 genera and 1000 species placed in three subfamilies, Eurytominae, Heimbrinae, and Rileynae, and most eurytomid species have been considered to be endophytic, either as phytophagous or as parasitoids of phytophagous insects (Noyes 2019).

Eurytoma is included in Eurytominae and the members are known to be plant feeders or entomophagous, idiobiont, solitary ectoparasitoids of insect larvae feeding within plant tissue (e.g. Varley 1937). They attack Coleoptera, gall-forming Hymenoptera (mostly Cynipidae), Diptera (especially Tephritidae) and Lepidoptera (e.g. Claridge 1959; Claridge and Askew 1960). A few species of *Eurytoma* exhibit a variety of parasitic modes, such as gregarious ectoparasitoids of Lepidoptera (e.g. Piel 1933), koinobiont endoparasitoids of a tephritid (Claridge 1961), hyperparasitoid on other ectoparasitoids (Fisher 1965), and predacious in multichambered galls on rose made by *Diplolepis* species (Cynipidae) (Blair 1944).

Eurytoma attacks various species of Cecidomyiidae (Table 11.1). Barnes (1935) and Sen (1954) reported that *Eurytoma aciculata* Ratzeburg and *Eurytoma saliciperdae* Mayr parasitized *Rabdophaga saliciperda* (Dufour) and its congeners on *Salix* spp. (Salicaceae). Gall midges of the genera *Lasioptera* and *Neolasioptera* are parasitized by at least four species of *Eurytoma* (Thompson 1955; Boucek 1977; Wilson and Heaton 1987; Zerova and Seryogina 2006). *Eurytoma dentata* Mayr and their congeners attack various species of *Asphondylia* in many countries (Tiwari 1974; Zerova and Seryogina 2006; Noyes 2019; Dorchin et al. 2014; Doğanlar and Elsayed 2015; Elsayed et al. 2015). *Eurytoma dentata* also attacks two species of *Contarinia* (Mayr 1878; Zerova and Seryogina 2006). *Eurytoma setitibia* Gahan parasitizes the Asian and African rice gall midges, *Orseolia oryzae* (Wood-Mason) and *Orseolia oryzivora* Harris & Gagné (Ogah and Nwilene 2014).

Table 11.1 Selected examples of *Eurytoma* species attacking cecidomyiids

Species of *Eurytoma*	Host gall midge	Host plant	Reference
E. aciculata Ratzeburg	*Rabdophaga* spp.	*Salix* spp. (Salicaceae)	Barnes (1935)
E. saliciperdae Mayr	*Rabdophaga saliciperda* (Dufour)	*Salix fragilis*, *S. alba* (Salicaceae)	Barnes (1935), Sen (1954)
E. setitibia Gahan	*Orseolia oryzae* (Wood-Mason)	*Oryza sativa* (Poaceae)	Ogah and Nwilene (2014)
	Orseolia oryzivora Harris & Gagné	*Oryza glaberrima* (Poaceae)	Ogah and Nwilene (2014)
E. aterrima (Schrank)	*Lasioptera rubi* (Schrank)	*Rubus idaeus* (Rosaceae)	Thompson (1955)
E. curculionum Mayr	*Lasioptera rubi* (Schrank)	*Rubus idaeus* (Rosaceae)	Boucek (1977)
E. rosae Nees	*Lasioptera rubi* (Schrank)	*Rubus idaeus* (Rosaceae)	Boucek (1977)
E. laserpitii Mayr	*Lasioptera eryngii* (Vallot)	*Eryngium campestre* (Apiaceae)	Zerova and Seryogina (2006)
Eurytoma sp.	*Neolasioptera brevis* Gagné	*Gleditsia triacanthos* (Fabaceae)	Wilson and Heaton (1987)
E. dentata Mayr	*Contarinia loti* (De Geer)	*Lotus corniculatus* (Fabaceae)	Mayr (1878)
	Contarinia medicaginis Kieffer	*Medicago sativa*, *M. falcata* (Fabaceae)	Zerova and Seryogina (2006)
	Asphondylia gennadii (Marchal)	Various host plants	Zerova and Seryogina (2006)
	Asphondylia lupini Silvestri	*Lupinus albus* (Fabaceae)	Rizzo and Massa (1998)
	Asphondylia miki Wachtl	*Medicago sativa* (Fabaceae)	Zerova and Seryogina (2006)
	Asphondylia punica Marchal	*Atriplex halimus* (Amaranthaceae)	Doğanlar & Elsayed et al. (2015)
	Asphondylia sarothamni (Loew)	*Cytisus scoparius* (Fabaceae)	Zerova and Seryogina (2006)
	Asphondylia trabuti Marchal	*Solanum nigrum* (Solanaceae)	Rizzo and Massa (1998)
	Asphondylia verbasci (Vallot)	*Verbascum* spp. (Scrophulariaceae)	Zerova and Seryogina (2006)
	Asphondylia spp.	-----	Noyes (2019)
E. nesiotes Crawford	*Asphondylia sesami* Felt	*Sesamum indicum* (Pedaliaceae)	Tiwari (1974)
E. nohurae Zerova	*Asphondylia* sp.	*Astragalus* sp. (Fabaceae)	Zerova and Seryogina (2006)
E. sp. nr *coleophorae* Zerova	*Asphondylia punica* Marchal	*Atriplex halimus* (Amaranthaceae)	Elsayed et al. (2015)
E. sp. nr *dentata* Mayr	*Asphondylia scopuli* Dorchin	*Atriplex lanfrancoi* (Amaranthaceae)	Dorchin et al. (2014)

In Japan, unidentified species of *Eurytoma* have been reported to attack *Asphondylia aucubae* Yukawa & Ohsaki, *Asphondylia baca* Monzen (Yukawa and Ohsaki 1988), *Asphondylia sphaera* Monzen (Yukawa and Miyamoto 1979), and *Asphondylia tojoi* Elsayed & Tokuda (Elsayed et al. 2018). Up to now, we have reared unidentified species of *Eurytoma* from galls induced by *Asphondylia itoi* Uechi & Yukawa, *A. yushimai*, and several unidentified species of *Asphondylia*, such as those induce fruit galls on the species of *Alpinia*, *Ardisia*, *Callicarpa*, and *Hedera*. In addition, Yukawa et al. (2012) reported that *Pitydiplosis puerariae* Yukawa Ikenaga & Sato was parasitized by *Eurytoma* sp.

It should be remarked here that *Eurytoma erythrinae* Gates & Delvare (Fig. 11.1d) was introduced from Africa to Hawaii and has been used as a biological control agent against *Quadrastichus erythrinae* Kim (Eulophidae) that infests by gall induction various species of *Erythrina* trees (Fabaceae) in many tropical and subtropical regions (Kim et al. 2004; Gates and Delvare 2008) including Okinawa and Amami, Japan (Uechi et al. 2007; Kanai et al. 2008). Because *E. erythrinae* had successfully reduced the degree of infestation in Hawaii (Kaufmana et al. 2020), Okinawa Prefecture introduced it from Hawaii to Okinawa and started experiments and field works for successful release of the natural enemy (e.g. Matsuo 2016).

11.3.5 Torymidae

The family Torymidae currently includes 68 genera and 986 species placed in two subfamilies, Megastigminae and Toryminae (Noyes 2019). The majority of species of the subfamily Toryminae are idiobiont ectoparasitoids of the inhabitants of plant galls.

Torymus (Fig. 11.1e) is included in Toryminae and contains at least 394 species worldwide (Grissell 1995; Matsuo and Yukawa 2009a, b; Matsuo 2010; Noyes 2019; Matsuo 2020). However, about 25% of them were described without males and host information because they were collected by net sweeping. The most of remaining species are known to be ectoparasitoids of larvae of Cecidomyiidae or Cynipidae (e.g. Moser 1965; Fulmek 1968; Grissell 1976, 1995; Murakami 1981; Graham and Gijswijt 1998; Matsuo et al. 2012; Noyes 2019), which is supported by the long ovipositor that enables most *Torymus* females to lay their eggs on concealed hosts (Graham and Gijswijt 1998). Several species are also known as parasitoids of Apoidea (Hymenoptera), Psyllidae, Cicadidae (Hemiptera) and Pyralidae (Lepidoptera) (e.g. Moser 1965; Fulmek 1968; Graham and Gijswijt 1998; Grissell 1976, 1995, 2007; Kazmi and Chauhan 2003; Noyes 2019). Phytophagous life histories, such as seed parasitoids, inquilines or gall inducers are reported for several species (e.g. Grissell 1995; LaSalle 2005; Matsuo et al. 2012; Noyes 2019). Zoophytophagy, in which a parasitoid larva consumes both host insect and surrounding plant tissues, is also found in this genus (Askew 1961; Murakami 1981).

At least 65% of Holarctic species of *Torymus* that attack cecidomyiids and cynipids are monophagous or oligophagous within a single host insect genus, while the remaining species are polyphagous across different host insect genera or families (Grissell 1976; Graham and Gijswijt 1998). However, the host range of such polyphagous species needs to be re-examined because a preliminary investigation of Japanese *Torymus* species by Matsuo (2020) indicates that morphologically similar species with different host insect records could be frequently mixed into one species.

Ashmead (1904a) described three species of *Torymus* from Japan, for the first time. Thereafter, various taxonomic studies have been published for Japanese species (e.g. Grissell 1995; Kamijo 1964, 1979, 1982; Yasumatsu and Kamijo 1979; Zerova et al. 2003; Grissell et al. 2004; Matsuo and Yukawa 2009b; Matsuo 2010; Matsuo et al. 2011; Janšta et al. 2018). Following the latest revision of *Torymus* (Matsuo 2020), 31 species of *Torymus* are now recognized from Japan. Among them, eight species are associated with cynipids including the chestnut gall wasp *Dryocosmus kuriphilus* Yasumatsu, one with a fruit fly (Diptera: Tephritidae), one is a seed feeder, and two are host unknown species. The remaining 19 species attack gall-inducing cecidomyiids (Table 11.2) but there was no strong association of the *Torymus* species with a particular cecidomyiid genus. Because some of the host gall midges are unidentified, their species identification is needed.

11.3.6 Pteromalidae

The family Pteromalidae consists of at least 3000 described species in about 640 genera worldwide. Most of them are parasitoids of other insects (Noyes 2019). They exhibit various parasitic modes of life, such as endo- or ectoparasitoids, koino- or idiobionts, and primary or hyperparasitoids.

The genus *Gastrancistrus* is endoparasitoid and contains at least 130 species worldwide (Graham 1969; Noyes 2019). Host information is unknown for about 75% of them. The remaining 25% have been recorded as parasitoids of Cecidomyiidae, Agromyzidae (Diptera), and Tenthredinidae (Hymenoptera) (Graham 1969; Herting 1977; Hansson 1987), except for *Gastrancistrus fuscicornis* Walker (Fig. 11.1f), which has been suggested to be a parasitoid of aphids (Graham 1969). In Japan, two unidentified species of *Gastrancistrus* have been reared from cecidomyiid galls (Yukawa 1971, as *Gastrancistrum*, misspelling) (Fig. 9.2c). One species parasitizes larvae of *P. neolitseae* that induce leaf galls on *N. sericea* and another species attacks larvae of *Daphnephila machilicola* Yukawa that induce leaf galls on *Machilus thunbergii* Siebold & Zuccarini (Lauraceae). They are solitary and univoltine, emerging in late April and May a little after emergence season of the host gall midges (Yukawa 1983).

A solitary endoparasitic species of pteromalid was reared in Japan from pupae of an aphidophagous gall midge, *Aphidoletes aphidimyza* (Rondani) (Diptera: Cecidomyiidae), which is an important biological control agent against aphids (see Chap. 13 for the details of *Aphidoletes*). Matsuo (2012) identified the pteromalid as

Table 11.2 List of *Torymus* species attacking cecidomyiids in Japan

Species of *Torymus*	Host cecidomyiids in Japan	Host plant of gall midges in Japan
T. achyranthii Matsuo	*Lasioptera achyranthii* Shinji	*Achyranthes bidentata* var. *japonica* (Amaranthaceae)
T. aciculatus Matsuo	Unidentified gall midge	*Neolitsea aciculata* (Lauraceae)
T. acutissimus Matsuo	Unidentified gall midge	*Quercus acutissima* (Fagaceae)
T. angustitemple Matsuo	*Celticecis japonica* Yukawa & Tsuda	*Celtis sinensis* (Cannabaceae)
T. azureus Boheman	Unidentified gall midge[a]	*Picea glehnii* (Pinaceae)
T. caudatus Boheman	Unidentified gall midge[b]	*Picea jezoensis* (Pinaceae)
T. celtidigalla Matsuo	*Celticecis japonica* Yukawa & Tsuda	*Celtis sinensis* (Cannabaceae)
T. celticolus Matsuo	*Celticecis japonica* Yukawa & Tsuda	*Celtis sinensis* (Cannabaceae)
T. ezomatsuanus Kamijo	Unidentified gall midge	*Picea glehnii* & *P. jezoensis* (Pinaceae)
T. fuscicornis (Walker)	Unidentified gall midge	*Betula platyphylla* var. *japonica* (Betulaceae)
T. hirtipennis Matsuo	*Schizomyia sasakii* (Monzen)	*Ilex crenata* (Aquifoliaceae)
T. macrops Matsuo	*Mikiola glandaria* Sato & Yukawa	*Fagus crenata* (Fagaceae)
T. minamii Matsuo	*Kiefferia pericarpiicola* (Bremi)	*Angelica* spp. & *Spuriopimpinella calycina* (Apiaceae)
T. nitidulus (Walker)	Unknown, possibly *Semdobia* spp.	*Betula platyphylla* var. *japonica* (Betulaceae)
T. rosariae Graham & Gijswijt	*Rabdophaga rosaria* (Loew)	*Salix integra* (Salicaceae)
T. rugosus Matsuo	*Dasineura* sp. (& *Diplolepis* sp.[c])	*Rosa rugosa* (Rosaceae)
T. sasae Matsuo	*Procystiphora uedai* Sato & Yukawa	*Sasa nipponica* (Poaceae)
T. sawadai Matsuo	*Asphondylia sphaera* Monzen	*Weigela hortensis* (Caprifoliaceae)
T. wanggyui Matsuo	*Macrodiplosis selenis* Kim & Yukawa	*Quercus serrata* (Fagaceae)

Summarized from Matsuo 2020

[a]*Kaltenbachiola strobi* (Winnertz) and *Plemeliella abietina* Seitner in Europe

[b]*K. strobi* in Europe

[c]a cynipid host

G. fuscicornis, a species that has been recorded from Europe (Graham 1969) and confirmed that *G. fuscicornis* is an endoparasitoid of *A. aphidimyza*.

11.3.7 Braconidae

In the family Braconidae, members of the tribe Braconini have been known as parasitoids of gall-inducing insects belonging to Cecidomyiidae, Curculionidae (Coleoptera), and Momphidae (Lepidoptera) (Quicke 1986). They are typical late attackers, idiobiont, polyphagous, and ectoparasitoids attacking older larvae or pupae. In Japan, three braconid species, *Campyloneurus asphondyliae* Watanabe, *Philomacroploea pleuralis* (Ashmead) and *Ipobracon scurra* Fischer, were known to attack gall midges (Watanabe 1940, 1952; Fischer 1980). However, Maeto (1991) combined *C. asphondyliae* with the genus *Bracon*, *P. pleuralis* with the genus *Testudobracon*, and synonymized *I. scurra* with *B. asphondyliae*. At the same time, four new species, *Bracon sunosei* Maeto, *Bracon tamabae* Maeto (Fig. 9.2d), *Simplicibracon curticaudis* Maeto (Fig. 11.1g), and *Testudobracon longicaudis* Maeto (Fig. 11.1h) were described. Maeto (1991) reared these braconids from at least eight identified and three unidentified species of gall midges. Later, Matsuo et al. (2016b) found many additional host gall midges. As a result, the six species of braconid are now known to parasitize 11 identified and 11 unidentified species of Asphondyliini (Diptera: Cecidomyiidae) in Japan (Table 11.2). It should be remarked here that all known host gall midges of the braconid parasitoids in Japan are members of the tribe Asphondyliini. However, *Bracon* spp. are known to attack gall midges belonging to tribes other than Asphondyliini, such as the African rice gall midge, *Orseolia oryzivora* Harris & Gagné (Pathak and Khan 1994) and *Lasioptera falcata* Felt and *Lasioptera bryoniae* Schiner on bitter gourd in India (Muthukumar et al. 2017). A species of Opiinae (Braconidae) also parasitizes *Horidiplosis ficicola* Gagné (new name for *Dyodiplosis fici* Rao) in India (Sallem 2015).

Analysis of cytochrome oxidase subunit I (COI) did not show any evidence of host races among the braconids (Matsuo et al. 2016b). Among the six braconids, *B. asphondyliae, B. sunosei, T. longicaudis* and *T. pleuralis* did not show any particular association with host plant species, gall shape or galled organ. They are polyphagous and possibly multivoltine using different host gall midges on various plant species in different seasons. In contrast, *B. tamabae* and *S. curticaudis* attack only univoltine gall midges that belong to genera other than *Asphondylia* and induce leaf or axillary bud galls on evergreen broad-leaved trees. They may be univoltine, ovipositing on mature larvae or pupae of the host gall midges inhabiting the leaf galls during the period from October to the following March, and emerging from the host galls in April and May. However, where the emerged adults spend summer before coming to the host plants in autumn is still unknown. If they are not univoltine, they may utilize alternative hosts from May to October. Further filed surveys are needed.

11.4 Manipulation of Galls by Parasitoids

11.4.1 Host Manipulation by Parasitoids Associated with Mobile Hosts

Many parasitoids have an ability to manipulate their hosts to increase their own fitness. Behavioral changes of aphids and lepidopteran larvae to avoid predation or hyperparasitism have been intensively studied (Brodeur and Boivin 2004; Seyahooei et al. 2009; Maure et al. 2013). For example, aphids that were parasitized by *Aphidius nigripes* Ashmead (Hymenoptera: Braconidae) leave the colony and hide in concealed microhabitats to avoid an attack of hyperparasitoids (Brodeur and McNeil 1989; Brodeur and McNeil 1992). Larvae of *Pieris brassicae* (Linnaeus) (Lepidoptera: Pieridae) that were parasitized by a primary parasitoid, *Cotesia glomerata* (Linnaeus) (Hymenoptera: Braconidae), change their behavior after *C. glomerata* larvae quit the host larvae. The host larvae protect cocoons of *C. glomerata* against hyperparasitoids by spinning silk webs over the cocoons (Brodeur and Vet 1994; Tanaka and Ohsaki 2006). Larvae of *Thyrinteina leucocerae* Rindge (Lepidoptera: Geometridae) parasitized by a braconid wasp *Glyptapanteles* sp. defend cocoons of the parasitoid against approaching predators with violent head swinging (Grosman et al. 2008; Janssen et al. 2010).

11.4.2 Host Manipulation by Parasitoids Associated with Endophytic or Immobile Hosts

In contrast, host manipulation by parasitoids associated with endophytic or immobile hosts has seldom been studied. Fujii et al. (2014) investigated the interactions between a leaf gall inducer *Masakimyia pustulae* Yukawa & Sunose (Diptera: Cecidomyiidae) (Fig. 2.6e, f) and an early parasitoid *Platygaster ingeniosus* Matsuo & Yamagishi (Hymenoptera: Platygastridae). By examining the relationship between the leaf gall thickness and percentage parasitism by *P. ingeniosus* and late parasitoids such as *Chrysonotomyia* sp. and *Pnigalio* sp. (Hymenoptera: Eulophidae), they tested the hypothesis that the early parasitoid manipulates host larvae to increase gall thickness to avoid the attacking by late parasitoids.

Galls parasitized by *P. ingeniosus* were mostly thicker than those unparasitized (Figs. 3 and 4 in Fujii et al. 2014). Both the upper and lower gall walls were significantly thicker in galls parasitized by *P. ingeniosus* than in those unparasitized (Fig. 5 in Fujii et al. 2014). These results suggested that galls parasitized by *P. ingeniosus* were thickened to avoid hyperparasitism by the late parasitoids. Hypertrophy of gall wall is well known to decrease the rate of parasitism (e.g. Tabuchi and Amano 2004; Marchosky and Craig 2004) probably by preventing the ovipositor of parasitoids from penetrating the thick gall wall. The percentage parasitism by the late parasitoids was significantly lower in galls parasitized by

P. ingeniosus than those unparasitized (Tables 5 and 6 in Fujii et al. 2014). These results strongly suggested that *P. ingeniosus* manipulated the behavior of *M. pustulae* larvae, possibly by enhancing their feeding activity, to thicken the gall wall to avoid hyperparasitism. An alternative possibility was that females of *P. ingeniosus* might be able to detect differences in host egg size, if any. They then would prefer to lay their eggs in large host eggs, which are expected to give rise gall midge larvae that induce thick galls.

11.4.3 Manipulation of Gall Tissues

In recent years, manipulation of gall tissues by a parasitoid was also found in the interaction between the gall wasp *Barbotinia oraniensis* (Barbotin) (Hymenoptera: Cynipidae) and its parasitoid *Parnips nigripes* (Barbotin) (Cynipidae) (Ronquist et al. 2018). Galls containing *B. oraniensis* parasitized by *P. nigripes* have distinctly thicker walls than those containing *B. oraniensis* alone. This mechanism hypothesized that *Barbotinia* larvae parasitized by *Parnips* might lose their ability to control accurately the development of the gall towards the end of their life, resulting in some of the gall tissue that should normally develop into nutritional tissue remaining undifferentiated. Another possible explanation was that *Parnips* larva manipulated its host, such that the wall of gall became thicker and provided better protection against hyperparasitoids (Ronquist et al. 2018).

Asphondylia tojoi (Diptera: Cecidomyiidae) and *Ceratoneura* sp. (Hymenoptera: Eulophidae) induce galls on leaf buds of *Schoepfia jasminodora* Siebold & Zuccarini (Schoepfiaceae) (Elsayed et al. 2018). Galls containing *Ceratoneura* sp. were significantly larger in diameter than those of *A. tojoi* alone and had hard walls without residues of the gall midge or its fungal mycelia. *Ceratoneura* sp. also controlled the development of gall tissues for large size of galls to avoid an attack of hyperparasitoids (Elsayed et al. 2018).

Thus, the aforementioned studies demonstrated well that host manipulation by early parasitoids associated with endophytic or immobile hosts has an adaptive significance for them to reduce the rate of hyperparasitism by late parasitoids.

11.5 Population Dynamics of Parasitoids

11.5.1 Hymenopteran Effects on Gall Midge Populations

There are many examples where hymenopteran parasitoids affect positively or negatively the population dynamics of various gall-inducing cecidomyiids, such as *Torymus nigritarsus* (Walker) (Torymidae) and *Mesopolobus diffinis* (Walker) (Pteromalidae) on the 2-years type populations of *Taxomyia taxi* (Inchbald) (Cameron and Redfern 1978; Redfern and Hunter 2005), a platygastrid on *M. pustulae*

(Sunose 1985), *Platygaster* (Platygastridae) on *Semudobia* species (Roskam 1986), *Aprostocetus diplosidis* Crawford (Eulophidae) on *Stenodiplosis sorghicola* (Coquillett) (Manzoli-Palma and da Teles 1990; Lampo 1994), *Platygaster oebalus* Walker on *Dasineura napi* (Loew) (Axelsen 1994), and a guild of five interacting parasitoids on *Rhopalomyia californica* Felt (Ehler 1985, 1986). In most studies, the abundance or strength of parasitoid effects has been expressed as percentage parasitism, and the actual number of parasitoid individuals has seldom been treated because of difficulties in counting parasitoid individuals in the field. However, if the number of galls from which parasitoids emerged is surveyed intensively, we can relatively easily obtain necessary data as to the individual number of parasitoids. In this Section, three such examples of population dynamics of parasitoid species are demonstrated.

11.5.2 Parasitoids Attacking Taxomyia taxi

Through the long-term study of population dynamics of the yew gall midge *T. taxi*, Redfern and Cameron (1978) showed that two hymenopteran parasitoid *T. nigritarsus* and *M. diffinis* attacked the gall midge. Then, Cameron and Redfern (1978) analyzed the data and demonstrated that the population densities of both parasitoids were dependent on the host density. In *T. nigritarsus*, this tendency was true especially for those generations attacking even year galls (whose density was usually lower than that of odd year galls). *Mesopolobus diffinis* showed a similar dependence on host density. This dependence was roughly proportional to the amount of available food resources, although the mean percentage parasitism was much lower in 2-years galls than in 1-year galls. See Sect. 8.3 for detailed effects of the two parasitoids on the yew gall midge populations based on the subsequent 35-year filed survey by Redfern and Hunter (2005).

11.5.3 Parasitoids Attacking Pseudasphondylia neolitseae

Two species, *Gastrancistrus* sp. (Pteromalidae) (Fig. 9.2c) and *B. tamabae* (Fig. 9.2d) have been known to parasitize larvae of *P. neolitseae* that induces leaf galls on *N. sericea* (e.g. Yukawa 1983; Matsuo et al. 2016b) (Figs. 2.1a, b and 9.2a). The former is a univoltine and possibly monophagous endoparasitoid attacking first instars, and the latter is a multivoltine and polyphagous ectoparasitoid, attacking full-grown larvae or pupae. J. Yukawa and his students of Kagoshima and Kyushu Universities studied population dynamics of *P. neolitseae* from 1970 to 1996 in an evergreen broad-leaved forest at Shiroyama, Kagoshima, Japan (see

Chaps. 2 and 8). On that occasion, the number of parasitoid adults emerged from the leaf galls was surveyed every March and April to evaluate their effects on the host gall midge population as top-down biological mortality factors. Usually, each exit hole retains pupal case after the gall midge emerged, but no pupal case was left on the exit hole of gall from which a parasitoid adult emerged. The diameter of exit hole made by *Gastrancistrus* sp. was distinctly larger than that by *B. tamabae*. Thus, the number of *P. neolitseae* and the two parasitoid species at the time of emergence was easily recognized.

The annual fluctuation of their numbers is shown in Fig. 11.2. The number of *Gastrancistrus* sp. and *B. tamabae* fluctuated largely from year to year, and their fluctuations are not always parallel with those expressed in percentage parasitism. This means that actual population numbers are more important to evaluate the effects of parasitoid species. Figure 11.3 indicates that density dependency to the number of host gall midge is distinct in *B. tamabae* but unclear in *Gastrancistrus* sp. In common sense, *Gastrancistrus* sp. should rely on the abundance of *P. neolitseae* because it is a univoltine and possibly monophagous endoparasitoid. However, density dependency on the host gall midge became obscure in *Gastrancistrus* sp. at the time of emergence probably because of hyperparasitism by *B. tamabae* later in the season.

11.5.4 Parasitoids Attacking **Asphondylia sphaera**

The following five species of hymenopteran ectoparasitoids are known to attack *A. sphaera* that induces fruit galls on *Ligustrum* spp. (Oleaceae): *B. asphondyliae, T. longicaudis*, *T. pleuralis* (Braconidae) (Table 11.3), *Ceratoneura indi* Girault (Eulophidae) and *Eurytoma* sp. (Eurytomidae) (Yukawa et al. 2016). *Bracon asphondyliae* is most prominent among them and polyphagous, attacking, *A. aucubae*, *A. baca*, *A. itoi*, *A. yushimai*, *Pseudasphondylia matatabi* (Yuasa & Kumazawa), and several other unidentified species of Asphondyliini in Japan (Table 11.3; Yukawa et al. 1981; Maeto 1991; Matsuo et al. 2016b).

J. Yukawa and his students of Kagoshima and Kyushu Universities studied population dynamics of *A. sphaera* from 1970 to 1996 in an evergreen broad-leaved forest at Shiroyama, Kagoshima, Japan (see Chap. 8). On that occasion, the number of *B. asphondyliae* adults emerged from the fruit galls was surveyed every April. Emergences of *B. asphondyliae* fluctuated largely between 0 and 328 from year to year (Fig. 11.4) and its percentage parasitism varied from 0 to 54.2% (Yukawa et al. 2016).

To detect density-dependent and independent forces operating on the *A. sphaera* population and to assess the relative strength of top-down and bottom-up effects, the method of Key-factor/key-stage analysis (Yamamura 1999) was used to analyze

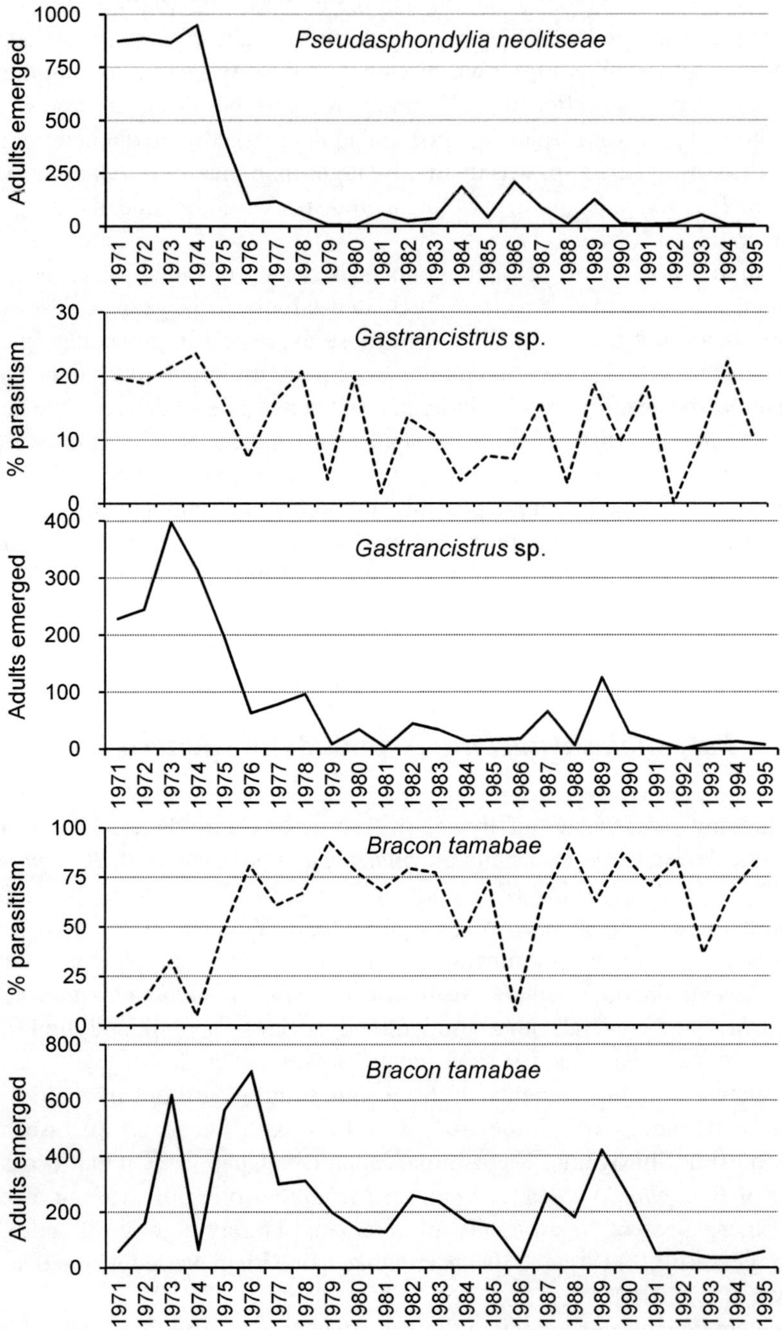

Fig. 11.2 Population dynamics of *Pseudasphondylia neolitseae* and its two parasitoid species, *Gastrancistrus* sp. and *Bracon tamabae* at the time of emergence in March and April

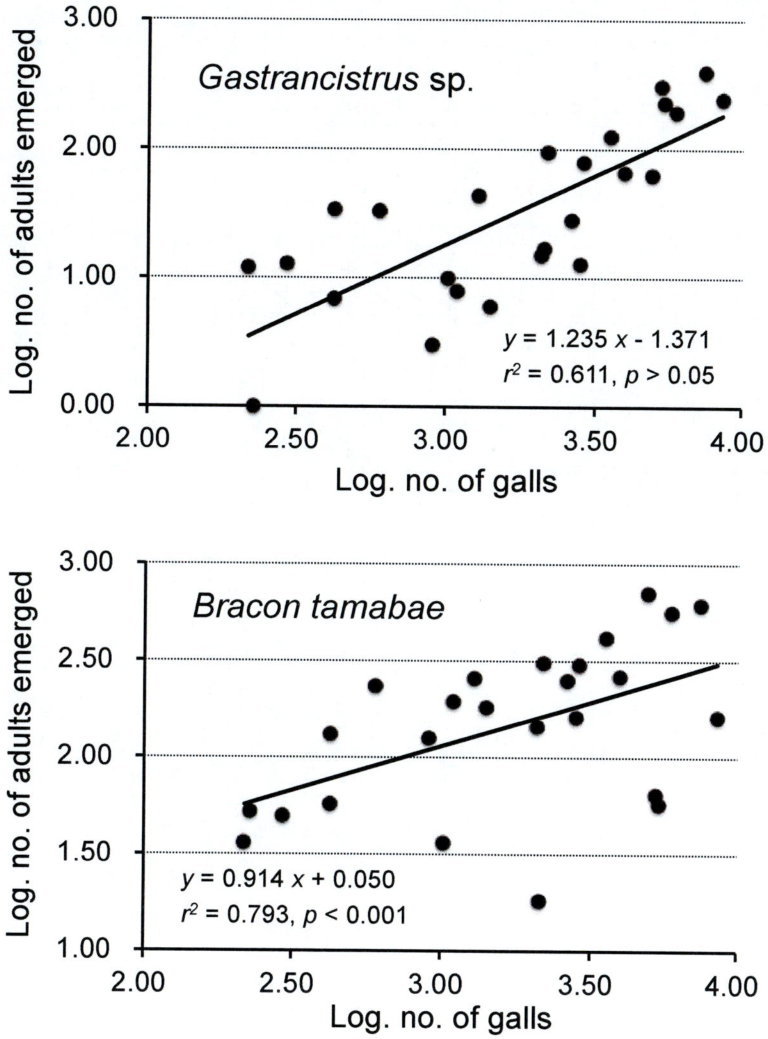

Fig. 11.3 Relationships between the log number of leaf galls induced by *Pseudasphondylia neolitseae* and that of *Gastrancistrus* sp. and *Bracon tamabae* adults emerged from the galls

long-term life table data. The result was that a top-down effect of *B. asphondyliae* was weakly density-dependent and contributed less to the total survival rate (Yukawa et al. 2016).

Table 11.3 Host insects, host plants of the host insects and distributional information for braconid parasitoids attacking Asphondyliini in Japan

Braconid species	Host gall midge	Associated plant (Family)	Galled part	Distribution[a]	Reference[b]
Bracon asphondyliae	*Asphondylia aucubae*	*Aucuba japonica* (Aucubaceae)	Fruit	JPN (HN, KY)	1, 2
	Asphondylia baca	*Ampelopsis glandulosa* var. *heterophylla* (Vitaceae)	Fruit	JPN (HN)	1, 2
	Asphondylia baca	*Weigela hortensis* (Caprifoliaceae)	Leaf bud	JPN (HN, KY)	1
	Asphondylia itoi	*Distylium racemosum* (Hamamelidaceae)	Fruit	JPN (KY)	10
	Asphondylia sphaera	*Ligustrum obtusifolium* (Oleaceae)	Fruit	JPN (KY)	1, 2
	Asphondylia sphaera	*Ligustrum obtusifolium* var. *pacificumi* (Oleaceae)	Fruit	JPN (IZ)	10
	Asphondylia yushimai	*Laurocerasus zippeliana* (Rosaceae)	Fruit	JPN (HN, KY)	1
	Asphondylia sp. 1	*Helwingia japonica* (Helwingiaceae)	Fruit	JPN (KY)	2
	Asphondylia sp. 2	*Hedera rhombea* (Araliaceae)	Flower bud	JPN (HN, KY)	1
	Asphondylia sp. 3	*Hedera rhombea* (Araliaceae)	Fruit	JPN (KY)	10
	Asphondylia sp. 4	*Pieris japonica* (Ericaceae)	Flower bud	JPN (KY)	10
	Asphondylia sp. 5	*Neoshirakia japonica* (Euphorbiaceae)	Leaf bud	JPN (HN)	10
	Pseudasphondylia matatabi	*Actinidia polygama* (Actinidiaceae)	Flower bud	JPN (HN, KY)	2
	Asphondyliini sp. 1	*Eupatorium makinoi* (Asteraceae)	Flower	JPN (HN)	10
	Asphondyliini sp. 2	*Jacobaea cannabifolia* (Asteraceae)	Flower or fruit	JPN (HN)	10
	Asphondyliini sp. 3	*Styrax japonica* var. *kotoensis* (Styracaceae)	Leaf	JPN (IZ)	10
Bracon sunosei	*Asphondylia baca*	*Ampelopsis glandulosa* var. *heterophylla* (Vitaceae)	Fruit	JPN (HK)	2
	Asphondylia baca	*Weigela hortensis* (Caprifoliaceae)	Leaf bud	JPN (HK)	2
	Asphondyliini sp. 4	*Hydrangea serrata* (Hydrangeaceae)	Fruit	JPN (HN)	10
	Asphondyliini sp. 5[c]	*Styrax japonica* (Styracaceae)	Leaf	JPN (KY)	10
Bracon tamabae	*Schizomyia sasakii*	*Ilex crenata* (Aquifoliaceae)	Axillary bud	JPN (HN)	3, 4
	Schizomyia soyogo	*Ilex integra* (Aquifoliaceae)	Axillary bud	JPN (HN)	3, 4
	Daphnephila machilicola	*Machilus thunbergii* (Lauraceae)	Leaf	JPN (KY)	2

	Illiciomyia yukawai	*Illicium anisatum* (Schisandraceae)	Leaf	JPN (HN, KY)	10
	Masakimyia pustulae[d]	*Euonymus japonicus* (Celastraceae)	Leaf	JPN (KY)	2
	Pseudasphondylia neolitseae	*Neolitsea sericea* (Lauraceae)	Leaf	JPN (HN, SK, KY, IZ)	2
Simplicibracon curticaudis	*Daphnephila machilicola*	*Machilus thunbergii* (Lauraceae)	Leaf	JPN (KY, IZ)	2
Testudobracon longicaudis	*Asphondylia aucubae*	*Aucuba japonica* (Aucubaceae)	Fruit	JPN (KY)	10
	Asphondylia baca	*A. glandulosa* var. *heterophylla*	Fruit	JPN (SK, KY)	2
	Asphondylia sphaera	*Toxicodendron sylvestre* (Anacardiaceae)	Flower bud or fruit	JPN (KY)	10
	Asphondylia sphaera	*Toxicodendron succedaneum* (Anacardiaceae)	Flower bud or fruit	JPN (KY)	10
	Asphondylia yushimai	*Laurocerasus zippeliana* (Rosaceae)	Fruit	JPN (KY)	2
	Unknown	Unknown	Unknown	JPN (KY)	2
Testudobracon pleuralis	*Asphondylia itoi*	*Distylium racemosum* (Hamamelidaceae)	Fruit	JPN (KY)	10
	Asphondylia sphaera	*Ligustrum obtusifolium* (Oleaceae)	Fruit	JPN (HN, KY)	2
	Asphondylia sphaera	*Ligustrum obtusifolium* var. *pacificumi* (Oleaceae)	Fruit	JPN (IZ)	10
	Asphondylia sphaera	*Toxicodendron sylvestre* (Anacardiaceae)	Flower bud or fruit	JPN (KY)	10
	Asphondylia sphaera	*Toxicodendron succedaneum* (Anacardiaceae)	Flower bud or fruit	JPN (KY)	10
	Asphondylia yushimai	*Glycine max* (Fabaceae)	Pod	JPN (HN, KY)	1, 2, 5
	Asphondylia yushimai	*Sophora flavescens* (Fabaceae)	Pod	JPN (KY)	10
	Asphondylia sp. 2	*Hedera rhombea* (Araliaceae)	Flower bud	JPN (KY)	10
	Asphondylia sp. 3	*Hedera rhombea* (Araliaceae)	Fruit	JPN (HN, KY)	10
	Asphondylia sp. 6	*Schoepfia jasminodora* (Schoepfiaceae)	Leaf bud	JPN (KY)	10

(continued)

Table 11.3 (continued)

Braconid species	Host gall midge	Associated plant (Family)	Galled part	Distribution[a]	Reference[b]
	Unknown	Unknown	Unknown	JPN (HN), TW, CH	2, 6, 7
	Etiella zinckenella (Lepidoptera)[d]	Unknown	Unknown	JPN (HN)	8

Modified from Matsuo et al. (2016b)

[a]Abbreviation of distributional areas: *JPN* Japan, *HK* Hokkaido, *HN* Honshu, *SK* Shikoku, *KY* Kyushu, *IZ* the Izu Islands, *TW* Taiwan, *CH* mainland China

[b]Reference: 1, Yukawa et al. (1981); 2, Maeto (1991); 3, Tabuchi and Amano (2003); 4, Tabuchi and Amano (2004); 5, Naito and Osaka (1959); 6, Ashmead 1906; 7, Wang et al. (2003); 8, Watanabe (1952); 10, Matsuo et al. (2016b)

[c]Asphondyliini sp. 5 induces galls on *Styrax japonica* but the gall shape is different from that induced by Asphondyliini sp. 3 on var. *kotoensis* of the same host plant

[d]These host insect records should be reexamined (Maeto 1991) and note that *M. pustulae* does not belong to Asphondyliini

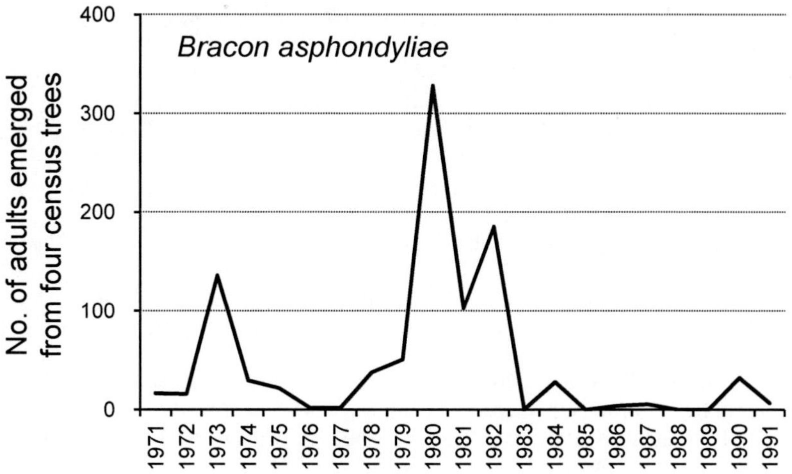

Fig. 11.4 Annual fluctuation in the number of *Bracon asphondyliae* adults at the time of emergence in April

References

Abe J, Ganaha-Kikumura T, Yukawa J (2011) Morphological features, distribution, prey mites, and life history traits of *Feltiella acarisuga* (Diptera: Cecidomyiidae) in Japan. Appl Entomol Zool 46:271–279

Ashmead WH (1904a) Descriptions of new Hymenoptera from Japan – I. J N Y Entomol Soc 1:65–84

Ashmead WH (1904b) Descriptions of new Hymenoptera from Japan II. J N Y Entomol Soc 12:146–165

Ashmead WH (1906) Descriptions of new Hymenoptera from Japan. Proc US Nat Mus 30:169–201

Askew RR (1961) On the Palaearctic species of *Syntomaspis* Förster (Hymenoptera, Torymidae). Entomol Month Mag 96:184–191

Askew RR (1975) The organisation of chalcid-dominated parasitoid communities centred upon endophytic hosts. In: Price PW (ed) Evolutionary strategies of parasitic insects and mites. Plenum Press, New York, pp 130–153

Axelsen JA (1994) Host–parasitoid interactions in an agricultural ecosystem: a computer simulation. Ecol Model 73:189–203

Barnes HF (1935) On the gall midges injurious to the cultivation of willows. II. The so called "shot hole" gall midges (*Rhabdophaga* spp.). Ann Appl Biol 22:86–105

Blair KG (1944) A note on the economy of the rose bedeguar gall, *Rhodites rosae* L. Proc South Lond Entomol Nat Hist Soc 1943–44:55–59

Boucek Z (1977) A faunistic review of the Yugoslavian Chalcidoidea (parasitic Hymenoptera). Acta Entomol Jugoslav 13 (Suppl):1–145

Brodeur J, Boivin G (2004) Functional ecology of immature parasitoids. Annu Rev Entomol 49:27–49

Brodeur J, McNeil JN (1989) Seasonal microhabitat selection by an endoparasitoid through adaptive modification of host behavior. Science 244:226–228

Brodeur J, McNeil JN (1992) Host behaviour modification by the endoparasitoid *Aphidius nigripes*: a strategy to reduce hyperparasitism. Ecol Entomol 17:97–144

Brodeur J, Vet LEM (1994) Usurpation of host behavior by a parasitic wasp. Anim Behav 48:187–192

Buhl PN (1994a) A new extralimital species of the *Platygaster* complex (Hymenoptera: Platygastridae) from Denmark and England. Reichenbachia 30:163–165

Buhl PN (1994b) Three new species of *Platygaster* Latreille, 1809, from Denmark (Hymenoptera: Platygastridae). Entomofauna 15:77–84

Buhl PN (2006a) Key to *Platygaster* (Hymenoptera: Platygastridae) from Denmark, with descriptions of new species. Steenstrupia 29:127–167

Buhl PN (2006b) New species of Platygastrinae from Canada (Hymenoptera: Platygastridae). Entomofauna 27:193–205

Buhl PN, Duso C (2008) *Platygaster robiniae* sp. (Hymenoptera: Platygastridae) parasitoid of *Obolodiplosis robiniae* (Diptera: Cecidomyiidae) in Europe. Ann Entomol Soc Am 101:297–300

Buhl PN, O'Connor JP, Ashe P (2010) A new reared species of *Aphanogmus* (Hymenoptera: Ceraphronidae) from Ireland. Entomol Month Mag 146:53–55

Cameron RAD, Redfern M (1978) Population dynamics of two hymenopteran parasites of the yew gall midge *Taxomyia taxi* (Inchbald). Ecol Entomol 3:265–272

Claridge MF (1959) A contribution of the biology and taxonomy of the British species of the genus *Eudecatoma* Ashmead (Hym., Eurytomidae). Trans Soc Br Entomol 13:149–168

Claridge MF (1961) Biological observations on some eurytomid (Hymenoptera: Chalcidoidea) parasites associated with compositae, and some taxonomic implications. Proc R Entomol Soc Lond A 36:153–158

Claridge MF, Askew RR (1960) Sibling species in the *Eurytoma rosae* group (Hym., Eurytomidae). Entomophaga 5:141–153

Dessart P (1992) Revision d'*Aphanogmus fulmeki* Szelényi, 1940 (Hymenoptera: Ceraphronoidea: Ceraphronidae) avec remarques biologiques. Bull Inst r sci nat Belg 62:83–91

Doğanlar M, Elsayed AK (2015) Parasitoids complex in summer populations of *Asphondylia punica* Marchal, 1897 (Diptera: Cecidomyiidae) on the Mediterranean saltbush, *Atriplex halimus* L. (Chenopodiaceae) in Egypt, with descriptions of new species from Eupelmidae and Eulophidae (Hymenoptera: Chalcidoidea). Munis Entomol Zool 10:75–85

Dorchin N, Mifsud D, Askew R (2014) Saltbush-associated *Asphondylia* species (Diptera: Cecidomyiidae) in the Mediterranean Basin and their chalcidoid parasitoids (Hymenoptera: Chalcidoidea). Zootaxa 3869:383–396

Ehler LE (1985) Species-dependent mortality in a parasite guild and its relevance to biological control. Environ Entomol 14:1–6

Ehler LE (1986) Distribution of progeny in two ineffective parasites of a gall midge (Diptera: Cecidomyiidae). Environ Entomol 15:1268–1271

Elsayed AK, Skuhravá M, Karam HH, Elminshawy A, Al-Eryan MA (2015) New records and new species of gall midges (Diptera: Cecidomyiidae) developing on Chenopodiaceae in Egypt. Zootaxa 3904:105–115

Elsayed AK, Matsuo K, Kim W, Uechi N, Yukawa J, Gyoutoku N, Tokuda M (2018) A new *Asphondylia* species (Diptera: Cecidomyiidae) and a eulophid wasp (Hymenoptera) inducing similar galls on leaf buds of *Schoepfia jasminodora* (Schoepfiaceae), with reference to their ecological traits and a description of the new gall midge. Entomol Sci 21:324–339

Evans GA, Dessart P, Glenn H (2005) Two new species of *Aphanogmus* (Hymenoptera: Ceraphronidae) of economic importance reared from *Cybocephalus nipponicus* (Coleoptera: Cybocephalidae). Zootaxa 1018:47–54

Fischer M (1980) Fünf neue Raupenwespen (Hymenoptera, Braconidae). Frustula Entomol, Nouva Series 1:147–160

Fisher JP (1965) A contribution to the biology of *Eurytoma curculionum* Mayr. (Hym., Eurytomidae). Entomophaga 10:317–318

Fujii T, Matsuo K, Abe Y, Yukawa J, Tokuda M (2014) An endoparasitoid avoids hyperparasitism by manipulating immobile host herbivore to modify host plant morphology. PLoS One 9: e102508.https://doi.org/10.1371/journal.pone.0102508

Fulmek L (1968) Parasitinsekten der Insektengallen Europas. Beiträge zur Entomologie 18:720–952

Ganaha-Kikumura T, Yukawa J, Tokuda M, Ohno S, Abe J (2012) Occurrence of two acarivorous species of the genus *Feltiella* (Diptera: Cecidomyiidae) in Okinawa, southern Japan, and redescription of *F. acarivora* (Zehntner). Appl Entomol Zool 46:319–329

Gates M, Delvare G (2008) A new species of *Eurytoma* (Hymenoptera: Eurytomidae) attacking *Quadrastichus* spp. (Hymenoptera: Eulophidae) galling *Erythrina* spp. (Fabaceae), with a summary of African *Eurytoma* biology and species checklist. Zootaxa 1751:1–24

Gauld ID, Bolton B (1988) The Hymenoptera. British museum of natural history. Oxford University Press, Oxford

Gauthier N, LaSalle J, Quicke DLJ, Godfray HCJ (2000) Phylogeny of Eulophidae (Hymenoptera; Chalcidoidea), with a reclassification of Eulophinae and the recognition that Elasmidae are derived eulophids. Syst Entomol 25:521–539

Gilkeson LA, McLean JP, Dessart P (1993) *Aphanogmus fulmeki* Ashmead (Hymenoptera: Ceraphronidae), a parasitoid of *Aphidoletes aphidimyza* Rondani (Diptera: Cecidomyiidae). Can Entomol 125:161–162

Graham MWR d V (1969) The Pteromalidae of North-Western Europe (Hymenoptera: Chalcidoidea). Bull Br Mus (Nat Hist) Suppl 16:1–908

Graham MWR d V, Gijswijt MJ (1998) Revision of the European species of *Torymus* Dalman (Hymenoptera: Torymidae). Zool Verh, Leiden 317:1–202

Grissell EE (1976) A revision of Western Nearctic species of *Torymus* Dalman (Hymenoptera: Torymidae). Univ Calif Pub Entomol 79:1–120

Grissell EE (1995) Toryminae (Hymenoptera: Chalcidoidea: Torymidae): A redefinition, generic classification and annotated world catalogue of species. Associated Publishers, Gainesville, FL

Grissell EE (2007) Torymidae (Hymenoptera: Chalcidoidea) associated with bees (Apoidea), with a list of chalcidoid bee parasitoids. J Hym Res 16:234–265

Grissell EE, Kamijo K, Hobbs KR (2004) *Torymus* Dalman (Torymidae: Hymenoptera) associated with coniferous cones, with descriptions of three new species. J Hymenopt Res 13:31–47

Grosman AH, Janssen A, de Brito EF, Cordeiro EG, Colares F, Fonseca JO, Lima ER, Pallini A, Sabelis MW (2008) Parasitoid increases survival of its pupae by inducing host to fight predators. PLoS One 3:e2276. https://doi.org/10.1371/journal.pone.0002276

Hansson C (1987) New records of Swedish Eulophidae and Pteromalidae (Hymenoptera: Chalcidoidea), with data on host species. Entomologisk Tidskrift 108:167–173

Herting B (1977) Hymenoptera. A catalogue of parasites and predators of terrestrial arthropods. Section A. Host or prey/enemy. Commonwealth Agricultural/Institute of Biological Control, Bureaux

Ikeda E (1999) A revision of the world species of *Sigmophora* Rondani (Hymenoptera: Eulophidae). Insecta Matsumurana 55:37–74

Ishii T (1950) New chalcid-parasites of cecidomyiid-flies injurious to the soy bean. Ôyô-Kontyû 6:141–143

Ishii T (1953) A report of the studies of the parasitic wasps of injurious insects. Bull Fac Agric, Tokyo Univ Agric Technol 1:1–10

Janssen A, Grosman AH, Corderio EG, de Brito EF, Fonseca J, Colares F, Lima ER, Pallini A, Sabelis MW (2010) Context-dependent fitness effects of behavioral manipulation by a parasitoid. Behav Ecol 21:33–36

Janšta P, Cruaud A, Delvare G, Genson G, Heraty J, Křížková B, Rasplus JY (2018) Torymidae (Hymenoptera: Chalcidoidea) revised: molecular phylogeny, circumscription and reclassification of the family with discussion of its biogeography and evolution of life-history traits. Cladistics 34:627–651

Johnson NF, Musetti L (2004) Catalog of systematic literature of the superfamily Ceraphronoidea (Hymenoptera). Contrib Am Entomol Inst 33:1–149

Kamijo K (1964) A new species of the genus *Diomorus* Walker (Hymenoptera: Torymidae). Insecta Matsumurana 27:16–17

Kamijo K (1979) Four new species of Torymidae from Japan, with notes on two known species. Akitu, New Series 24:1–11

Kamijo K (1982) Two new species of *Torymus* (Hymenoptera: Torymidae) reared from *Dryocosmus kuriphilus* (Hymenoptera: Cynipidae) in China and Korea. Kontyû 50:505–510

Kanai K, Matsuhira K, Uechi N, Yukawa J (2008) Invasion of the Amami Islands, Kagoshima, Japan by *Quadrastichus erythrinae* (Hymenoptera: Eulophidae). Jpn J Appl Entomol Zool 52:151–154. (In Japanese with English abstract)

Kaufmana LV, Yalemarb J, Wrighta MG (2020) Classical biological control of the erythrina gall wasp, *Quadrastichus erythrinae*, in Hawaii: conserving an endangered habitat. Biol Control 142. https://doi.org/10.1016/j.biocontrol.2019.104161

Kazmi SI, Chauhan N (2003) Chalcidoid parasitoids (Hymenoptera: Chalcidoidea) of *Hypsipyla robusta* (Lepidoptera: Pyralidae), a pest of cedars and mahogany. Orient Insects 37:261–275

Kim IK, Delvare G, LaSalle J (2004) A new species of *Quadrastichus* (Hymenoptera: Eulophidae): a gall-inducing pest on *Erythrina* spp. (Fabaceae). J Hymenopt Res 13:243–249

Kim IK, Parl JD, Shin SC, Park IK (2011) Prolonged embryonic stage and synchronized life-history of *Platygaster robiniae* (Hymenoptera: Platygastridae), a parasitoid of *Obolodiplosis robiniae* (Diptera: Cecidomyiidae). Biol Control 57:24–30

Kim W, Matsunaga K, Gyoutoku N, Matsuo K, Minami T, Yukawa J (2015) Life history strategy and adult and larval behavior of *Macrodiplosis selenis* (Diptera: Cecidomyiidae), a species that induces leaf-margin fold galls on deciduous *Quercus* (Fagaceae). Entomol Sci 18:470–478

Lampo M (1994) The importance of refuges in the interaction between *Contarinia sorghicola* and its parasitic wasp *Aprostocetus diplosidis*. J Anim Ecol 63:176–186

LaSalle J (2005) Biology of gall inducers and evolution of gall induction in Chalcidoidea (Hymenoptera: Eulophidae, Eurytomidae, Pteromalidae, Tanaostigmatidae, Torymidae). In: Raman A, Schaefer CW, Withers TM (eds) Biology, ecology, and evolution of gall-inducing arthropods. Science Publishers, Inc, Enfield, NH, pp 507–537

Leiby RW, Hill CC (1924) The polyembryonic development of *Platygaster vernalis*. J Agric Res 28:829–839

Maeto K (1991) Braconid parasitoids (Hymenoptera) of the gall-making Cecidomyiidae (Diptera) in Japan. Jpn J Entomol 59:295–313

Manzoli-Palma MF, da Teles MC (1990) Ocorrência de inimigos naturais de *Contarinia sorghicola* (Coq., 1898) (Diptera: Cecidomyiidae) na regiao de Ribeirao Preto. Rev Bras Entomol 34:589–594

Marchosky JR, Craig PT (2004) Gall size-dependent survival for *Asphondylia atriplicis* (Diptera: Cecidomyiidae) on *Atriplex canescens*. Environ Entomol 33:709–719

Matsuo K (2010) A redescription of *Torymus sapporoensis* Ashmead and description of a new species of *Torymus* Dalman (Hymenoptera: Torymidae) parasitizing *Paratephritis fukaii* (Diptera: Tephritidae) in Japan. Zootaxa 2410:53–58

Matsuo K (2012) An endoparasitic species of Pteromalidae (Hymenoptera: Chalcidoidea) attacking *Aphidoletes aphidimyza* (Diptera: Cecidomyiidae). Appl Entomol Zool 47:373–378

Matsuo K (2016) Damage of *Erythrina* trees caused by an invasive pest, *Quadrastichus erythrinae*, (Hymenoptera; Eulophidae) and measures toward a successful control. Nat Insects 51:9–11. (In Japanese)

Matsuo K (2020) A revision of Japanese *Torymus* Dalman (Hymenoptera: Torymidae). Zootaxa 4758:401–441

Matsuo K, Yukawa J (2009a) Host insect records and distributional information for *Torymus* species (Hymenoptera: Torymidae) in Japan. Jpn J Environ Entomol Zool 20:133–135. (In Japanese with English summary)

Matsuo K, Yukawa J (2009b) Two new species of *Torymus* Dalman, 1820 (Hymenoptera: Torymidae) parasitizing *Celticecis japonica* Yukawa & Tsuda, 1987 (Diptera: Cecidomyiidae) that induces leaf galls on *Celtis* species (Ulmaceae) in Japan. Entomol Sci 12:261–269

Matsuo K, Yara K, Kagoshima K, Tuda M, Moriya S (2011) Finding of *Torymus koreanus* (Hymenoptera: Torymidae) attacking *Dryocosmus kuriphilus* (Hymenoptera: Cynipidae) in Japan. Entomol Sci 14:100–102

Matsuo K, Yang MM, Tung GS, Tokuda M, Yukawa J (2012) Descriptions of a new and redescriptions of two known species of *Torymus* (Hymenoptera: Torymidae) in Taiwan with a key to Taiwanese species. Zootaxa 3409:47–57

Matsuo K, Uechi N, Tokuda M, Yukawa J (2013) Taxonomic position and ecological traits of *Sigmophora tricolor* (Hymenoptera: Eulophidae), a parasitoid of *Asphondylia yushimai* (Diptera: Cecidomyiidae), and description of a new species of *Sigmophora*. Appl Entomol Zool 48:355–363

Matsuo K, Ganaha-Kikumura T, Ohno S, Yukawa J (2016a) Description of a new species of *Aphanogmus* Thomson (Hymenoptera, Ceraphronidae) that parasitizes acarivorous gall midges of *Feltiella* (Diptera, Cecidomyiidae) in Japan. ZooKeys 596:77–85

Matsuo K, Uechi N, Tokuda M, Maeto K, Yukawa J (2016b) Host range of braconid species (Hymenoptera: Braconidae) that attack Asphondyliini (Diptera: Cecidomyiidae) in Japan. Entomol Sci 19:3–8

Matsuo K, Fujii T, Tokuda M, Ganaha-Kikumura T, Yukawa K, Yamagishi K (2018) Descriptions of two new species of *Platygaster* Latreille that attack gall midges (Diptera: Cecidomyiidae) with notes on their biology (Hymenoptera: Platygastridae). ZooKeys 754:113–125

Maure F, Daoust PS, Brodeur J, Mitta G, Thomas F (2013) Diversity and evolution of bodyguard manipulation. J Exp Biol 216:36–42

Mayr G (1878) Arten der Chalcidier Gattung *Eurytoma* durch Zucht erhalten. Verh Zool-Bot Ges Wien 28:297–334

Mikó I, Deans AR (2009) *Masner*, a new genus of Ceraphronidae (Hymenoptera: Ceraphronoidea) described using controlled vocabularies. ZooKeys 20:127–153

Moser JC (1965) The interrelationships of three gall makers and their natural enemies, on hackberry (*Celtis occidentalis* L.). N Y State Mus Sci Ser 402:1–95

Murakami Y (1981) The parasitoids of *Dryocosmus kuriphilus* Yasumatsu (Hymenoptera: Cynipidae) in Japan and the introduction of a promising natural enemy from China (Hymenoptera: Chalcidoidea). J Fac Agric Kyushu Univ 25:167–174

Muthukumar M, Sridharan S, Kennedy JS, Jeyakumar P, Arumugam T (2017) Biology and natural parasitization of gall fly *Lasioptera falcata* Felt and *Lasioptera bryoniae* Schiner infesting bitter gourd. J Entomol Zool Stud 5:1635–1639

Naito A, Osaka K (1959) Occurrence of the soy been pod gall midge and its hymenopterous parasites in Kanto District, and notes on the parasites. Jpn J Appl Entomol Zool 3:91–98. (In Japanese with English summary)

Noyes JS (2019) Universal Chalcidoidea database. World Wide Web electronic publication. Available from: http://www.nhm.ac.uk/chalcidoids. Accessed 10 Apr 2020

Oatman ER (1985) *Aphanogmus floridanus* (Hymenoptera: Ceraphronidae), a primary parasite of *Feltiella acarivora* (Diptera: Cecidomyiidae), a predator of spider mites on strawberry. Proc Entomol Soc Wash 87:714–716

Ogah EO, Nwilene FE (2014) Biological control of African Rice gall midge (*Orseolia oryzivora*, Harris and Gagné) in Nigeria: a review. Ann Res Rev Biol 4:2995–3006

Pathak MD, Khan ZR (1994) Insect pests of Rice. International Rice Research Institute, Manila

Piel O (1933) *Monema flavescens* Wkr. and its parasites (Lepidoptera, Heterogeneidae). Lingnan Sci J 12(Supplement):173–202

Polaszak A, Dessart P (1996) Taxonomic problems in the *Aphanogmus hakonensis* species complex (Hymenoptera: Ceraphronidae), common hyperparasitoids in biocontrol programs against lepidopterous pests in the tropics. Bull Entomol Res 86:419–422

Quicke DLJ (1986) Seven new genera and species of Braconinae (Hymenoptera, Braconidae) from Australia and Indonesia. Entomol Month Mag 122:9–29

Redfern M, Cameron RAD (1978) Population dynamics of the yew gall midge *Taxomyia taxi* (Inchbald) (Diptera: Cecidomyiidae). Ecol Entomol 3:251–263

Redfern M, Hunter M (2005) Time tells: long-term patterns in the population dynamics of the yew gall midge, *Taxomyia taxi* (Cecidomyiidae), over 35 years. Ecol Entomol 30:86–95

Rizzo MC, Massa B (1998) On two Italian gall midges (Diptera: Cecidomyiidae) and their parasitoids. Entomologica, Bari, 32:121–131

Ronquist F, Nylander JAA, Vårdal H, Nieves-Aldrey JL (2018) Life history of *Parnips* and the evolutionary origin of gall wasps. J Hymenopt Res 65:91–110

Roskam JC (1986) Biosystematics of insects living in female birch catkins. IV. Egg-larval parasitoids of the genera *Platygaster* Latreille and *Metaclisis* Förster (Hymenoptera, Platygastridae). Tijdschr Entomol 129:125–140

Sallem UA (2015) Hymenopteran parasitoids associated with plant galls. Thesis, Department of Zoology, University of Calicut

Segoli M, Harari AR, Rosenheim JA, Bouskila A, Keasar T (2010) The evolution of polyembryony in parasitoid wasps. J Evol Biol 23:1807–1819

Sen P (1954) On the bionomics and post-embryonic development of *Eurytoma saliciperdae* Mayr, a supposed chalcidoid parasite of the cecidomyiid or gall midge *Rhabdophaga saliciperda* (Duf.). Rec Indian Mus 52:137–150

Seyahooei MA, Kraaijeveld-Smit FJL, Kraaijeveld K, Crooijmans JBM, Van Dooren TJM, Van Alphen JJM (2009) Closely related parasitoids induce different pupation and foraging responses in *Drosophila* larvae. Oikos 118:1148–1157

Sharkey MJ, Carpenter JM, Vilhelmsen L, Heraty J, Liljeblad J, Dowling APG, Schulmeister S, Murray D, Deans AR, Ronquist F, Krogmann L, Wheeler WC (2012) Phylogenetic relationships among superfamilies of Hymenoptera. Cladistics 28:80–112

Stireman JO III, Nason JD, Heard SB, Seehawer JM (2006) Cascading host-associated genetic differentiation in parasitoids of phytophagous insects. Proc R Soc B Biol Sci 273:523–530

Sunose T (1985) Population regulation of the euonymus gall midge *Masakimyia pustulae* Yukawa and Sunose (Diptera: Cecidomyiidae) by hymenopterous parasitoids. Res Popul Ecol 27:287–300

Tabuchi K, Amano H (2003) Polymodal emergence pattern and parasitoid composition of *Asteralobia sasakii* (Monzen) (Diptera: Cecidomyiidae) on *Ilex crenata* and *I. integra* (Aquifoliaceae). Appl Entomol Zool 38:493–500

Tabuchi K, Amano H (2004) Impact of differential parasitoid attack on the number of chambers in multilocular galls of two closely related gall midges (Diptera: Cecidomyiidae). Evol Ecol Res 6:695–707

Tanaka S, Ohsaki N (2006) Behavioral manipulation of host caterpillars by the primary parasitoid wasp *Cotesia glomerata* (L.) to construct defensive webs against hyperparasitism. Ecol Res 21:570–577

Thompson WR (1955) Section 2. Host parasite catalogue, part 3. Hosts of the Hymenoptera (Calliceratid to Evaniid). In: Commonwealth agricultural Bureaux, A catalogue of the parasites and predators of insect pests. The Commonwealth Institute of Biological Control, Ottawa, ON, pp 191–332

Tiwari NK (1974) *Eurytoma nesiotes* Crawford (Hym. Chalcidoidea) an external parasite of the gall midge *Asphondylia sesami* Felt. Z Angew Entomol 77:169–194

Tokuda M (2012) Biology of Asphondyliini (Diptera: Cecidomyiidae). Entomol Sci 15:361–383

Tokuda M, Nohara M, Yukawa J, Usuba S, Yukinari M (2004) *Oxycephalomyia*, gen. Nov., and life history strategy of *O. styraci* comb. nov. (Diptera: Cecidomyiidae) on *Styrax japonicus* (Styracaceae). Entomol Sci 7:51–62

Uechi N, Uesato T, Yukawa J (2007) Detection of an invasive gall-inducing pest, *Quadrastichus erythrinae* (Hymenoptera: Eulophidae), causing damage to *Erythrina variegata* L. (Fabaceae) in Okinawa prefecture, Japan. Entomol Sci 10:209–212

Varley GC (1937) Description of the eggs and larvae of four species of chalcidoid Hymenoptera parasitic on the knapweed gall-fly. Proc R Entomol Soc Lond B 6:122–130. 5 figs
Vlug HJ (1995) Catalogue of the Platygastridae (Platygastroidea) of the world. Hymenopterorum Catalogues 19:1–168
Wang YP, Chen XX, He JH (2003) The genus *Testudobracon* in eastern China and the description of three new species (Hymenoptera: Braconidae: Braconinae). Eur J Entomol 100:411–416
Watanabe C (1940) A new braconid parasite of the gall midge of *Actinidia polygama* Maximowicz (Hymenoptera: Braconidae). Insecta Matsumurana 14:138–140
Watanabe C (1952) Synonymy and other notes on Braconidae (Hymenoptera). Insecta Matsumurana 18:41–42
Wilson LF, Heaton GC (1987) Life history, damage, and gall development of the gall midge, *Neolasioptera brevis* (Diptera: Cecidomyiidae), injurious to honeylocust in Michigan. Great Lake Entomol 20:111–118
Yamagishi K (1980) Platygastrid parasites of willow gall midges in Japan. Esakia 15:161–175
Yamamura K (1999) Key-factor/key-stage analysis for life table data. Ecology 80:533–537
Yasumatsu K, Kamijo K (1979) Chalcidoid parasites of *Dryocosmus kuriphilus* Yasumatsu (Cynipidae) in Japan, with descriptions of five new species (Hymenoptera). Esakia 14:93–111
Yefremova ZA, Yegorenkova EN (2009) A review of the subfamily Tetrastichinae (Hymenoptera: Eulophidae) in Yemen, with descriptions of new species. Fauna Arabia 24:169–210
Yoshida N, Hirashima Y (1979) Systematic studies on proctotrupoid and chalcidoid parasites of gall midges injurious to *Pinus and Cryptomeria* in Japan and Korea. Esakia 14:113–133
Yukawa J (1971) A revision of the Japanese gall midges (Diptera: Cecidomyiidae). Mem Fac Agric Kagoshima Univ 8:1–203
Yukawa J (1983) Arthropod community centred upon the neolitsea leaf gall midge, *Pseudasphondylia neolitseae* Yukawa (Diptera, Cecidomyiidae) and its host plant, *Neolitsea sericea* (Blume) Koidz. (Lauraceae). Mem Fac Agric Kagoshima Univ 19:89–96
Yukawa J (1987) Life history strategies of univoltine gall-making Cecidomyiidae (Diptera) in Japan. Phytophaga 1:121–139
Yukawa J, Miyamoto K (1979) Redescription of *Asphondylia sphaera* Monzen (Diptera, Cecidomyiidae), with notes on its bionomics. Mem Fac Agric Kagoshima Univ 15:99–106
Yukawa J, Ohsaki N (1988) Separation of the aucuba fruit midge, *Asphondylia aucubae* sp. nov. from the ampelopsis fruit midge, *Asphondylia baca* Monzen (Diptera, Cecidomyiidae). Kontyû 56:365–376
Yukawa J, Ohsako S, Ikenaga H (1981) Parasite complex of the Japanese species of the genus *Asphondylia* (Diptera: Cecidomyiidae), including the soybean pod gall midge. Proc Assoc Plant Prot Kyushu 27:113–115. (In Japanese with English summary)
Yukawa J, Ikenaga H, Sato S, Tokuda M, Ganaha-Kikumura T, Uechi N, Matsuo K, Mishima M, Tung GH, Paik JC, Ren BQ, Don XO (2012) Description and ecological traits of a new species of *Pitydiplosis* (Diptera: Cecidomyiidae) that induces leaf galls on *Pueraria* (Fabaceae) in East Asia, with a possible diversification scenario of intraspecific groups. Entomol Sci 15:81–98
Yukawa J, Miyamoto K, Yamaguchi T, Takesaki K, Uechi N, Matsuo K (2016) Key factor/key stage analysis of long-term life table data for a fruit gall midge, *Asphondylia sphaera* (Diptera: Cecidomyiidae). Ecol Entomol 41:516–526
Zerova MD, Seryogina LY (2006) A review of the Palaearctic species of the genus *Eurytoma*, belonging to the *E. robusta* species-group (Hymenoptera: Eurytomidae), with description of two new species. Entomol Rev 86:695–705
Zerova MD, Seryogina LY, Stetsenko IT (2003) Torymid chalcid wasps (Hymenoptera: Chalcidoidea: Torymidae) of tribe Torymini of the Ukrainian Fauna. Vestnik zoologii Suppl 17:3–83

Part VI
Applied Fields

Chapter 12
Invasive Species

Makoto Tokuda and Nami Uechi

Abstract Invasions by gall-inducing insects are serious global concerns. In gall midges, *Contarinia maculipennis*, which was originated in Southeast Asia, has invaded various places including Hawaii, Korea, China, and Japan. This species has a broad host range including at least eight plant families. Several species of *Procontarinia* infesting mango have spread to various places. *Dasineura oxycoccana* associated with blueberry and *Obolodiplosis robiniae* with black locust *Robinia pseudoacacia* are both native to North America, but invaded Europe and Asian countries. *Dasineura gleditchiae* on honey locust *Gleditsia triacanthos* (Fabaceae) is also North American origin and has colonized Europe, Australia, and South America. The swede midge *Contarinia nasturtii* that is distributed widely in Europe and southwestern Asia entered North America and infests Brassicaceae.

Keywords *Contarinia* · *Dasineura* · Invasive pest · *Obolodiplosis* · *Procontarinia*

12.1 Introduction

Following the increase of international trades and globalization trends, biological invasions have become serious concerns worldwide (Kiritani 1998; Hurley et al. 2016). Invasion by alien species severely affects natural and agricultural ecosystems (e.g. Sax and Gaines 2008; Bellard et al. 2016; Paini et al. 2016).

Many gall inducers including gall midges have been reported to invade various regions, and some became important pests in invaded areas (Skuhravá et al. 2010; Csóka et al. 2017). For example, recent invasion of Asian countries and USA by

M. Tokuda (✉)
Laboratory of Systems Ecology, Faculty of Agriculture, Saga University, Saga, Japan
e-mail: tokudam@cc.saga-u.ac.jp

N. Uechi
Institute of Plant Protection, National Agriculture and Food Research Organization, Ibaraki, Japan
e-mail: uechinmi@affrc.go.jp

J. Yukawa, M. Tokuda (eds.), *Biology of Gall Midges*, Entomology Monographs,
https://doi.org/10.1007/978-981-33-6534-6_12

Quadrastichus erythrinae Kim (Hymenoptera: Eulophidae) has been causing heavy damage to *Erythrina* trees (Fabaceae) (Rubinoff et al. 2010; Csóka et al. 2017). *Dryocosmus kuriphilus* Yasumatsu (Hymenoptera: Cynipidae) invaded European countries and seriously harms chestnuts there (EFSA Panel on Plant Health 2010; Csóka et al. 2017).

Many gall midges have frequently been intercepted under plant quarantine inspection at sea- and airports in Japan and other countries (e.g. Tokuda et al. 2002; Iwaizumi et al. 2007). Some species became invasive pests elsewhere in the world (e.g. Tokuda et al. 2019). In this chapter, we refer to some of the invasive gall midge pests associated with various beneficial plant species. First, we introduce *Contarinia maculipennis* Felt, which has an extraordinary wide host range and is known especially as an important pest of orchids (Orchidaceae). Second, we review invasive species of *Procontarinia* infesting mango *Mangifera indica* Linnaeus (Anacardiaceae). Third, we focus on other invasive gall midges such as *Dasineura oxycoccana* (Johnson) on blueberry, *Obolodiplosis robiniae* (Haldeman) on *Robinia*, *Dasineura gleditchiae* Osten Sacken on honey locust, and *Contarinia nasturtii* (Kieffer) on brassicaceous plants.

12.2 Orchid Flower Bud Gall Midge, *Contarinia maculipennis*

12.2.1 Host Range

Contarinia maculipennis was originally described based on specimens infesting flower buds of hibiscus (Malvaceae) in Hawaii (Felt 1933). Jensen (1950) confirmed its wide host range by rearing experiments, and Gagné (1995) observed morphological similarities among adults emerged from several host plants. Later, Uechi et al. (2003) compared the cytochrome oxidase subunit I gene in mitochondrial DNA of *C. maculipennis* collected from various host plants in Hawaii, Japan and Thailand, and confirmed the polyphagy of *C. maculipennis*. It is now known to infest various plant species across at least eight families, namely, Amaranthaceae, Apocynaceae, Brassicaceae, Cucurbitaceae, Malvaceae, Oleaceae, Orchidaceae, and Solanaceae (Jensen 1946, 1950; Gagné 1995; Uechi et al. 2003; Uechi et al. 2007b; Uechi et al. 2011). Thereafter, Plant Protection Office (2018) in Japan found *C. maculipennis* from several orchid species other than *Dendrobium* spp. such as *Mokara* (an intergeneric hybrid of Arachnis and Ascocentrum and Vanda) and *Aranda* (an intergeneric hybrid of Arachnis and Vanda).

12.2.2 Life History

According to the observation in Hawaii by Jensen (1950) (Fig. 12.1), females of *C. maculipennis* insert their ovipositor into the host flower bud and lay their eggs, which hatch within 5–7 days. Larvae mature within 3 days, then drop to the ground and enter the soil, where they pupate. Pupal period is approximately 9–15 days. Adults normally emerge about 14 days after the larvae enter the soil. Thus, *C. maculipennis* is multivoltine and requires 24–28 days to complete one generation in Hawaii.

In some circumstances the larvae remain in the soil for additional several days without pupating. Lack of adequate soil moisture apparently is one of the factors responsible for retarding or preventing pupation (Jensen 1950).

12.2.3 Damages

Contarinia maculipennis attacks flowers of orchids such as *Dendrobium phalaenopsis* Fitzgerald (Orchidaceae) in Southeast Asia, and *Plumeria rubra* Linnaeus (Apocynaceae), *Hibiscus rosa-sinensis* Linnaeus (Malvaceae), etc. in Hawaii. The infested flower buds usually remain closed and eventually drop to the ground (Fig. 12.2a, b). Even when lightly infested flower buds manage to open, the petals are conspicuously damaged, so the commercial value is reduced to nothing (Tokuda et al. 2002). In 2005, Uechi et al. (2007b) detected *C. maculipennis* infesting bitter gourd *Momordica charantia* Linnaeus (Cucurbitaceae) that is one of the most important cash crops in Okinawa (Fig. 12.2c). Thereafter, Uechi et al. (2011) found that *C. maculipennis* infests the flower buds of *Pseuderanthemum laxiflorum* (A. Gray) Hubbard ex Baillon (Amaranthaceae) and *Jasminum sambac* (Linnaeus)

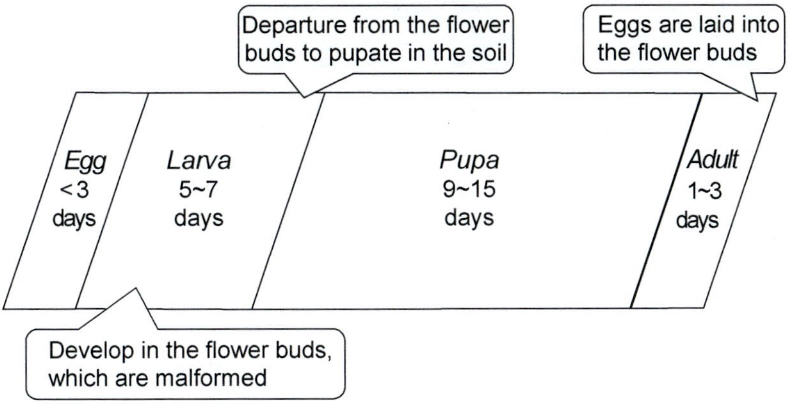

Fig. 12.1 Schematic representation of *Contarinia maculipennis* life cycle. The drawing was made based on the description in Jensen (1950)

Fig. 12.2 Some of invasive gall midges and their damages. (**a**), a flower bud of *Dendrobium phalaenopsis* damaged by *Contarinia maculipennis*; (**b**), flower buds of *Plumeria rubra* damaged by *C. maculipennis*; (**c**), a female of *C. maculipennis* ovipositing into a flower bud of *Momordica charantia*; (**d**), a mango leaf damaged by *Procontarinia mangicola*; (**e**), a leaf bud of blueberry damaged by *Dasineura oxycoccana*; (**f**), an adult of *Obolodiplosis robiniae* and its leaf margin roll gall on *Robinia pseudoacacia*; (**g**), mature larvae and cocoons of *Lasioptera yoichiensis* in the damaged stem of tomato; (**h**), blacking of tomato fruit bottom

Aiton (Oleaceae) in Okinawa, and *Dendrobium* spp. (Orchidaceae) in Mie Prefecture, Japan.

In India, *C. maculipennis* has attained the status of a major pest of jasmine, *J. sambac*, causing severe economic loss by reducing the marketable quality of the flowers (Merlin Kamala and Kennedy 2018). When larvae enter the buds at the base of the corollas, the base of the buds swells and shrivels. It causes deformed and pink discolored buds and blossoms. In case of severe infestations, buds dry prematurely leading to bud drop or blossom drop, thereby the marketable quantity of the flowers are greatly reduced (Merlin Kamala and Kennedy 2018).

12.2.4 Distribution

Contarinia maculipennis was first recorded from Hawaii (Felt 1933) but Gagné (1995) suggested that it originally came from Southeast Asia because the same species was found in Florida on orchids imported from Thailand. It is now known to occur also in Vietnam and Taiwan (Gagné and Jaschhof 2017). Along with the import and/or export of cut flowers, *C. maculipennis* expanded its distribution range to Hawaii (Gagné 1995), Japan (Tokuda et al. 2002; Uechi et al. 2003; Uechi et al. 2007b; Uechi et al. 2011), and Guadeloupe FWI (Gagné and Jaschhof 2017). Thereafter, it was reported from South Korea (Kang et al. 2010), China (Zhou et al. 2015), Sri Lanka (Dias et al. 2017), and India (Merlin Kamala 2018). It once occurred in California and Florida, USA, but was not established (Gagné and Jaschhof 2017). In the Netherlands, cut flowers of *Dendrobium* orchid with *C. maculipennis* from Thailand was intercepted around 2000 (Tokuda et al. 2002). It has been repeatedly intercepted from plants imported from Southeast Asia at sea- or airports in Japan, the Netherlands, and the USA (Tokuda et al. 2002; Uechi et al. 2003).

In 1989, a *Contarinia* species was found heavily infesting the flower buds of cultivated orchids in several greenhouses in Nago, Okinawa Island, Japan (Yasuda and Uehara 1994). By 1993 this gall midge was successively found in green houses of other places on Okinawa Island (Yasuda and Uehara 1994). Based on morphological features, this gall midge was identified as *C. maculipennis* (Tokuda et al. 2002). Thereafter, its occurrence was recorded from Miyazaki, Fukuoka and Mie Prefectures, Honshu, Japan (Uechi et al. 2011).

12.2.5 Control Measures

As always, plant quarantine inspection is most important to prevent pest species entering the country. All who are responsible for orchid growing, international trading, and inspection, are requested to pay much attention to the dangers of accidental international and internal transfer (Tokuda et al. 2002). Among cecidomyiid pests, *C. maculipennis* has been most frequently found under plant quarantine inspection at Japanese sea- and airports during the period from 2000 to 2005 (Iwaizumi et al. 2007). This tendency is still continuing until today (Plant Protection Station 2018). At the local level, people should refrain from bringing orchid seedlings or young plants into the country from the areas where the pests have been occurring. Because mature gall midge larvae leave the galled flower buds or flowers to drop to the ground for pupation, they could be easily transported with soil or other substrates covering the roots of *Dendrobium* plants in nurseries (Tokuda et al. 2002).

Insecticides would be one of the most practical control measures. In Japan, *C. maculipennis* occurring in the orchid greenhouses seemed to have developed resistance to various insecticides such as organophosphorus, synthetic pyrethroid, carbamate, and other major IGR chemicals (Tokuda et al. 2002). In China, six insecticides against the larva of *C. maculipennis* were tested (Zhou et al. 2015). As results of indoor toxicity test and field efficacy, acetamiprid is the most effective, the effect has reached nearly 80% after 3 days of spraying, and spinetoram and avermectins followed acetamiprid (Zhou et al. 2015).

Natural enemies have been considered. In India, the efficacy of six entomophagous fungi to *C. maculipennis* infesting jasmine was evaluated under in vitro and field conditions (Merlin Kamala 2018). This study revealed that *Metarhizium anisopliae* (Metschnikoff) Sorokin (Hypocreales: Clavicipitaceae) in three rounds of spray at 14 days interval was quite effective in reducing the damage by *C. maculipennis* and increasing the flower yield of jasmine. *Beauveria bassiana* (Balsamo-Crivelli) Vuillemin (Hypocreales: Cordycipitaceae) and *Lecanicillium lecanii* R. Zare & W. Gams (Hypocreales: Cordycipitaceae) also proved their efficacy in the next level in managing blossom midge of jasmine (Merlin Kamala 2018).

Two endoparasitoid species belonging to the genus *Synopeas* (Hymenoptera: Platygastridae) were found in Okinawa, Japan from *C. maculipennis* larvae in the flower buds of *Dendrobium* and *Momordica* (Uechi et al. 2007a). However, their ecological traits have not yet been studied. Because female sex pheromones of a congener *C. nasturtii* were identified (Hillbur et al. 2005; Boddum et al. 2009) and applied in field experiments (Samietz et al. 2012), similar monitoring or control techniques can be applied to *C. maculipennis*.

12.3 Mango Gall Midges

12.3.1 Procontarinia mangicola

Gall midges of the genus *Procontarinia* are exclusively associated with mango, *M. indica* (Gagné and Jaschhof 2017). Among 16 species of *Procontarinia* (Gagné and Jaschhof 2017; Jiao et al. 2018), *Procontarinia mangicola* (Shi), *Procontarinia mangiferae* (Felt), and *Procontarinia matteiana* Kieffer & Cecconi are known as invasive species.

Procontarinia mangicola was originally described from southern China (Guangxi Zhuang Autonomous Region) (Shi 1980), and has been reported to invade Guam (Harris and Schreiner 1992), the Southwest Islands of Japan including

Okinawa and Amami Islands (Uechi et al. 2002; Yukawa et al. 2004; Yamaguchi 2012), and the Bonin Islands (Tokuda et al. 2019). This gall midge induces circular blister galls on mango leaves (Fig. 12.2d) and its heavy infestation causes early leaf falling as well as anthracnose infection on mature leaves (Harris and Schreiner 1992; Uechi et al. 2002). Harris and Schreiner (1992) added that high humidity improved larval and pupal survival, and more galls appeared. Uechi et al. (2002) found that the production of inflorescences was inhibited on heavily infested shoots.

12.3.2 Procontarinia mangiferae

Procontarinia mangiferae is distributed in India and southern China (Guandong and Hainan) and has been reported to immigrate to various localities such as Réunion, Iran, Brazil, and the Lesser Antilles including Guadeloupe, St. Lucia, St. Vincent, and Trinidad (Amouroux et al. 2013; Gagné and Jaschhof 2017). Unlike almost all other *Procontarinia* species inducing galls on host leaves (Raman et al. 2009; Jiao et al. 2018), *P. mangiferae* induces galls not only on leaves but also on stems and inflorescences, which directly affect fruit production of mango (Raman et al. 2009; Amouroux et al. 2013). In Pakistan, the application of bifenthrin or neem seed kernel extracts to tree canopies was effective to kill *P. mangiferae* larvae and raking over the ground for disturbing pupation and emergence of gall midges reduced gall midge adult densities (Rehman et al. 2016). At least nine species of hymenopteran parasitoids seem to attack *P. mangiferae* in the native distribution range, but the information is still scattered, and further studies are needed (Singh 2018).

12.3.3 Procontarinia matteiana

Procontarinia matteiana is reported to have invaded Oman, Mauritius, Kenya, Réunion, Italy and South Africa (Augustyn et al. 2013; Henri et al. 2015; Gagné and Jaschhof 2017). Some resistant cultivars of mango are known to deter the development of galls, which causes larval mortality of *P. matteiana* (Augustyn et al. 2013). Although a parasitoid *Chrysonotomyia pulcherimma* Kerrich (Hymenoptera: Eulophidae) is known from its native distribution range, the lack of natural enemies seems to cause the outbreak of *P. matteiana* in invaded areas (Augustyn et al. 2013). The parasitoid has been non-intentionally introduced into South Africa, possibly with its host gall midge and mango trees (Morgan et al. 2017). At the field scale, parasitism rates by *C. pulcherimma* was higher in areas closer to natural vegetations, possibly because of the presence of non-host resources including nectar and pollen provided by natural vegetations (Morgan et al. 2017). Recently, Saxena (2020) reported that leaf damage and gall intensity were related positively with sunshine and negatively with temperature, relative humidity, wind velocity, and rainfall.

12.4 Other Invasive Pests

12.4.1 Dasineura oxycoccana

The blueberry gall midge *D. oxycoccana* is one of the most important pests of blueberry cultivation causing severe yield reduction (Fig. 12.2e) (e.g. Lyrene and Payne 1992; Sampson et al. 2002; Liburd and Roubos 2010). This species is native to North America but invaded Europe and has rapidly spread there (e.g. Collins et al. 2010; Gagné and Jaschhof 2017; Survilienė and Kazlauskaitė 2019). In Asia, this gall midge has found in South Korea (Kim et al. 2015; Lee et al. 2016) and Japan (Yoshida et al. 2017; Tokuda et al. 2019). Although gall midges associated with blueberry and cranberry had been regarded as *D. oxycoccana*, they were clarified to have genetically diverged from each other, suggesting the presence of cryptic species (Mathur et al. 2012; Fitzpatrick et al. 2013). Several hymenopteran parasitoids such as *Aprostocetus* and *Quadrastichus* (Eulophidae) are known to attack *D. oxycoccana* in USA, and an application of malathion is effective to control the larvae (Sampson et al. 2002).

12.4.2 Obolodiplosis robiniae *and* Dasineura gleditchiae

Obolodiplosis robiniae and its host, black locust *Robinia pseudoacacia* Linnaeus (Fabaceae) (Fig. 12.2f), are native to northeastern parts of USA, but the gall midge has spread in other parts of North America following the planting of its host as ornamental trees (Csóka et al. 2017). In the 2000's, *O. robiniae* has expanded its range to East Asia including Japan (Kodoi et al. 2003), South Korea (Woo et al. 2003), and mainland China (Yang et al. 2006) and to Europe, first to Italy (Duso and Skuhravá 2002) and is now distributed in many countries (Yao et al. 2015; Csóka et al. 2017).

Dasineura gleditchiae induces leaflet galls on honey locust *Gleditsia triacanthos* Linnaeus (Fabaceae), which is native to North America but introduced to Europe, Australia, and South America (Csóka et al. 2017). The gall midge was first found in Europe in 1975 (Nijveldt 1980) and is now widely distributed there (Csóka et al. 2017).

12.4.3 Contarinia nasturtii

Swede midge *C. nasturtii* is widely distributed in Europe and southwestern Asia but was found in Canada (Ontario) in 2000 and in USA (New York) in 2004. This species feeds on various crops of Brassicaceae (Hallett and Heal 2001; Kikkert et al. 2006), and brassicaceous weeds seem to serve as a reservoir for the gall midge (Chen et al. 2009). In spite of intensive surveys in 2004 and 2005, no parasitoids were obtained from larvae of *C. nasturtii* in Canada (Corlay et al. 2007). In USA, Wu et al. (2006) investigated efficacy of various insecticides applied with foliar spray to the gall midge eggs and larvae, and then detected several effective ones.

12.4.4 Lasioptera yoichiensis

Recently two new species, *Lasioptera tomaticola* Yukawa & Harris from the Mediterranean Region (Greece, Romania and Turkey) and *Lasioptera yoichiensis* Yukawa & Kim from Japan (Hokkaido) were described (Yukawa et al. 2020). Morphological characters and genetic data indicated that the two species were distinct. Both of them damage tomato *Solanum lycopersicum* Linnaeus (Solanaceae: South American origin) (Fig. 12.2g, h) (Perdikis et al. 2011; Hashimoto et al. 2018). *Lasioptera tomaticola* also infests cucumber *Cucumis sativus* Linnaeus (Cucurbitaceae: Asian origin) in the Mediterranean Region but it is morphologically and genetically different from known *Lasioptera* species on cucurbitaceous plants in the Palaearctic and Oriental Regions. Yukawa et al. (2020) first confirmed that no *Lasioptera* species have previously been found on solanaceous plants in the Palaearctic Region even though many entomologists have been working on gall midges. It is also natural that farmers would not overlook such conspicuous damages on tomato by the gall midge for a long period of time before the spring of 2001 (Perdikis et al. 2011). Nevertheless, they could not conclude that *L. tomaticola* is not a species of Palaearctic origin, because of the existence of at least six haplotypes in the *L. tomaticola* populations indicating that it has been in the Mediterranean Region for a considerable time before its infestations were noticed.

In contrast, Yukawa et al. (2020) considered that *L. yoichiensis* was accidentally introduced from overseas to Yoichi, Hokkaido, Japan because of several reasons as follows: (1) its distribution range is restricted only to Yoichi and its adjacent areas in Hokkaido, Japan; (2) the existence of only one haplotype in the *L. yoichiensis* population indicates its recent arrival in Hokkaido; (3) there is no report of its occurrence from agricultural experiment stations in other prefectures than Hokkaido; and (4) no *Lasioptera* species has been found on wild plants of Solanaceae in Japan (Yukawa et al. 2014). Further field surveys are needed to establish appropriate control measures against the newly invaded pest gall midge and to prevent its range expansion to other parts of Japan.

References

Amouroux P, Normand F, Nibouche S, Delatte H (2013) Invasive mango blossom gall midge, *Procontarinia mangiferae* (Felt) (Diptera: Cecidomyiidae) in Reunion Island: ecological plasticity, permanent and structured populations. Biol Invasions 15:1677–1693

Augustyn WA, Du Plooy W, Botha BM, van Wilpe E (2013) Infestation of *Mangifera indica* by the mango gall fly, *Procontarinia matteiana*, (Kieffer & Cecconi) (Diptera: Cecidomyiidae). Afr Entomol 21:79–88

Bellard C, Genovesi P, Jaschke JM (2016) Global patterns in threats to vertebrates by biological invasions. Proc R Soc B Biol Sci 283:2015–2454

Boddum T, Skals N, Wiren M, Baur R, Rauscher S, Hillbur Y (2009) Optimization of the pheromone blend of the swede midge, *Contarinia nasturtii*, for monitoring. Pest Manag Sci 65:851–856

Chen N, Shelton AM, Wang P, Hoepting CA, Kain WG, Brainard DC (2009) Occurrence of the new invasive insect *Contarinia nasturtii* (Diptera: Cecidomyiidae) on cruciferous weeds. J Econ Entomol 102:115–120

Collins DW, Harris KM, Glover RH (2010) *Dasineura oxycoccana* (Johnson, 1899) (Diptera: Cecidomyiidae), a pest of cultivated blueberries, new to Britain. Dipter Dig 17:53–59

Corlay F, Boivin G, Bélair G (2007) Efficiency of natural enemies against the swede midge *Contarinia nasturtii* (Diptera: Cecidomyiidae), a new invasive species in North America. Biol Control 43:195–201

Csóka G, Stone GN, Melika G (2017) Non-native gall-inducing insects on forest trees: a global review. Biol Invasions 19:3161–3181

Dias MA, Amarasinghe LD, Jayalath WGH, Attanayake RN (2017) *Contarinia maculipennis* as an emerging threat to *Dendrobium* in Sri Lanka - a case study. In: Abstract of international research symposium on pure and applied sciences, 2017, p. 14

Duso C, Skuhravá M (2002) First record of *Obolodiplosis robiniae* (Haldeman) (Diptera: Cecidomyiidae) galling leaves of *Robinia pseudoacasia* L. (Fabaceae) in Italy and Europe. Frustula entomol 25:117–122

EFSA Panel on Plant Health (2010) Risk assessment of the oriental chestnut gall wasp, *Dryocosmus kuriphilus* for the EU territory and identification and evaluation of risk management options. EFSA J 8:1619

Felt EP (1933) A hibiscus bud midge new to Hawaii. Proc Hawaiian Entomol Soc 79:87–89

Fitzpatrick SM, Gries R, Khaskin G, Peach DAH, Iwanski J, Gries G (2013) Populations of the gall midge *Dasineura oxycoccana* on cranberry and blueberry produce and respond to different sex pheromones. J Chem Ecol 39:37–49

Gagné RJ (1995) *Contarinia maculipennis* (Diptera: Cecidomyiidae), a polyphagous pest newly reported for North America. Bull Entomol Res 85:209–214

Gagné RJ, Jaschhof M (2017) A catalog of the Cecidomyiidae (Diptera) of the World, 4th edn. Digital. [Cited 10 May 2020] Available from URL: https://www.ars.usda.gov/ARSUserFiles/80420580/Gagne_2017_World_Cat_4th_ed.pdf

Hallett RH, Heal JD (2001) First Nearctic record of the swede midge (Diptera: Cecidomyiidae), a pest of cruciferous crops from Europe. Can Entomol 133:713–715

Harris KM, Schreiner H (1992) A new species of gall midge (Diptera: Cecidomyiidae) attacking mango foliage in Guam, with observations on its pest status and biology. Bull Entomol Res 82:41–48

Hashimoto N, Sumino A, Iwasaki A, Furudate T, Nishimura K, Takada M, Kim W, Yukawa J (2018) First record of *Lasioptera* sp. (Diptera: Cecidomyiidae) infesting tomato in Hokkaido, Japan. Jpn J Appl Entomol Zool 32:193–197. (In Japanese with English abstract)

Henri DC, Jones O, Tsiattalos A, Thébault E, Seymour CL, van Veen FJF (2015) Natural vegetation benefits synergistic control of the three main insect and pathogen pests of a fruit crop in southern Africa. J Appl Ecol 52:1092–1101

Hillbur Y, Celander M, Baur R, Rauscher S, Haftmann J, Franke S, Francke W (2005) Identification of the sex pheromone of the swede midge, *Contarinia nasturtii*. J Chem Ecol 31:1807–1828

Hurley BP, Garnas J, Wingfield MJ, Branco M, Richardson DM, Slippers B (2016) Increasing numbers and intercontinental spread of invasive insects on eucalypts. Biol Invasions 18:921–933

Iwaizumi R, Tokuda M, Yukawa J (2007) Identification of gall midges (Diptera: Cecidomyiidae) intercepted under plant quarantine inspection at Japanese sea- and airports from 2000 to 2005. Appl Entomol Zool 42:231–240

Jensen DD (1946) The identity and host plants of blossom midge in Hawaii (Diptera: Cecidomyiidae: *Contarinia*). Proc Hawaiian Entomol Soc 12:525–534

Jensen DD (1950) Notes on the life history and ecology of blossom midge, *Contarinia lycopersici* Felt (Diptera: Cecidomyiidae). Proc Hawaiian Entomol Soc 14:91–100

Jiao KL, Wang H, Wei DW et al (2018) A new species of *Procontarinia* (Diptera: Cecidomyiidae) damaging fruit of mango, *Mangifera indica* (Anacardiaceae), in China. Zootaxa 4413:368–376

Kang TJ, Lee HS, Oh SH, Cho MR, Kim HH, Jung JA (2010) Occurrence and damages of *Contarinia maculipennis* Felt (Diptera: Cecidomyiidae) on *Dendrobium phalaenopsis* (Asparagales: Orchidaceae). Korean J Appl Entomol 49:371–374. (in Korean with English abstract)

Kikkert JR, Hoepting CA, Wu QJ, Wang P, Baur R, Shelton AM (2006) Detection of *Contarinia nasturtii* (Diptera; Cecidomyiidae) in New York, a new pest of cruciferous plants in the United States. J Econ Entomol 99:1310–1315

Kim H, Rodriguez-Saona C, Kwon DH, park S, Kang TJ, Kim SJ, Hong KJ, Lee HS (2015) Development and characterization of 12 microsatellite loci from the blueberry gall midge *Dasineura oxycoccana* (Diptera: Cecidomyiidae). Appl Entomol Zool 50:415–418

Kiritani K (1998) Exotic insects in Japan. Entomol Sci 1:291–298

Kodoi F, Lee HS, Uechi Yukawa J (2003) Occurrence of *Obolodiplosis robiniae* (Diptera: Cecidomyiidae) in Japan and South Korea. Esakia 43:35–41

Lee W, Lee Y, Kim S, Lee JH, Lee H, Lee S, Hong KJ (2016) Current status of exotic insect pests in Korea: comparing border interception and incursion during 1996–2014. J Asia-Pac Entomol 19:1095–1101

Liburd OE, Roubos CR (2010) Blueberry gall midge: a key pest of rabbiteye blueberries. Proc Florida State Hortic Soc 123:23–25

Lyrene PM, Payne JA (1992) Blueberry gall midge: a pest on rabbiteye blueberry in Florida. Proc Florida State Hortic Soc 105:297–300

Mathur S, Cook MA, Sinclair BJ, Fitzpatrick SM (2012) DNA barcodes suggest cryptic speciation in *Dasineura oxycoccana* (Diptera: Cecidomyiidae) on cranberry, *Vaccinium macrocarpon*, and blueberry, *V. corymbosum*. Fla Entomol 95:387–394

Merlin Kamala I (2018) Incidence of blossom midge, *Contarinia maculipennis* Felt on Jasmine (*Jasminum sambac* L.) and its natural enemies in Tamil Nadu. Indian J Ecol 45:367–372

Merlin Kamala I, Kennedy JS (2018) Evaluation of entomopathogens against blossom midge, *Contarinia maculipennis* Felt in *Jasmine* (*Jasminum sambac* L). J Biol Control 32:121–127

Morgan WH, Thébault E, Seymour CL, van Veen FJF (2017) Density dependence and environmental factors affect population stability of an agricultural pest and its specialist parasitoid. BioControl 62:175–184

Nijveldt WC (1980) Nieuwe galmuggen voor de Nederlandse fauna (VII). Entomologische berichten 40:53–56

Paini DR, Sheppard AW, Cook DC, De Barro PJ, Worner SP, Thomas MB (2016) Global threat to agriculture from invasive species. Proc Nat Acad Sci USA 113:7575–7579

Perdikis D, Lykouressis D, Paraskevopoulos A, Harris KM (2011) A new insect pest, *Lasioptera* sp. (Diptera: Cecidomyiidae), on tomato and cucumber crops in glasshouses in Greece. Bull OEPP 41:442–444

Plant Protection Station (2018) Annual report on quarantine disease and pest list in 2018. Ministry of Agriculture, Forestry and Fisheries. http://www.pps.go.jp/TokeiWWW/Pages/report/download.xhtml;jsessionid=CDFA69F9A077FD78667CC84ED7CABDC1?indexId=002887. (In Japanese)

Raman A, Burckhardt D, Harris KM (2009) Biology and adaptive radiation in the gall-inducing Cecidomyiidae (Insecta Diptera) and Calophyidae (Insecta Hemiptera) on *Mangifera indica* (Anacardiaceae) in the Indian subcontinent. Trop Zool 22:27–56

Rehman HM, Mahmood R, Razaq M, Reddy GV (2016) Damage patterns, monitoring, and management of *Procontarinia mangiferae* (Diptera: Cecidomyiidae) in Pakistan. J Econ Entomol 109:2446–2453

Rubinoff D, Holland BS, Shibata A, Messing RH, Wright MG (2010) Rapid invasion despite lack of genetic variation in the erythrina gall wasp (*Quadrastichus erythrinae* Kim). Pac Sci 64:23–31

Samietz J, Baur R, Hillbur Y (2012) Potential of synthetic sex pheromone blend for mating disruption of the swede midge, *Contarinia nasturtii*. J Chem Ecol 38:1171–1177

Sampson BJ, Stringer SJ, Spiers JM (2002) Integrated pest management for *Dasineura oxycoccana* (Diptera: Cecidomyiidae) in blueberry. Ecol Entomol 31:339–347

Sax DF, Gaines SD (2008) Species invasions and extinction: the future of native biodiversity on islands. Proc Nat Acad Sci USA 105:11490–11497

Saxena SK (2020) Impact of major abiotic factors and crop phenology on abundance of mango leaf gall midge *Procontarinia matteiana* Kieffer and Cecconi. Int J Res Anal Rev 7:4–9

Shi DS (1980) A new species of gall midges affecting young leaf of mango (Diptera: Cecidomyiidae). Entomotaxonomia 2:131–134. (In Chinese with English summary)

Singh S (2018) A new species of *Aprostocetus* Westwood (Hymenoptera: Eulophidae), parasitizing mango leaf gall midge (Diptera: Cecidomyiidae), from India. J Asia Pac Entomol 21:553–559

Skuhravá M, Martinez M, Roques A (2010) Diptera chapter 10. BioRisk 4:553–602

Survilienė E, Kazlauskaitė S (2019) First report of *Dasineura oxycoccana* in Lithuania – short communication. Plant Prot Sci 55:218–221

Tokuda M, Yukawa J, Yasuda K, Iwaizumi R (2002) Occurrence of *Contarinia maculipennis* (Diptera: Cecidomyiidae) infesting flower buds of *Dendrobium phalaenopsis* (Orchidaceae) in greenhouses on Okinawa Island, Japan. Appl Entomol Zool 37:583–587

Tokuda M, Uechi N, Yukawa J (2019) Invasive pest species of gall-inducing cecidomyiids (Diptera) in Japan. Form Entomol 38:33–41

Uechi N, Kawamura F, Tokuda M, Yukawa J (2002) A mango pest, *Procontarinia mangicola* comb. nov. (Diptera: Cecidomyiidae), recently found in Okinawa, Japan. Appl Entomol Zool 37:589–593

Uechi N, Tokuda M, Yukawa J, Kawamura F, Teramoto KK, Harris KM (2003) Confirmation by DNA analysis that *Contarinia maculipennis* (Diptera: Cecidomyiidae) is a polyphagous pest of orchids and other unrelated cultivated plants. Bull Entomol Res 93:545–551

Uechi N, Ooishi T, Yasuda K, Yamagishi K, Taniguchi M, Yukawa J (2007a) Discovery of 2 platygastrid species parasitizing *Contarinia maculipennis* (Diptera: Cecidomyiidae) from Okinawa, Japan. Kyushu Plant Prot Res 53:107–110. (In Japanese with English summary)

Uechi N, Yasuda K, Gyoutoku N, Yukawa J (2007b) Further detection of an invasive gall midge, *Contarinia maculipennis* (Diptera: Cecidomyiidae), on bitter gourd in Okinawa and on orchids in Fukuoka and Miyazaki, Japan, with urgent warning against careless importation of orchids. Appl Entomol Zool 42:277–283

Uechi N, Yukawa J, Tokuda M, Ganaha-Kikumura T, Taniguchi M (2011) New information on host plants and distribution ranges of an invasive gall midge, *Contarinia maculipennis* (Diptera: Cecidomyiidae), and its congeners in Japan. Appl Entomol Zool 46:383–389

Woo K, Choe HJ, Kim H (2003) A report of the occurrence of yellow locust midge *Obolodiplosis robiniae* (Haldeman, 1987) from Korea. Korean J Appl Entomol 42:77–79. (In Korean with English summary)

Wu Q, Zhao J, Taylor AG, Shelton AM (2006) Evaluation of insecticides and application methods against *Contarinia nasturtii* (Diptera: Cecidomyiidae), a new invasive insect pest in the United States. J Econ Entomol 99:117–122

Yamaguchi T (2012) Insect pests of the mango plant in Japan, and their managements. Plant Prot 66:612–618. (In Japanese)

Yang Z, Qiao X, Bu W, Yao Y, Xiao Y, Han Y (2006) First discovery of an important invasive insect pest, *Obolodiplosis robiniae* (Diptera: Cecidomyiidae) in China. Acta Entomol Sin 49:1050–1053. (In Chinese with English summary)

Yao Y, Zhao W, Shang X (2015) Development of polymorphic microsatellite markers of *Obolodiplosis robiniae* (Haldeman) (Diptera: Cecidomyiidae), a north American pest invading Asia. J Insect Sci 15:127

Yasuda K, Uehara K (1994) Occurrence of orchid pests in Okinawa. Proc Assoc Plant Prot Kyushu 40:134–136. (In Japanese)

Yoshida K, Elsayed AK, Matsuura H, Horigome M, Aizawa T, Tokuda M (2017) Occurrence of *Dasineura oxycoccana* (Diptera: Cecidomyiidae) infesting cultivated blueberry in Japan. Jpn J Appl Entomol Zool 61:25–28. (In Japanese with English abstract)

Yukawa J, Uechi N, Tokuda M, Kawamura F (2004) Recent invasion of Okinawa by gall midges infesting orchid flower buds and mango leaves. Plant Prot 58:216–219. pl. (In Japanese)

Yukawa J, Tokuda M, Yamagishi K (2014) Host plant ranges and distribution records of identified and unidentified species of the genus *Lasioptera* (Diptera: Cecidomyiidae) in Japan. Esakia 54:1–15

Yukawa J, Harris KM, Kim W (2020) Descriptions of two new species of the genus *Lasioptera* (Diptera: Cecidomyiidae) that infest tomato in the Mediterranean region and Hokkaido, Japan. Appl Entomol Zool 55:129–140

Zhou H, Li PZ, Chen SM, Lin MG (2015) Indoor toxicity test and field efficacy of six insecticides against the larva of *Contarinia maculipennis* Felt. J South Agric 46:260–264. (In Chinese with English abstract)

Chapter 13
Beneficial Species

Makoto Tokuda, Junichiro Abe, and Tomoko Ganaha-Kikumura

Abstract Several gall midges are effective biological control agents against various pest species. Predatory species such as *Aphidoletes aphidimyza* feeding on aphids and *Feltiella acarisuga* on spider mites are important natural enemies in agricultural ecosystems. Some phytophagous cecidomyiids are used to control invasive plants. In this chapter, we summarize the information and present status of gall midges used as biological control agents worldwide. Recent studies have emphasized the importance of dipterans as pollinators in natural environments. We also review the current knowledge about pollinating gall midges and their associations with various plant families.

Keywords *Aphidoletes* · Biological control · *Feltiella* · Natural enemy · Pollinator

13.1 Introduction

Up to the present, various natural enemies of arthropods and weeds have been used as biological control agents to regulate pest populations (e.g. van Lenteren et al. 2006; Jennings et al. 2017). Some gall midges are predators or parasitoids of arthropods including important pests such as spider mites, aphids, mealybugs, and psyllids (e.g. Harris 1968; Abe et al. 2011d; Fujita et al. 2019; see also below). They are considered as effective biological control agents against the pests. Overgrowth of

M. Tokuda (✉)
Laboratory of Systems Ecology, Faculty of Agriculture, Saga University, Saga, Japan
e-mail: tokudam@cc.saga-u.ac.jp

J. Abe
Western Region Agricultural Research Center, National Agriculture Research Organization, Hiroshima, Japan
e-mail: abejun@affrc.go.jp

T. Ganaha-Kikumura
Okinawa Prefectural Central Wholesale Market, Okinawa, Japan
e-mail: kikumurt@pref.okinawa.lg.jp

J. Yukawa, M. Tokuda (eds.), *Biology of Gall Midges*, Entomology Monographs,
https://doi.org/10.1007/978-981-33-6534-6_13

exotic plants is a recent global concern and many attempts to use specialist herbivores as their biological control agents have been conducted (e.g. Schwarzländer et al. 2018). Some phytophagous gall midges are adopted to control these invasive plants as classical biological control agents. In this chapter, we summarize the biological information and present status of predatory gall midges of *Aphidoletes* that feed on aphids, and *Feltiella* on spider mites, and phytophagous cecidomyiids inducing galls on invasive plants.

Insect pollinators facilitate reproduction of flowering plants and thus they have a critical role in supporting the diversity of land plants (Ollerton 2017). Flower-visiting dipterans are one of the most common pollinators for various plants (e.g. Larson et al. 2001; Raguso 2020). Recent studies emphasize especially the importance of Sciaroidea (Diptera), including gall midges, as pollinators (e.g. Ollerton et al. 2017; Mochizuki and Kawakita 2018). In the basal clade of angiosperms, known as ANITA grade, dipterans are dominant pollinators, suggesting a long-lasting relationship between flowering plants and dipterans (Thien et al. 2009). In this chapter, we also introduce gall midges known as pollinators of various plant families.

13.2 *Aphidoletes aphidimyza* as Biological Control Agents against Aphids

13.2.1 Aphidoletes aphidimyza *and its Congeners*

The genus *Aphidoletes* belongs to the tribe Aphidoletini of the subfamily Cecidomyiinae (Harris 1968) and includes four identified species, *Aphidoletes aphidimyza* (Rondani), *Aphidoletes abietis* (Kieffer), *Aphidoletes thompsoni* (Möhn), and *Aphidoletes urticaria* (Kieffer). They are morphologically remarkably similar to each other, but the male terminalia serve for species identification (Harris 1973). *Aphidoletes aphidimyza* is distinguished from three other species by having a distally enlarged and cordate hypoproct; the enlarged portion of hypoproct is densely pubescent.

Aphidoletes aphidimyza is a cosmopolitan species. Adult body length is about 2.5 mm, wing length being 1.5–1.8 mm and 1.6–2.0 mm for male and female, respectively (Rondani 1847). Egg is orange in color, oval, about 0.3 mm in long axis and about 0.1 mm in short axis, and unrecognizable by naked eyes. The newly hatched larva is only 0.3 mm long and mature larva is 2–3 mm long (Fig. 13.1a). The color of the larvae varies from light orange to red, depending on the food source (Markkula and Tittanen 1985).

Fig. 13.1 Mature larvae of predacious gall midges. (**a**), a larva of *Aphidoletes aphidimyza* attacking an aphid; (**b**), a larva of *Feltiella* sp.

13.2.2 Molecular Data for Japanese and European Aphidoletes aphidimyza

There are no clear diagnostic morphological differences between Japanese and European specimens of *A. aphidimyza* (Yukawa et al. 1998). Therefore, Shirota et al. (1999) analyzed their mitochondrial cytochrome oxidase subunit I (COI) gene and found that there were clear genetic differences (about 10% of 315 bp) between Japanese and Dutch populations. Thus, the partial COI gene of *A. aphidimyza* can be useful to monitor the introgression of European traits into Japanese strains. According to Yukawa et al. (2008), experiments of cross mating between the two populations revealed that about half of the eggs could not give rise to F_1 progeny, suggesting that the two populations have started to diversify into different species, and if so, genetic contamination would hardly occur. However, ecological interference still exists between them.

13.2.3 Distribution

Aphidoletes aphidimyza is widely distributed in the world, being recorded from Canada, Chile, China, Egypt, Hawaii, India, Israel, Japan, Russia, Sudan, Tunisia, Turkey, USA, and many European countries (Gagné 1971; Harris 1973; Meadow et al. 1985; Kulp et al. 1989; Grover et al. 1988; Zhao and Cheng 1990; Halima-Kamel and Hamouda 1993; Gagné and Jaschhof 2017). In Japan, the occurrence of an aphidophagous species was recorded, for the first time, by Ninomiya (1959), who tentatively identified it as a North American species, *Aphidoletes meridionalis* Felt, which has since been synonymized with *A. aphidimyza* by Gagné (1971). Subsequently, Yukawa (1971) confirmed the presence of *A. aphidimyza* in Kagoshima Prefecture, and Yukawa et al. (1998) revealed that *A. aphidimyza* is distributed commonly and abundantly in Japan, except Okinawa, the most southern prefecture. The three other Palearctic and Nearctic congeners, *A. abietis*, *A. thompsoni*, and *A. urticaria* have never been found in Japan.

13.2.4 Prey Range

Larvae of *A. aphidimyza* feed on a wide variety of aphids, at least 61 species having been known as prey from all over the world (Nijveldt 1969; Harris 1973). Most aphid species belong to the subfamily Aphidinae, some to Chaitoporinae and Callaphidinae, and a few to other subfamilies.

Yukawa et al. (1998) confirmed in Japan that *A. aphidimyza* has a prey range across at least 19 aphid species and that the larvae derived from *A. aphidimyza* populations imported from the Netherlands as a biological control agent developed to adults by feeding on seven Japanese aphid species. When all these aphid species are regarded as prey, the number of prey aphid species totals at least 80 in the world (Yukawa et al. 1998). Among these aphid species, economically important horticultural pests, such as *Aphis gossypii* Glover, *Myzus persicae* Sulzer, *Aulacorthum solani* Kaltenbach, and *Macrosiphum euphorbiae* Thomas, are included. Of course, many more aphid species will be detected as prey in the future, in particular when aphids on wild host plants are searched more extensively. The known prey ranges of *A. thompsoni* and *A. urticaria* largely overlap with that of *A. aphidimyza*, but *A. abietis* is known to prey exclusively on adelgids on conifers (Harris 1973).

13.2.5 Biology

Eggs are laid singly or in clusters of up to 40 eggs and they are usually placed on plants, within or nearby aphid colonies (Boulanger et al. 2018). The egg stage lasts for 48–72 h at a constant temperature of 25 °C. Immediately after hatching, first instars seek out and attack aphids. The larvae usually attack their prey by piercing a leg or another body joint (Bouchard et al. 1981). The attacked aphid is paralyzed with the toxic saliva injected by *A. aphidimyza* larva. The number of aphids killed by one larva is depending on aphid species, size, and density. In general, one larva needs about 30 small or 10 large aphids to complete development (Boulanger et al. 2018). However, larvae of *A. aphidimyza* have been known to kill more aphids than they actually need to fulfill the nutritional needs. This behavior has been called as 'overkill' (Harris 1973). After maturity, larvae crawl down the stem of plant or drop to the ground. They then burrow down to a depth of about in the top few mm to 30 mm of the soil and spin cocoons. The larvae pupate within a few days in the cocoons and adults emerge after 7–10 days at constant temperature of 25 °C (Boulanger et al. 2018). Before emergence, pupae come out of the cocoons and move up to the soil surface. Adults emerge from the pupae during the day, and mating and oviposition occur between sunset and sunrise. At the time of mating, males and females hang on abandoned spider-webs or other substrates in a face-to-face position (Hauser 1986). Adults are not predaceous and both males and females take aphid honeydew as energy source (Harris 1973). The life span of males is always shorter than that of females. Adult life span was significantly extended if the adults were supplied with honeydew or sugar solution (Kuo 1985). After mating, female flies to find aphid colonies. In Japan, the adults first appear around May in the field and the larvae can be seen near aphid colonies up to the middle of December in southern areas. Larvae drop to the soil in December and diapause in the cocoons on the ground. Pupation takes place in the following spring.

13.2.6 Use as a Biological Control Agent

In many countries, *A. aphidimyza* has been used as a biological control agent against aphids in greenhouses (e.g. Meadow et al. 1985; Morse and Croft 1987; Nijveldt 1988; Solarska 2004; Yano et al. 2008; Abe et al. 2011b; Jandricic et al. 2016). *Aphis gossypii* or *M. persicae* on fruit vegetables, such as sweet peppers and cucumbers, have mainly been targeted to control (Boulanger et al. 2018). In Japan, this species was registered as a biological control agent in 1998. Thereafter, the mass-produced cocoons imported from Europe had been commercially available until 2012. After the registration, opportunities for using *A. aphidimyza* increased especially in the greenhouses of fruit vegetables in the temperate region. However, striking effects of this species were seldom reported in Japan.

In order to improve methods of releasing the cocoons, Yukawa et al. (2008) investigated the detailed behavioral and ecological traits of imported individuals of *A. aphidimyza*. They first examined the relationship between the successful emergence rate of *A. aphidimyza* and the depth of vermiculate layer in the commercial bottle, with which mass-reared cocoons were carried, because the following two release methods were recommended by the producer: (1) open the cap of commercial bottle and leave it in a greenhouse, and (2) divide the contents of bottle into several containers with 10 cm depth and leave them in the greenhouse. Under these conditions, however, 55% of individuals could not successfully emerge, and most individuals failed to emerge if the depth of vermiculate was more than 5 cm. Based on the results, Yukawa et al. (2008) recommended that cocoons and vermiculite substrates must be divided into many shallow dishes with 1 cm depth.

Yukawa et al. (2008) also proposed to provide substrates for mating near the release point since *A. aphidimyza* mates by hanging on abandoned spider-webs or other substrates (Hauser 1986). In their experiments, *A. aphidimyza* females derived from a container provided with substrates could oviposit normally while those from a container without hanging substrates laid no eggs.

Banker plant system has been developed in many countries to use *A. aphidimyza* more effectively (Hansen 1983; Bennison 1992; Bennison and Corless 1993; Kim and Kim 2004). In Japan, the banker plant method has developed especially in greenhouse fruit vegetables (Yano et al. 2009; Nagasaka et al. 2010). Abe et al. (2011c) reared four species of cereal aphids, *Melanaphis sacchari* (Zehntner), *Rhopalosiphum maidis* (Fitch), *Rhopalosiphum padi* (Linnaeus), and *Schizaphis graminum* (Rondani) with barley or sorghum leaf fragments and investigated their abilities of development and reproduction to select the most suitable combination of banker plant and alternative prey. They concluded that sorghum and *M. sacchari* could be a suitable combination of banker plant and alternative prey, particularly under warm conditions. Then, they actually introduced *A. aphidimyza* with this combination of banker plant system into sweet pepper greenhouses from spring to autumn (Abe et al. 2011c). In these greenhouses, populations of *A. aphidimyza* were maintained longer with higher densities than in those without this system, and the occurrences of pest aphids, such as *A. gossypii* and *M. persicae*, were successfully suppressed.

Jandricic et al. (2016) reported that females of *A. aphidimyza* lay more eggs in the colonies of *M. persicae* on plant meristems than those of *Aulacorthum solani* (Kaltenbach) on flowers or lower leaves. They noticed the necessity for careful monitoring of the less-preferred pest species and its relative position on the plant.

13.3 *Feltiella* spp. as Biological Control Agents against Mites

13.3.1 Spider Mites

Many spider mites (Acarina: Tetranychidae) are economically important pests of various crops, vegetables, ornamentals, and fruit trees worldwide (e.g. Jeppson et al. 1975; Bolland et al. 1998). Their tiny body size makes the detection of initial occurrence difficult. When we noticed, damaged leaves caused by feeding often became stippled with tiny white dots, resulted occasionally in the retard of plant growth. Because of their short life cycles and high reproductive potential, the mites rapidly develop resistance to many compounds (e.g. Cranham and Helle 1985; Stumpf and Nauen 2001). For example, the twospotted spider mite *Tetranychus urticae* Koch is ranked in the top 10 resistant arthropods in the world, developing more than 90 active ingredients (Whalon et al. 2008). Under such situations, we need to desert our principles depending solely on chemical control and to use natural enemies for successful integrated management of spider mites.

13.3.2 Species of Feltiella *Feeding on Spider Mites*

Some predacious gall midges are natural enemies of various pest arthropods. Larvae of the genus *Feltiella* (Diptera: Cecidomyiidae) have been known to attack tetranychid mites (Gagné and Jaschhof 2017). The genus is cosmopolitan and was formerly described under many different names that are now treated as synonyms. Gagné (1995) revised and redefined *Feltiella* and reduced the number of nominal species from 24 to 8. Later, Gagné (2004) synonymized the genus *Acaroletes* with *Feltiella*, adding two species to *Feltiella*.

Among 10 nominal species of the genus *Feltiella, Feltiella acarisuga* (Vallot) is regarded as an important natural enemy against spider mites including *T. urticae* (e.g. Gillespie et al. 1998) and *Tetranychus cinnabarinus* (Boisduval) (Mo and Liu 2006). Pupae of *F. acarisuga* have been commercially available as a biological control agent provided by several producers and suppliers in Europe, North and South America since 1990 (Lenteren 2012). Opit et al. (1997) found that *F. acarisuga* prefers spider mites to a predacious mite species, *Phytoseiulus persimilis* Athias-Henriot (Acarina: Phytoseiidae), one of the key natural enemies.

In Japan, the occurrence of an acarivorous species of Cecidomyiidae was first recorded by Hotta (1916), who tentatively identified it as *Arthrocnodax* sp. Because his description was inadequate and he did not leave any specimens, we could not identify that species. Kawano (1969) reported a *Feltiella* species from Kagoshima City, southern Japan, and Yukawa (1971) provided drawings of the male terminalia and male fifth flagellomere of the *Feltiella* sp. but left it unidentified. Gagné (2004) treated the Japanese *Feltiella* sp. as *F. acarisuga* based on the drawings in Yukawa (1971) and Abe et al. (2011a) identified acarivorous gall midges collected from all over Japan as *F. acarisuga* based on morphological features. After that, Ganaha-Kikumura et al. (2012) found another acarivorous species in various localities of Okinawa Prefecture, southern Japan and identified it as *Feltiella acarivora* (Zehntner) based on morphological features and molecular sequencing data. Up to date, *Feltiella* species have been found on 43 plant species (including varieties and subspecies) across 35 genera belonging to 19 families throughout our intensive filed surveys in Okinawa (Table 13.1). *Feltiella acarisuga* and *F. acarivora* were often found co-existing on a single plant. Ganaha-Kikumura et al. (2012) also suggested the existence of one or two additional species of *Feltiella*, including one similar to *F. acarivora* but they left them unidentified because of inadequate male specimens for identification.

13.3.3 Identification of the Two Japanese **Feltiella** *Species*

Up to date, 10 nominal species of *Feltiella* have been recognized; *F. acarisuga, F. acarivora, Feltiella curtistylus* Gagné, *Feltiella insularis* (Felt), *Feltiella kanchanjungaensis* (Grover), *Feltiella ligulata* Gagné, *Feltiella occidentalis* (Felt), *Feltiella pini* (Felt), *Feltiella reducta* (Felt), *Feltiella tetranychi* (Kieffer) (Gagné and Jaschhof 2017). They can be identified to the species level based mainly on male terminalia (see Gagné 1995 for a key to male *Feltiella*). *Feltiella acarisuga* is distinguished from other congeners by the combination of the following characters: tarsal claws toothed on forelegs and simple on mid and hind legs; hypoproct entire,

Table 13.1 List of plant species on which *Feltiella* species and their prey mites were found in Okinawa

Plant family	Important crops (indicated with bold style) or wild plant species
Adoxaceae	*Sambucus chinensis* var. *formosana*
Anacardiaceae	***Mangifera indica***
Annonaceae	***Annona squamosa, A. cherimola*** **×** ***A. squamosa***
Araliaceae	*Hedera* spp.
Asteraceae	***Arctium lappa***, *Ixeridium dentatum* subsp. *dentatum*
Caricaceae	***Carica papaya***[a]
Convolvulaceae	***Ipomoea batatas***, *I. biflora*
Cucurbitaceae	***Citrullus lanatus***, ***Luffa aegyptica***
Euphorbiaceae	*Jatropha integerrima*, *Mallotus japonicus*[a], ***Manihot esculenta***, *Melanolepis multiglandulosa*[a]
Fabaceae	*Bauhinia purpurea*[a], *Canavalia* spp., *Erythrina variegate*[a], *Lablab purpurea, Mucuna macrocarpa, M. pruriens,* ***Phaseolus vulgaris***[a], ***Psophocarpus tetragonolobus***, *Pueraria montana*[a], ***Vigna angularis***, *V. marina,* ***V. unguiculata***
Lamiaceae	*Clerodendrum trichotomum*
Moraceae	*Broussonetia papyrifera*[a], *Morus australis*
Mulvaceae	***Corchorus olitorius***, *Hibiscus mutabilis*
Oxalidaceae	*Oxalis corniculata*
Passifloraceae	***Passiflora edulis***
Phyllanthaceae	*Antidesma pentandrum*
Rosaceae	***Fragaria*** **×** ***ananassa***[a], *Prunus persica*
Solanaceae	*Solanum americanum*, ***S. lycopersicum***, ***S. melongena***, *S. nigrum*
Zingiberaceae	*Hedychium coccineum*

[a]Plant species on which *Feltiella* species were found on at least three occasions

longer than cerci; aedeagus broad, elongate with blunt apex, much longer than cerci and hypoproct, gradually narrowing to rounded apex; gonocoxite with mesobasal lobe. *Feltiella acarivora* is distinguished from *F. acarisuga* and other congeners by the combination of the following characters: hypoproct a little shorter than aedeagus, parallel sided basally, slightly tapering to a blunt apex; apex of hypoproct with an apical ligule; gonocoxite with a spiny mediobasal lobe. In addition, circumfila of female flagellomeres are appressed in *F. acarivora*, whereas these of *F. acarisuga* have slightly looping circumfila (Gagné 1995; Abe et al. 2011a).

Immature stages of *Feltiella* as well as other genera of the tribe Lestodiplosini are hardly identified to the species level based on morphological features because previous descriptions have been rarely provided with morphological characters of immature stages (Fig. 13.1b). Applied entomologists frequently need an immediate species identification of cecidomyiid larvae feeding on spider mites in the field. In order to contribute to such a request, molecular data for each identified species are actually helpful.

A region of mitochondrial DNA COI was analyzed for the two Japanese species, *F. acarisuga* and *F. acarivora*, to confirm the morphological species identifications

(Ganaha-Kikumura et al. 2012) and the sequencing data were registered in GenBank etc. for future larval identification. Both the two maximum parsimonious trees and neighbor-joining tree obtained from the analysis had distinct two clades which were supported by a 100% bootstrap value, respectively; one was the clade of *F. acarisuga* and the other was that of *F. acarivora*.

Seven haplotypes were recognized among individuals of *F. acarisuga* and were divided into two subclades, roughly representing their collecting sites. The minimum sequence divergence was 1/814 bp (0.123%) between them. In the clade of *F. acarivora*, two haplotypes were found, between which the sequence divergence was only 1/814 bp (0.123%) (Ganaha-Kikumura et al. 2012). A slight sequence divergence among individuals of *F. acarivora*, together with absence records from the mainland of Japan (Abe et al. 2011a), may indicate that Okinawa is the northern distributional limit of *F. acarivora* in Japan. To confirm this, we need further specimens from various localities in Asia including its type locality, Indonesia, for future phylogenetic and biogeographic studies. If sufficient molecular sequencing data are available in the future, we will be able to identify *Feltiella* species based solely on larval specimens.

13.3.4 Distribution Range, Prey Mites and Seasonal Occurrence

Information on known distribution ranges and prey mites of 10 identified species of *Feltiella* are summarized in Tables 13.2 and 13.3 based mainly on Gagné and Jaschhof (2017). *Feltiella acarisuga* is most widely distributed in the world except for the Neotropical Region. Known distribution ranges of the other species are restricted to particular regions or countries probably because of less extensive and intensive field surveys.

Many species of Tetranychidae (Acarina) have been reported as preys of *F. acarisuga*, *F. acarivora*, *F. insularis*, *F. occidentalis* and *F. pini* (Table 13.3). In the table, all 10 out of 11 species of prey mites fed by *F. acarisuga* have been found in Japan (Abe et al. 2011a; Ganaha-Kikumura et al. 2012).

In Okinawa, both *F. acarisuga* and *F. acarivora* were found in the filed all the year round (Table 13.4). A small peak appeared in August, but the data were not distinctly different from those in December. An unidentified species of *Feltiella* was found on only one occasion in October.

13.3.5 Biology of Feltiella acarisuga

Feltiella acarisuga is multivoltine and seems to overwinter at any developmental stage. Eggs of *F. acarisuga* are laid in or near the colonies of tetranychid mites and

Table 13.2 Known distribution range of *Feltiella* species (based mainly on Gagné and Jaschhof 2017)

Species of *Feltiella*	Distribution range
acarisuga	Australia, Europe, India, Iran, Israel, Japan, Korean Peninsula, Sri Lanka, Taiwan, Nearctic, New Zealand, N. America,
acarivora	Australia, Indonesia, Japan
curtistylus	Brazil, USA
insularis	Argentina, Colombia, Jamaica, Puerto Rico, Trinidad, USA
kanchanjungaensis	India
ligulata	Cape Verde Is.
occidentalis	Japan, USA
pini	Australia, Bahamas, Nicaragua, N. America,
reducta	USA
tetranychi[a]	Bulgaria, Czechoslovakia, Germany, Ukraine

[a]This species is possibly synonymous with *F. acarisuga* (Gagné and Jaschhof 2017)

larvae feed on eggs, nymphs, and adults of prey mites. *Feltiella* larvae prey on all stages of spider mites, generally preferring eggs, and young nymphs (Sharaf 1984). The larvae of *F. acarisuga* consume on an average 175.4 eggs of *T. cinnabarinus* (Mo and Liu 2006), 185.4 eggs of *T. urticae* (Mo and Liu 2007), at least 250 eggs of *T. kanzawai* (Nakagawa 1986), and 32.3 adults of *Tetranychus* sp. (Kawano 1969) during the whole larval stage. Female larvae of *F. acarisuga* consume spider mite eggs 18.6% more than male larvae (Mo and Liu 2007). Xiao et al. (2013) showed that the larvae of *F. acarisuga* fed on 50 eggs of *T. urticae* per a day, which was more numerous than 25.6 or 15.1, respectively, by two other biological control agents, *Neoseiulus californicus* (McGregor) and *Amblyseius swirskii* (Anthias-Henriot) (Acarina: Phytoseiidae). Mature larvae pupate in cocoons. Cocoons are usually spun along veins mainly on the undersurface of leaves infested by mites. When veins are not prominent, cocoons are spun in hollows or curls of leaves.

Feltiella acarisuga develops and reproduces at temperatures ranging from 15 to 30 °C (Ho and Chen 1998; Gillespie et al. 2000). Days required for the development from egg to adult vary with temperatures, prey mites or host plants; 10 days at 27 °C and 34 days at 15 °C, preying on *T. urticae* on cucumber (Gillespie et al. 2000), 9 days at 28 °C and 33 days at 15 °C on *T. kanzawai* on eggplant (Ho and Chen 1998), 16 days at 27 °C on *T. cinnabarinus* (Mo and Liu 2006) and 17 days at 27 °C on *T. urticae* on tomato (Mo and Liu 2007). However, the developmental period becomes shorter in higher relative humidity than in lower (Gillespie et al. 2000; Agamy and Gomaa 2002), hence the aforementioned developmental data might be influenced not only by temperatures, prey mites or host plants but also by humidity conditions in the respective experiments.

Females of *F. acarisuga* preying on *T. cinnabarinus* on tomato laid an average of 32 eggs (Mo and Liu 2006) and those preying on *T urticae* on tomato laid 33 eggs (Mo and Liu 2007). Ho and Chen (1998) also reported that *F. acarisuga* preying on *T. kanzawai* on eggplant laid eight eggs at 15 °C and 17 eggs at 20 °C on average.

Table 13.3 Known prey mites of *Feltiella* species

	F. acarisuga	*F. acarivora*	*F. curtistylus*	*F. insularis*	*F. kanchanjungaensis*	*F. ligulata*	*F. occidentalis*	*F. pini*	*F. reducta*[a]	*F. tetranychi*[b]
Tetranychidae										
Eotetranychus asiaticus Ehara		△								
Euetranychus africanus (Tucker)		□								
Oligonychus biharensis (Hirst)	□	□								
Oligonychus coffeae (Nietner)		△								
Oligonychus exsccator (Zehntner)		◎								
Tetranychus cinnabarinus (Boisduval)	■									
Tetranychus evansi Baker & Pritchard	●		◎				◎			
Tetranychus kanzawai Kishida	●	□								
Tetranuchus neocaledonicus Andre	□	□								
Tetranychus okinawanus Ehara	□	□								

(continued)

Table 13.3 (continued)

	F. acarisuga	*F. acarivora*	*F. curtistylus*	*F. insularis*	*F. kanchanjungaensis*	*F. ligulata*	*F. occidentalis*	*F. pini*	*F. reducta*[a]	*F. tetranychi*[b]
Tetranychus parakanzawai Ehara	□	□								
Tetranychus piercei McGregor	●	□								
Tetranychus pueraricola Ehara	△	□								
Tetranychus telarius (Linnaeus)										◎
Tetranychus urticae Koch	●	□		○						
Tetranychus sp.	◎			◎		◎	◎	◎		
Tetranychus spp.	◎	◎								
Other tetranychid species	◎			◎			◎	◎		
Eriophyoidea										
Acaphylla theavagrans Kadono	◎									

References ◎: Gagné (1995) or Gagné and Jaschhof (2017), ●: Abe et al. (2011a), □: Ganaha-Kikumura et al. (2012), ■: Mo and Liu (2006), ○: Cédola (2002), △: Current data (Prey mites were identified by S. Ohno)

[a]Information on prey mites is not available for this species

[b]This species is possibly synonymous with *F. acarisuga* (Gagné and Jaschhof 2017)

Table 13.4 Seasonal occurrence of *F. acarisuga* and *F. acarivora*

Month	Number of sites surveyed in 2007–2010	Number of sites where larvae or cocoons of *Feltiella* species were found (% in parenthesis)		
		F. acarisuga	*F. acarivora*	*Feltiella* sp.
Jan.	79	3 (3.8)	1 (1.3)	0 (0.0)
Feb.	205	4 (2.0)	8 (3.9)	0 (0.0)
Mar.	111	2 (1.8)	0 (0.0)	0 (0.0)
Apr.	97	1 (1.0)	0 (0.0)	0 (0.0)
May	194	3 (1.5)	1 (0.5)	0 (0.0)
Jun.	124	2 (1.6)	4 (3.2)	0 (0.0)
Jul.	154	6 (3.9)	7 (4.5)	0 (0.0)
Aug.	277	16 (5.8)	15 (5.4)	0 (0.0)
Sep.	57	3 (5.3)	1 (1.8)	0 (0.0)
Oct.	121	3 (2.5)	5 (4.1)	1 (0.8)
Nov.	115	0 (0.0)	3 (2.6)	0 (0.0)
Dec.	167	7 (4.2)	10 (6.0)	0 (0.0)

13.3.6 Parasitoids

Aphanogmus floridanus Ashmead, *Aphanogmus fulmeki* Szelényi and *Aphanogmus parvulus* Roberti (Hymenoptera: Ceraphronidae) have been known to emerge from cocoons of *F. acarisuga* (Roberti 1954; Dessart 1992; Osborne et al. 2002).

13.3.7 Biological Control of Spider Mites

Feltiella acarisuga has been effectively used for biological control of spider mites on tomato, pepper, cucumber, strawberry and various ornamental crops (Wardlow and Tobin 1990; Helyer 1993; Gillespie and Quiring 1997; Gillespie et al. 1998; Oatman et al. 1985; Meesters et al. 1998). In some experiments, spider mites were successfully controlled by releasing 1000 individuals of *F. acarisuga* per ha per week (Wardlow and Tobin 1990; Helyer 1993; Gillespie and Quiring 1997; Gillespie et al. 1998). Multiple releases of *F. acarisuga* seem to be required for the establishment of its population (Gillespie et al. 1998). However, the use of predatory gall midges as biological control agents has been limited by the difficulty in maintaining their populations in greenhouses when target pests are absent. Xiao et al. (2011) demonstrated that the cultivation of corn as a banker plant would be effective to control *T. urticae* that infests green bean in greenhouses. Spider mites have frequently been managed by the combination of *F. acarisuga* and a predacious mite *P. persimilis* (Griffiths 1999).

13.4 Gall Midges as Biological Control Agents against Invasive Plants

13.4.1 Gall Midges that Control Invasive Plants

Gall midges infesting invasive plants are considered as potentially effective biological control agents. Among gall midges associated with Australian *Acacia* spp. (Fabaceae), *Asphondylia* and *Dasineura* species attacking reproductive organs are expected to suppress host seed production, and thus are considered as potential biological control agents of invasive *Acacia* plants in South Africa (Adair et al. 2000; Kolesik et al. 2005, 2010). *Dasineura dielsi* Rübsaamen and *Dasineura rubiformis* Kolesik, native to Australia, were released in 2001 and 2006, respectively, in South Africa to control them. According to Impson et al. (2011), *D. dielsi* extensively damaged rooikrans *Acacia cyclops* A. Cunningham ex G. Don and *D. rubiformis* considerably did black wattle *Acacia mearnsii* De Wild there. In addition, *Hemiasphondylia prosopidis* (Cockerell) and an undescribed congener native to USA are considered to be potential biological control agents against mesquite *Prosopis* spp. (Fabaceae) in South Africa (Zachariades et al. 2011; Park and Thompson 2019).

The broad-leaved paperbark *Melaleuca quinquenervia* (Cavanilles) Blake (Myrtaceae) native to Australia and New Caledonia became an invasive weed in wetland parts of Florida, USA. Under a classical biological control program, *Lophodiplosis trifida* Gagné and several other herbivores were released in Florida to control *M. quinquenervia* populations (Pratt et al. 2013; Tipping et al. 2017). The gall midge effectively suppressed the growth of *M. quinquenervia* saplings (Tipping et al. 2016). In addition, *Contarinia maleleucae* Kolesik infesting vegetable buds of *M. quinquenervia* is considered as another potential biological agent against the plant (Kolesik et al. 2018).

Mitodiplosis graminis Kieffer is associated with pyp grass *Ehrharta villosa* Schultes (Poaceae), which is native to South Africa and became an environmental weed in Australia and New Zealand. This gall midge is potentially a biological control agent against the plant in introduced areas (Kolesik and Wood 2019).

Pampas grass, *Cortaderia selloana* (JA et JH Schultes) Ascherson et Graebner (Poaceae), which is native to south America, was introduced into Europe in the nineteenth century, and subsequently to other places such as California, New Zealand, and Japan (e.g. Harradine 1991; Lambrinos 2001; Shimizu et al. 2001). *Spanolepis selloanae* Gagné of which larvae feed on developing ovaries of *C. selloana* was described in 2020 as a new species of Cecidomyiidae (Fagúndez et al. 2020). Its effects on the seed production have been studied in Spain since 2017, and the larvae of *S. selloanae* were found to feed on the ovaries with a mean seed depletion of 74% in the studied population (Fagúndez et al. 2020). Based on the field surveys, they considered that *S. selloanae* is a potential candidate agent for the effective biological control against pampas grass in southern Europe.

13.5 Pollinators

13.5.1 Pollinators of Schisandraceae

Some species of Schisandraceae are known to exhibit "flower-heating" phenomena, which are considered to promote pollination by Coleoptera and Diptera (Yuan et al. 2008; Luo et al. 2010). The pollination of *Schisandra henryi* Clark is exclusively dependent on females of a gall midge *Megommata* sp. (later identified to be *Resseliella* sp.) Although flowers of this plant do not emit either fragrance or thermal signals, the gall midge females visit them to feed on pollen grain to facilitate their ovary maturation and oviposition (Yuan et al. 2007). In contrast, flowers of *Kadsura longipedunculata* Finet & Gagnepain produce odors and heat in nocturnal periods, and females of *Resseliella kadsurae* Yukawa, Sato & Xu visit male flowers to eat pollens (Yuan et al. 2008; Yukawa et al. 2011). In both plants, female flowers seem not to offer any rewards but deceive gall midge females for achieving pollination (Yuan et al. 2007, 2008). Luo et al. (2017) reported that larvae of *Resseliella* species pollinating *Kadsura* develop in resin-filled chambers of host flowers.

At least some species of Chinese *Illicium* (Schisandraceae) also rely on gall midges for pollination (Luo et al. 2010). Females of *Clinodiplosis* sp. oviposit onto flowers and their larvae develop in the flowers. Flower heating of *Illicium* plants may have functions to facilitate the larval development in flowers (Luo et al. 2010).

Based on a phylogenetic analysis using molecular clocks, Schisandraceae and pollinating gall midges were inferred to have interacted at least from the Early Miocene (Luo et al. 2018).

13.5.2 Pollinators of Siparunaceae

Feil (1992) surveyed reproductive ecology of various species of *Siparuna* (Siparunaceae) in Ecuador and recognized that gall midges act as their main pollinators. Nocturnal gall midge species of *Asynapta* and several other species visited male flowers, possibly being attracted by odors, and laid eggs into flowers. During the oviposition, pollens attach to the gall midge body. The gall midges are also attracted by female flowers, but eggs were mainly laid on male flowers probably because the structure of male flower is more suitable for oviposition (Feil 1992). Renner et al. (1997) reported that gall midge females try to insert ovipositors into the pore and the floral roof of female flowers, but the pore is smaller than male ones and is closed by the dense styles, which prevent the gall midge from successfully ovipositing into female flowers. The contact of pollen-carrying females with styles during unsuccessful oviposition attempts seems to ensure pollination in *Siparuna* plants (Renner et al. 1997).

13.5.3 Pollinators of Malvaceae

Pollinators of cacao tree *Theobroma cacao* Linnaeus (Malvaceae) vary with localities but they are generally small dipterans especially Ceratopogonidae and Cecidomyiidae (Claus et al. 2018; de Schawe et al. 2018). Young (1985) reported that, in Costa Rica, the main pollinators of *T. cacao* are diurnal gall midges belonging to *Clinodiplosis* and *Mycodiplosis*, of which adults consume floral nectar and trap pollen grains.

13.5.4 Pollinators of Moraceae

Several tropical fruit trees of the genus *Artocarpus* (Moraceae) are pollinated by gall midges. In chempedak *Artocarpus integer* Merrill, two gall midge species (scientific names are under consideration) lay their eggs onto flowering staminate inflorescences, and at that time pollen grains cover the female bodies (Sakai et al. 2000; Sakai 2002). Females are also attracted to pistillate inflorescences by odor, and passively transfer pollen grains. Larvae feed on mycelia infecting the inflorescence (Sakai et al. 2000; Sakai 2002).

Similarly, Gardner et al. (2018) surveyed pollinators of jackfruit, *Artocarpus heterophyllus* Lamarck (Moraceae), and revealed that female adults of gall midge *Clinodiplosis ultracrepidata* Gagné visit male and female inflorescences of jackfruit. They laid eggs in fungus-infected male inflorescences and their larvae fed on the fungus. The gall midge is attracted by volatiles emitted from male and female inflorescences of jackfruit (Gardner et al. 2018).

13.5.5 Pollinators of Asparagaceae

Gall midge pollinations are also reported in monocots. *Aspidistra xuansonensis* N. Vislobokov (Asparagaceae) is pollinated by females of unidentified species of Cecidomyiidi (Vislobokov et al. 2014). The females lay their eggs on anthers and larvae develop in the flowers.

References

Abe J, Ganaha-Kikumura T, Yukawa J (2011a) Morphological features, distribution, prey mites, and life history traits of *Feltiella acarisuga* (Diptera: Cecidomyiidae) in Japan. Appl Entomol Zool 46:271–279

Abe J, Kumakura H, Yano E (2011b) Biological control of aphids in sweet pepper greenhouses using the banker plant system for aphidophagous gall midge, *Aphidoletes aphidimyza* (Rondani) (Diptera: Cecidomyiidae). Ann Rep Kansai Plant Prot Soc 53:37–46

Abe J, Mitsunaga T, Kumakura H, Yano E (2011c) Comparative studies on development and reproduction of four cereal aphid species reared on sorghum or barley to evaluate as alternative prey for banker plant system. Jpn J Appl Entomol Zool 55:227–239. (In Japanese with English summary)

Abe J, Sato S, Yukawa J (2011d) Descriptions of two new endoparasitic cecidomyiids (Diptera: Cecidomyiidae) from Japan. Appl Entomol Zool 46:15–25

Adair RJ, Neser S, Kolesik P (2000) Australian seed-preventing gall midges (Diptera: Cecidomyiidae) as potential biological control agents for invasive *Acacia* spp. in South Africa. Spencer NR (ed) Proceedings of the X international symposium on biological control of weeds. Bozeman, MT, pp 605–613

Agamy EA, Gomaa WO (2002) Biological studies and food consumption on *Feltiella acarisuga* (Vallot) (Diptera: Cecidomyiidae) a predator of the spider mite, *Tetranychus urticae* (Koch) (Acari: Tetranychidae). Egypt J Biol Pest Control 12:87–89

Bennison JA (1992) Biological control of aphids on cucumbers: use of open rearing systems or "banker plants" to aid establishment of *Aphidius matricariae* and *Aphidoletes aphidimyza*. Mededelingen van de Faculteit Landbouwwetenschappen, Universiteit Gent 57:457–466

Bennison JA, Corless SP (1993) Biological control of aphids on cucumbers: further development of open rearing units or "banker plants" to aid establishment of aphid natural enemies. IOBC/WPRS Bull 16:5–8

Bolland HR, Gutierrez J, Flechtmann CHW (1998) World catalogue of the spider mite family (Acari: Tetranychidae). Brill Academic Publishers, Leiden

Bouchard D, Tourneur JC, Paradis RO (1981) Bio-écologie d'*Aphidoletes aphidimyza* (Rondani) (Diptera: Cecidomyiidae) prédateur du puceron du pommier, *Aphis pomi* DeGeer (Homoptera: Aphididae). Ann de la Société Entomologique du Québec 26:119–130

Boulanger FX, Jandricic S, Bolckmans K, Wäckers FL, Pekas A (2018) Optimizing aphid biocontrol with the predator *Aphidoletes aphidimyza*, based on biology and ecology. Pest Manag Sci 75:1479–1493

Cédola CV (2002) First record in Argentine of *Feltiella insularis* (Cecidomyiidae), predatory gall midge of *Tetranychus urticae* (Acari: Tetranychidae). Rev Soc Entomol Argent 61:45–46

Claus G, Vanhove W, Van Damme P, Smagghe G (2018) Challenges in cocoa pollination: the case of Côte d'Ivoire. In: Mokwala PW (ed) Pollination in plants. IntecOpen, London

Cranham JE, Helle W (1985) Pesticide resistance in Tetranychidae. In: Helle W, Sabelis MW (eds) Spider mites: their biology, natural enemies and control. Elsvier, Amsterdam, pp 405–421

de Schawe CC, Kessler M, Hensen I, Tscharntke T (2018) Abundance and diversity of flower visitors on wild and cultivated cacao (*Theobroma cacao* L.) in Bolivia. Agrofor Syst 92:117–125

Dessart P (1992) Revision d'*Aphanogmus fulmeki* Szelényi, 1940 (Hymenoptera: Ceraphronoidea; Ceraphronidae) avec remarques biologiques. Bull Inst r sci nat Belg 62:83–91

Fagúndez J, Gagné RJ, Vila M (2020) A new gall midge species (Diptera, Cecidomyiidae) as a potential candidate for biological control of the invasive plant *Cortaderia selloana* (Poaceae). Phytoparasitica. https://doi.org/10.1007/s12600-020-00844-1

Feil JP (1992) Reproductive ecology of dioecious *Siparuna* (Monimiaceae) in Ecuador – a case of gall midge pollination. Bot J Linn Soc 110:171–203

Fujita S, Adachi S, Elsayed AK, Shiraishi S, Tokuda M (2019) Seasonal occurrence and host range of a predatory gall midge *Diadiplosis hirticornis* (Diptera: Cecidomyiidae), a native natural enemy of *Planococcus kraunhiae* (Hemiptera: Pseudococcidae). Appl Entomol Zool 54:197–201

Gagné RJ (1971) The genus *Aphidoletes* Kieffer (Diptera: Cecidomyiinae) in North America. Entomol News 82:177–181

Gagné RJ (1995) Revision of tetranychid (Acarina) mite predators of the genus *Feltiella* (Diptera: Cecidomyiidae). Ann Entomol Soc Am 88:16–30

Gagné RJ (2004) A catalog of the Cecidomyiidae (Diptera) of the world. Mem Entomol Soc Washington 25:1–408

Gagné RJ, Jaschhof M (2017) A catalog of the Cecidomyiidae (Diptera) of the world. 4th edn. Digital. [Cited 10 May 2020] Available from URL: https://www.ars.usda.gov/ARSUserFiles/80420580/Gagne_2017_World_Cat_4th_ed.pdf

Ganaha-Kikumura T, Yukawa J, Tokuda M, Ohno S, Abe J (2012) Occurrence of two acarivorous species of the genus *Feltiella* (Diptera: Cecidomyiidae) in Okinawa, southern Japan, and redescription of *F. acarivora* (Zehntner). Appl Entomol Zool 47:319–329

Gardner EM, Gagné RJ, Kendra PE, Montgomery WS, Raguso RA, McNeil TT, Zerega NJC (2018) A flower in fruit's clothing: pollination of jackfruit (*Artocarpus heterophyllus*, Moraceae) by a new species of gall midge, *Clinodiplosis ultracrepidata* sp. nov. (Diptera: Cecidomyiidae). Int J Plant Sci 179:350–367

Gillespie DR, Quiring DJM (1997) Collection and selection of natural enemies of twospotted spider mites for biological control. J Entomol Soc Br Columbia 94:7–11

Gillespie DR, Roitberg B, Basalyga E, Johnstone M, Opit G, Rodgers J, Sawyer N (1998) Biology and application of *Feltiella acarisuga* (Vallot) (Diptera: Cecidomyiidae) for biological control of twospotted spider mites on greenhouse vegetable crops. Pacific agri-food research centre (Agassiz) technical report, No 145. Agriculture and Agri-Food Canada

Gillespie DR, Opit G, Roitberg B (2000) Effects of temperature and relative humidity on development, reproduction and predation in *Feltiella acarisuga* (Vallot) (Diptera: Cecidomyiidae). Biol Control 17:132–138

Griffiths DA (1999) Biological control of mites. In: Albajes R, Gullino MA, van Lenteren JC, Elad Y (eds) Integrated pest and disease management in greenhouse crops. Kluwer Academic Publishers, Dordrecht, pp 217–234

Grover P, Ved M, Kashyap V (1988) Biology of *Aphidoletes aphidimyza* Rondani. Cecidologia Internationale 9:65–82

Halima-Kamel MBB, Hamouda MHB (1993) Les pucerons des cultures protegees et leurs enemies en Tunisie. Tropicultura 11:50–53

Hansen LS (1983) Introduction of *Aphidoletes aphidimyza* (Rond.) (Diptera: Cecidomyiidae) from an open rearing unit for the control of aphids in glasshouses. IOBC/WPRS Bull 6:146–150

Harradine AR (1991) The impact of pampas grass as weeds in southern Australia. Plant Prot Q 6:111–115

Harris KM (1968) A systematic revision and biological review of the cecidomyiid predators (Diptera: Cecidomyiidae) on world Coccoidea (Hemiptera-Homoptera). Trans R Entomol Soc London 119:401–494

Harris KM (1973) Aphidophagous Cecidomyiidae (Diptera): taxonomy, biology and assessments of field populations. Bull Entomol Res 63:305–325

Hauser I (1986) Untersuchungen zu Verhalten und Lebensweise der rauberischen Gallmucke *Aphidoletes aphidimyza* (Rond.) (Diptera: Cecidomyiidae). Dieertation, Göttingen

Helyer N (1993) *Verticillium lecanii* for control of aphids and thrips on cucumber. IOBC/WPRS Bull 16:63–66

Ho CC, Chen WH (1998) Life history, food consumption, and seasonal occurrence of *Feltiella minut* (Diptera: Cecidomyiidae) on eggplant. Chin J Entomol 18:27–37

Hotta M (1916) On the natural enemies of red spider mites. Insect World 20:231–232. (In Japanese)

Impson FAC, Kleinjan CA, Hoffmann JH, Post JA, Wood AR (2011) Biological control of Australian *Acacia* species and *Paraserianthes lophantha* (Willd.) Nielsen (Mimosaceae) in South Africa. Afr Entomol 19:186–207

Jandricic SE, Wraight SP, Gillespie DR, Sanderson JP (2016) Biological control outcomes using the generalist aphid predator *Aphidoletes aphidimyza* under multi-prey conditions. Insects 2016:75. https://doi.org/10.3390/insects7040075

Jennings DE, Duan JJ, Follett PA (2017) Environmental impacts of arthropod biological control: an ecological perspective. In: Coll M, Wajnberg E (eds) Environmental pest management: challenges for agronomists, ecologists, economists and policymakers. John Wiley & Sons, Oxford, pp 105–129
Jeppson LR, Keifer HH, Baker EW (1975) Mites injurious to economic plants. University of California Press, Berkeley
Kawano K (1969) Life history of *Feltiella* sp. (Diptera: Cecidomyiidae). Kontyû 37:127–133. (In Japanese with English summary)
Kim YH, Kim JH (2004) Biological control of *Aphis gossypii* using barley banker plants in greenhouse grown oriental melon. Proc Calif Conf Biol Control IV:124–126
Kolesik P, Wood AR (2019) Redescription of *Mitodiplosis graminis* (Diptera: Cecidomyiidae), a gall midge inhibiting the flowering of pyp grass *Ehrharta villosa* (Poaceae) in South Africa. Zootaxa 4614:173–179
Kolesik P, Adair RJ, Eick G (2005) Nine new species of *Dasineura* (Diptera: Cecidomyiidae) from flowers of Australian *Acacia* (Mimosaceae). Syst Entomol 30:454–479
Kolesik P, Adair RJ, Eick G (2010) Six new species of *Asphondylia* (Diptera: Cecidomyiidae) damaging flower buds and fruit of Australian *Acacia* (Mimosaceae). Syst Entomol 35:250–467
Kolesik P, Brown BT, Purcell MF (2018) A new species of gall midge (Diptera: Cecidomyiidae) feeding on vegetative buds of *Melaleuca quinquenervia* trees in Australia. Austral Entomol 57:387–391
Kulp D, Fortmann M, Hommes M, Plate HP (1989) Die rauberische Gallmucke *Aphidoletes aphidimyza* (Rondani) (Diptera: Cecidomyiidae): Ein bedeutender Blattelauspradator, Nachschlagewerk zur Systematik, Verbreitung, Biologie, Zucht und Anwedung. Mitteilungen aus der Biologischen Bundesanstalt fur Land- und Forstwirtschft, Berlin-Dahlem
Kuo HL (1985) Some bionomics of the predacious aphid midge, *Aphidoletes aphidimyza* (Rond.) (Diptera: Cecidomyiidae), and the possibility of using the rose grain aphid, *Metopolophium dirhodum* (Wlk.), as an alternative prey in an open rearing unit in greenhouses. Proceeding of a meeting of the EC experts group/Herakliom, pp 24–26
Lambrinos JG (2001) The expansion history of a sexual and asexual species of *Cortaderia* in California, USA. J Ecol 89:88–98
Larson BMH, Kevan PG, Inouye DW (2001) Flies and flowers: taxonomic diversity of anthophiles and pollinators. Can Entomol 133:439–465
Lenteren JC (2012) The state of commercial augmentative biological control: plenty of natural enemies, but a frustrating lack of uptake. BioControl 57:1–20
Luo SX, Chaw SM, Zhang D, Renner SS (2010) Flower heating following anthesis and the evolution of gall midge pollination in Schisandraceae. Am J Bot 97:1220–1228
Luo SX, Liu TT, Cui F, Yang ZY, Hu XY, Renner SS (2017) Coevolution with pollinating resin midges led to resin-filled nurseries in the androecia, gynoecia, and tepals of *Kadsura* (Schisandraceae). Ann Bot 120:653–664
Luo SX, Zhang LJ, Yuan S, Ma ZH, Zhang DX, Renner SS (2018) The largest early-diverging angiosperm family is mostly pollinated by ovipositing insects and so are most surviving lineages of early angiosperms. Proc R Soc B Biol Sci 285:20172365. https://doi.org/10.1098/rspb.2017.2365
Markkula M, Tittanem K (1985) Biology of the midge *Aphidoletes* and its potential for biological control. Blandford Press, UK, pp 74–81
Meadow RH, Kelly WC, Shelton AM (1985) Evaluation of *Aphidoletes aphidimyza* (Dip.: Cecidomyiidae) for control of *Myzus persicae* (Hom.: Aphididae) in greenhouse and field experiments in the United States. Entomophaga 30:385–392
Meesters P, Sterk G, Latet G (1998) Aspects of integrated production of raspberries and strawberries in Belgium. IOBC/WPRS Bull 21:45–50
Mo TL, Liu TX (2006) Biology, life table and predation of *Feltiella acarisuga* (Diptera: Cecidomyiidae) feeding on *Tetranychus cinnabarinus* eggs (Acari: Tetranychidae). Biol Control 39:418–426

Mo TL, Liu TX (2007) Predation and life table of *Feltiella acarisuga* (Diptera: Cecidomyiidae) preying on eggs of *Tetranychus urticae* (Acari: Tetranychidae). Environ Entomol 36:369–375

Mochizuki K, Kawakita A (2018) Pollination by fungus gnats and associated floral characteristics in five families of the Japanese flora. Ann Bot 121:651–663

Morse JG, Croft BA (1987) Biological control of *Aphis pomi* (Hom.: Aphididae) of *Aphidoletes aphidimyza* (Dip.: Cecidomyiidae); a predator-prey model. Entomophaga 32:339–356

Nagasaka K, Takahashi N, Okabayashi T, Abe J, Oya S (2010) Development of a practical banker plant system for aphid control in commercial greenhouse crops in Japan. Bull Nat Agric Res Center 15:1–50

Nakagawa T (1986) Effect of temperature on the prey consumption of *Feltiella* sp. as a predator of *Tetranychus kanzawai* Kishida. Proc Assoc Plant Prot Kyushu 32:214–217

Nijveldt WC (1969) Gall midges of economic importance, vol 8. Miscellaneous/Crosby Lockwood, London

Nijveldt WC (1988) Cecidomyiidae. In: Minks AK, Harrewijn P (eds) Aphids their biology, natural enemies and control, vol 2B. Elsevier, Amsterdam, pp 271–277

Ninomiya E (1959) *Aphidoletes meridionalis* Felt? In: Esaki T, Ishii T, Kawada A, Shirai T, Yusawa H (eds) Illustrated insect larvae of Japan. Hokuryukan, Tokyo, p 656 (In Japanese)

Oatman ER, Badgley ME, Platner GR (1985) Predators of the two-spotted spider mite on strawberry. Calif Agric 39:9–12

Ollerton J (2017) Pollinator diversity: distribution, ecological function, and conservation. Annu Rev Ecol Evol Syst 48:353–376

Ollerton J, Dötterl S, Ghorpadé K, Heiduk A, Liede-Schumann S, Masinde S, Meve U, Peter CI, Prieto-Beníez S, Punekar S, Thulin M, Whittington A (2017) Diversity of Diptera families that pollinate *Ceropegia* (Apocynaceae) trap flowers: An update in light of new data and phylogenetic analyses. Flora 234:233–244

Opit GP, Roitberg B, Gillespie DR (1997) The functional response and prey preference of *Feltiella acarisuga* (Vallot) (Diptera: Cecidomyiidae) for two of its prey: male and female twospotted spider mites, *Tetranychus urticae* Koch (Acari: Tetranychidae). Can Entomol 129:221–227

Osborne RS, Leppla NC, Osborne LS (2002) Predatory gall midge (unofficial common name), *Feltiella acarisuga* (Vallot) (Insecta: Diptera: Cecidomyiidae). Univ Florida IFAS Ext, EENY 269:1–4

Park I, Thompson DC (2019) Delineating the *Asphondylia prosopidis* complex (Diptera: Cecidomyiidae): possible biological control agents of honey mesquite. Environ Entomol 48:1042–1048

Pratt PD, Rayamajhi MB, Tipping PW, Center TD, Wright SA, Purcell M (2013) Establishment, population increase, spread, and ecological host range of *Lophodiplosis trifida* (Diptera: Cecidomyiidae), a biological control agent of the invasive tree *Melaleuca quinquenervia* (Myrtales: Myrtaceae). Environ Entomol 42:925–935

Raguso RA (2020) Don't forget the flies: dipteran diversity and its consequences for floral ecology and evolution. Appl Entomol Zool 55:1–7

Renner SS, Schwarbach AE, Lohmann L (1997) Phylogenetic position and floral function of *Siparuna* (Siparunaceae: Laurales). Int J Plant Sci 158:S89–S98

Roberti D (1954) I simbionti degli acari fitofagi. I *Therodiplosis persicae* Kieffer. Bollettino del Laboratorio di Entomologia Agraria Filippo Silvestri 13:285–302

Rondani C (1847) Osservazioni sopra parecchie specie di esapodi afidicidi e sui loro nemici. Società Agraria e Accademia delle Scienze dell'Istituto di Bologna, Nuovi Annali delle Scienze Naturali e Rendiconto 8(337–351):432–448

Sakai S (2002) A review of brood-site pollination mutualism: plants providing breeding sites for their pollinators. J Plant Res 115:161–168

Sakai S, Kato M, Nagamasu H (2000) *Artocarpus* (Moraceae) – gall midge pollination mutualism mediated by a male-flower parasitic fungus. Am J Bot 87:440–445

Schwarzländer M, Hinz HL, Winston RL, Day MD (2018) Biological control of weeds: an analysis of introductions, rates of establishment and estimates of success, worldwide. BioControl 63:319–331

Sharaf NS (1984) Studies on natural enemies of tetranychid mites infesting eggplant in the Jordan Valley. Zeitschrift fur Angewandte Entomologie-J Appl Entomol 98:527–533

Shimizu N, Hirota J, Morita H (2001) 600 species of plant invader. Zenkoku Nôson Kyôiku Kyôkai, Tokyo. (In Japanese)

Shirota Y, Iituka K, Asano J, Abe J, Yukawa J (1999) Intraspecific variations of mitochondrial cytochrome oxidase I sequence in an aphidophagous species, *Aphidoletes aphidimyza* (Diptera: Cecidomyiidae). Entomol Sci 2:209–215

Solarska E (2004) The use of *Aphidius colemani* and *Aphidoletes aphidimyza* to control damson-hop aphid (*Phorodon humuli* Schrank) on hop. J Plant Prot Res 44:85–90

Stumpf N, Nauen R (2001) Cross-resistance, inheritance, and biochemistry of mitochondrial electron transport inhibitor-acaricide resistance in *Tetranychus urticae* (Acari: Tetranychidae). J Econ Entomol 94:1577–1583

Thien LB, Bernhardt P, Devall MS, Chen Z, Luo Y, Fan J, Yuan L, Williams JH (2009) Pollination biology of basal angiosperms (ANITA grade). Am J Bot 96:166–182

Tipping PT, Martin MR, Gettys LA (2016) A gall-forming biological control agent suppresses vegetative growth of an invasive tree. Biocontrol Sci Tech 26:1586–1589

Tipping PW, Martin MR, Gettys LA (2017) Biological control increases the susceptibility of *Melaleuca quinquenervia* to fire. Biocontrol Sci Tech 27:1014–1017

van Lenteren JC, Bale J, Bigler F, Hokkanen HMT, Loomans AJM (2006) Assessing risks of releasing exotic biological control agents of arthropod pests. Annu Rev Entomol 51:609–634

Vislobokov NA, Galinskaya TV, Degtjareva GV, Valiejo-Roman CM, Samigullin TH, Kuznetsov AN, Sokoloff DD (2014) Pollination of Vietnamese *Aspidistra xuansonensis* (Asparagaceae) by female Cecidomyiidi flies: Larvae of pollinator feed on fertile pollen in anthers of anthetic bisexual flowers. Am J Bot 101:1519–1531

Wardlow LR, Tobin A (1990) Potential new additions to the armory of natural enemies for protected tomatoes. IOBC/WPRS Bull 13:225–227

Whalon ME, Mota-Sanchez RM, Hollingworth RM, Duynslager L (2008) Arthropods resistant to pesticides database (ARPD). https://www.pesticideresistance.org

Xiao Y, Osborne LS, Chen J, Mckenzie C, Houben K (2011) Evaluation of corn plant as potential banker plant for supporting predatory gall midge, *Feltiella acarisuga* (Diptera: Cecidomyiidae) against *Tetranychus urticae* (Acari: Tetranychidae) in greenhouse vegetable production. Crop Prot 30:1635–1642

Xiao Y, Osborne LS, Chen J, Mckenzie CL (2013) Functional responses and prey-stage preferences of a predatory gall midge and two predacious mites with twospotted spider mites, *Tetranychus urticae*, as host. J Insect Sci 13. https://doi.org/10.1673/0.31.013.0801

Yano E, Nishikawa S, Yamane M (2008) Biology of the aphidophagous gall midge, *Aphidoletes aphidimyza*, and its use for biological control. Mem Fac Agric Kinki University 41:1–10

Yano E, Nishikawa S, Yamane M, Abe J (2009) Development of the use of the banker plant system to control aphids in protected culture in Japan. IOBC/WPRS Bull 68:195–198

Young AM (1985) Studies of cecidomyiid midges (Diptera: Cecidomyiidae) as cocoa pollinators (*Theobroma cacao* L.) in Central America. Proc Entomol Soc Wash 87:49–79

Yuan LC, Luo YB, Thien LB, Fan JH, Xu HL, Chen ZD (2007) Pollination of *Schisandra henryi* (Schisandraceae) by female, pollen-eating *Megommata* species (Cecidomyiidae, Diptera) in South-central China. Ann Bot 99:451–460

Yuan LC, Luo YB, Thien LB, Fan JH, Xu HL, Yukawa J, Chen ZD (2008) Pollination of *Kadsura longipedunculata* (Schisandraceae), a monoecious basal angiosperm, by female, pollen-eating *Megommata* sp. (Cecidomyiidae: Diptera) in China. Biol J Linn Soc 93:523–536

Yukawa J (1971) A revision of the Japanese gall midges (Diptera: Cecidomyiidae). Mem Fac Agric Kagoshima University 8:1–203

Yukawa J, Yamaguchi D, Mizota K, Setokuchi O (1998) Distribution and host range of an aphidophagous species of Cecidomyiidae, *Aphidoletes aphidimyza* (Diptera), in Japan. Jpn J Appl Entomol Zool 33:185–193

Yukawa J, Abe J, Mizota K (2008) Improvement in the practical use of an aphidophagous gall midge, *Aphidoletes aphidimyza* (Diptera: Cecidomyiidae), in greenhouses. In: Mason PG, Gillespie DR, Vincent C (eds) Proceedings of the Third International Symposium on Biological Control of Arthropods, Christchurch, New Zealand. USDA Forest Service, Morgantown, WV, pp 77–87

Yukawa J, Sato S, Xu HL, Tokuda M (2011) Description of a new species of the genus *Resseliella* (Diptera: Cecidomyiidae), a pollinator of *Kadsura longipedunculata* (Schisandraceae) in China, with comments on its flower-visiting habit. Entomol Sci 14:297–303

Zachariades C, Hoffmann JH, Roberts AP (2011) Biological control of mesquite (*Prosopis* species) (Fabaceae) in South Africa. Afr Entomol 19:402–415

Zhao JH, Cheng HK (1990) Effect of several sorts of insecticide on *Aphidoletes aphidimyza* (Dip.: Cecidomyiidae). Chin J Biol Control 6:185. (In Chinese)

Name Index

J. Yukawa, M. Tokuda (eds.), *Biology of Gall Midges*, Entomology Monographs,
https://doi.org/10.1007/978-981-33-6534-6

Subject Index

J. Yukawa, M. Tokuda (eds.), *Biology of Gall Midges*, Entomology Monographs,
https://doi.org/10.1007/978-981-33-6534-6